NCCER's Instructor Res

MW01659723

This Web site is passcode-protected and is only meant for instructor use. The access code below provides you with access to all levels of the craft area. In order for you to access other craft areas, you will need to have an access code that is specific to that craft. Be sure to check back with the IRC on a regular basis for any updates to the exam banks or the TestGen software.

This site contains the following resources for instructor use:

TestGen – This product contains software to create the secure paper-based module exams that trainees must pass in order to acquire NCCER credentials. Instructors may use this software to scramble the order of exam questions and create additional questions as desired. Note that according to the NCCER Accreditation Policies and Guidelines, existing questions may ***not*** be removed or edited from any module examination. Doing so may seriously jeopardize either the accreditation status of the training program sponsor or any recognition of trainees, instructors, and trainers through the NCCER National Registry.

Performance Profile Sheets – Included on the Web site is a downloadable document containing all the hands-on tasks that are necessary for trainees to complete to receive credentials from NCCER.

Getting started is easy!

1. Visit **www.nccerirc.com.** You will need a valid email address that you check on a regular basis as well as the access code that is on this sheet. You are allowed to redeem this access code only once.
2. Select the appropriate craft area by either using the drop down menu or by selecting "Login" next to the book cover.
3. Login using your username and password every time you return to **www.nccerirc.com.**
4. If you are a new TestGen user, you will need to download the software. Go to: **www.pearsoned.com/testgen** and use your same user name and password from above.

Note: Additional resources may be added to this site as they become available.
For 24/7 technical support, please call **1-888-433-8435** or go to www.247pearsoned.com.

Access Code : HEO 1

PESHEQ-DROLL-TIPSY-CHIVY-HAUNT-PORES

STORE IN A SECURE AREA!

PEARSON

nccer

Heavy Equipment Operations

Level One

Annotated Instructor's Guide

Second Edition

PEARSON

Boston Columbus Indianapolis New York San Francisco Upper Saddle River
Amsterdam Cape Town Dubai London Madrid Milan Munich Paris Montreal Toronto
Delhi Mexico City São Paulo Sydney Hong Kong Seoul Singapore Taipei Tokyo

National Center for Construction Education and Research
President: Don Whyte
Director of Product Development: Daniele Stacey
Heavy Equipment Operations Project Manager: Patty Bird
Production Manager: Tim Davis
Quality Assurance Coordinator: Debie Ness
Desktop Publishing Coordinator: James McKay
Production Specialist: Heather Griffith-Gatson
Editor: Chris Wilson

Writing and development services provided by Topaz Publications, Liverpool, NY
Lead Writer/Project Manager: Roy Parker
Desktop Publisher: Joanne Hart
Art Director: Megan Paye
Permissions Editors: Andrea LaBarge, Alison Richmond
Writers: Roy Parker, Thomas Burke, Pat Vidler

Pearson Education, Inc.
Editorial Director: Vernon R. Anthony
Executive Editor: Alli Gentile
Senior Product Manager: Lori Cowen
Operations Supervisor: Deidra M. Skahill
Art Director: Jayne Conte
Director of Marketing: David Gesell
Executive Marketing Manager: Derril Trakalo
Marketing Manager: Brian Hoehl
Marketing Coordinator: Crystal Gonzalez
Composition: NCCER
Printer/Binder: Document Technology Resources, Fredericksburg, VA
Cover Printer: Document Technology Resources, Fredericksburg, VA
Text Fonts: Palatino and Univers
Cover Photo: Courtesey of Caterpillar, Inc.

Credits and acknowledgments for content borrowed from other sources and reproduced, with permission, in this textbook appear at the end of each module.

---

10 9 8 7 6 5 4 3 2

PEARSON

Perfect bound ISBN-13: 978-0-13-292166-4
ISBN-10: 0-13-292166-9

Preface

To the Instructor

Heavy equipment operators work on a wide variety of projects, including regular construction buildings, roads, bridges, mining, and timber operations, just to name a few. New construction and infrastructure projects will continue to increase the demand for qualified operators. The skills qualified operators provide are vital for clearing sites, moving materials, or any earth moving operations.

As a heavy equipment operations instructor, you play a key role in training qualified operators by sharing your knowledge and expertise. This book is a tool for you to use as you train tomorrow's heavy equipment operators.

New with *HEO Level One*

NCCER is proud to release the newest edition of *Heavy Equipment Operations* in full color with updates to the curriculum that will engage trainees and give them the best training possible. *Introduction to Earthmoving*, which was in Level 2, has been moved to Level 1 before *Grades* to provide a better understanding of earth moving techniques before explaining the grading process. Safety is always top priority, so additional safety measures have been added throughout the text. Information on trenching equipment has also been added to the modules, as well as information on tractor puller scrapers; both are types of equipment found on many sites. The module on grades has been environmentally enhanced with additional coverage of erosion and erosion control.

We invite you to visit the NCCER website at **www.nccer.org** for information on the latest product releases and training, as well as online versions of the *Cornerstone* newsletter and Pearson's NCCER product catalog.

Your feedback is welcome. You may email your comments to **curriculum@nccer.org** or send general comments and inquiries to **info@nccer.org**.

NCCER Curricula

NCCER is a not-for-profit 501(c)(3) education foundation established in 1996 by the world's largest and most progressive construction companies and national construction associations. It was founded to address the severe workforce shortage facing the industry and to develop a standardized training process and curricula. Today, NCCER is supported by hundreds of leading construction and maintenance companies, manufacturers, and national associations. The NCCER Curricula was developed by NCCER in partnership with Pearson, the world's largest educational publisher.

Some features of the NCCER Curricula are as follows:

- An industry-proven record of success
- Curricula developed by the industry for the industry
- National standardization providing portability of learned job skills and educational credits
- Compliance with the Office of Apprenticeship requirements for related classroom training *(CFR 29:29)*
- Well-illustrated, up-to-date, and practical information

NCCER also maintains a National Registry that provides transcripts, certificates, and wallet cards to individuals who have successfully completed modules of NCCER's Curricula. *Training programs must be delivered by an NCCER Accredited Training Sponsor in order to receive these credentials.*

Special Features

In an effort to provide a comprehensive, user-friendly training resource, we have incorporated many different features for your use. Whether you are a visual or hands-on learner, this book will provide you with the proper tools to get started in heavy equipment operations.

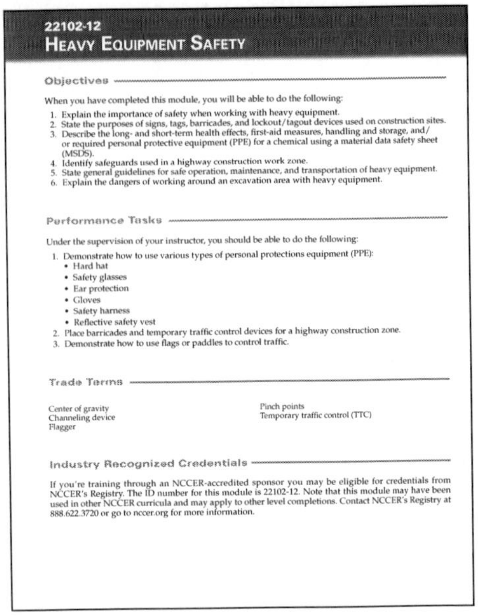
22102-12
HEAVY EQUIPMENT SAFETY

Objectives

When you have completed this module, you will be able to do the following:

1. Explain the importance of safety when working with heavy equipment.
2. State the purposes of signs, tags, barricades, and lockout/tagout devices used on construction sites.
3. Describe the long- and short-term health effects, first-aid measures, handling and storage, and/or required personal protective equipment (PPE) for a chemical using a material data safety sheet (MSDS).
4. Identify safeguards used in a highway construction work zone.
5. State general guidelines for safe operation, maintenance, and transportation of heavy equipment.
6. Explain the dangers of working around an excavation area with heavy equipment.

Performance Tasks

Under the supervision of your instructor, you should be able to do the following:

1. Demonstrate how to use various types of personal protections equipment (PPE):
   - Hard hat
   - Safety glasses
   - Ear protection
   - Gloves
   - Safety harness
   - Reflective safety vest
2. Place barricades and temporary traffic control devices for a highway construction zone.
3. Demonstrate how to use flags or paddles to control traffic.

Trade Terms

Center of gravity
Channeling device
Flagger
Pinch points
Temporary traffic control (TTC)

Industry Recognized Credentials

If you're training through an NCCER-accredited sponsor you may be eligible for credentials from NCCER's Registry. The ID number for this module is 22102-12. Note that this module may have been used in other NCCER curricula and may apply to other level completions. Contact NCCER's Registry at 888.622.3720 or go to nccer.org for more information.

Introduction

This page is found at the beginning of each module and lists the Objectives, Performance Tasks, Trade Terms, and Required Trainee Materials for that module. The Objectives list the skills and knowledge you will need in order to complete the module successfully. The Performance Tasks give you an opportunity to apply your knowledge to the real-world duties that heavy equipment operators perform. The list of Trade Terms identifies important terms you will need to know by the end of the module. Required Trainee Materials list the materials and supplies needed for the module.

On Site

On Site features provide a head start for those entering heavy equipment operations by presenting technical tips and professional practices from operators in a variety of disciplines. The On Site features often include real-life scenarios similar to those you might encounter on the job site.

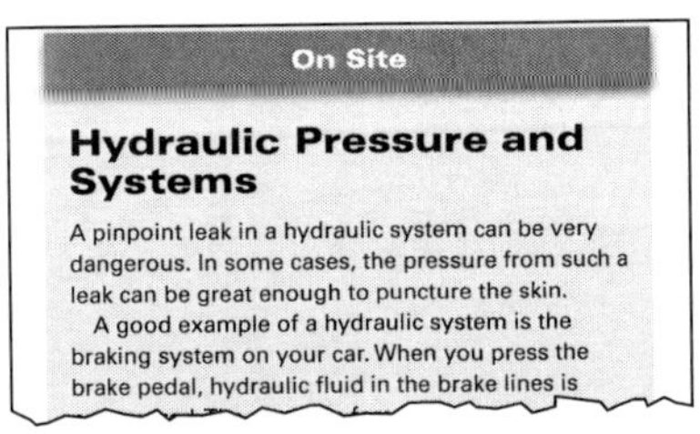
On Site

Hydraulic Pressure and Systems

A pinpoint leak in a hydraulic system can be very dangerous. In some cases, the pressure from such a leak can be great enough to puncture the skin.

A good example of a hydraulic system is the braking system on your car. When you press the brake pedal, hydraulic fluid in the brake lines is

Color Illustrations and Photographs

Full-color illustrations and photographs are used throughout each module to provide vivid detail. These figures highlight important concepts from the text and provide clarity for complex instructions. Each figure reference is denoted in the text in *italic type* for easy reference.

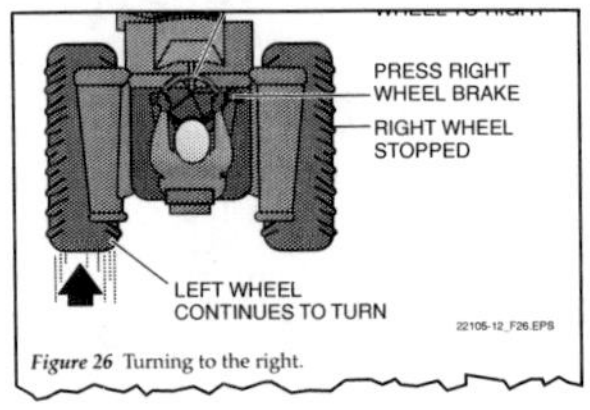

Figure 26 Turning to the right.

Notes, Cautions, and Warnings

Safety features are set off from the main text in highlighted boxes and are organized into three categories based on the potential danger of the issue being addressed. Notes simply provide additional information on the topic area. Cautions alert you of a danger that does not present potential injury but may cause damage to equipment. Warnings stress a potentially dangerous situation that may cause injury to you or a co-worker.

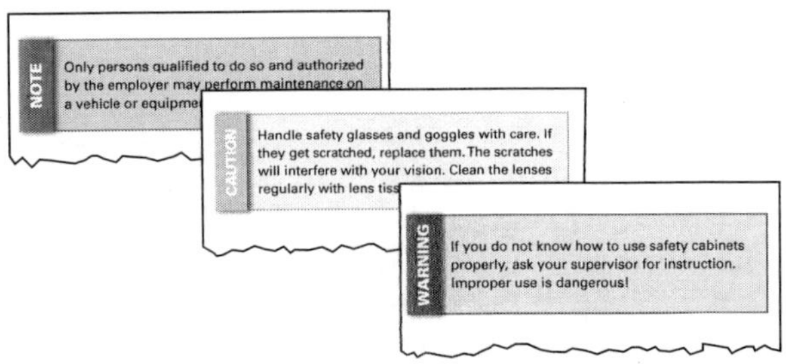

Going Green

Going Green looks at ways to preserve the environment, save energy, and make good choices regarding the health of the planet. Through the introduction of new construction practices and products, you will see how the "greening of America" has already taken root.

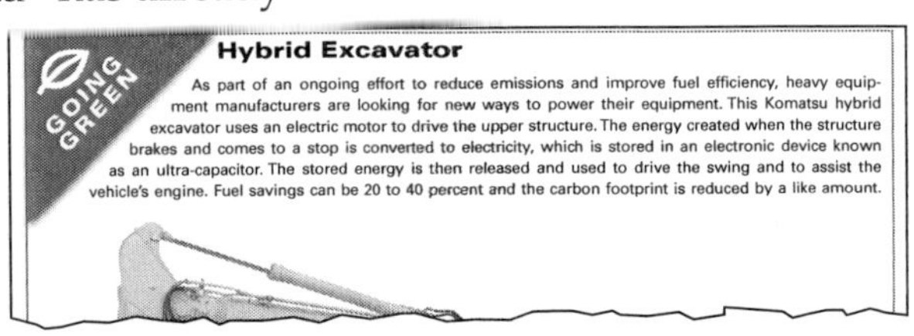
GOING GREEN

Hybrid Excavator

As part of an ongoing effort to reduce emissions and improve fuel efficiency, heavy equipment manufacturers are looking for new ways to power their equipment. This Komatsu hybrid excavator uses an electric motor to drive the upper structure. The energy created when the structure brakes and comes to a stop is converted to electricity, which is stored in an electronic device known as an ultra-capacitor. The stored energy is then released and used to drive the swing and to assist the vehicle's engine. Fuel savings can be 20 to 40 percent and the carbon footprint is reduced by a like amount.

Case History

Case History features emphasize the importance of safety by citing examples of the costly (and often devastating) consequences of ignoring best practices or OSHA regulations.

Case History

A 32-year-old construction surveyor, who had been on the job for only nine days, was doing a grade check on a roadway when the grader operator spotted a manhole cover that he could not go over without damaging it. The grader operator stopped as he approached the manhole cover, then checked his side mirrors for anything that might be behind him. Seeing nothing, he started to back up. The surveyor was bending down with his back to the grader to tie a ribbon on a stake. A co-worker saw the surveyor and heard the grader's back-up alarm. He yelled for the victim to move but it was too late. The surveyor died from multiple injuries.

The Bottom Line: Stay alert. Heavy equipment is dangerous to work around.

Source: The National Institute for Occupational Safety and Health (NIOSH)

Did You Know?

The Did You Know? features offer hints, tips, and other helpful bits of information from the trade.

Did You Know?

Safety Training

The general principle is that unsafe and unhealthful working conditions on mine sites fall under the Mine Safety and Health Administration (MSHA), and all else comes under the authority of the Occupational Safety and Health Administration (OSHA). Except for a few industry-specific requirements, the regulations are the same. Safety training is required for all activities. Never operate tools, machinery, or equipment without prior training.

Step-by-Step Instructions

Step-by-step instructions are used throughout to guide you through technical procedures and tasks from start to finish. These steps show you not only how to perform a task but also how to do it safely and efficiently.

5.6.0 Attaching a PTO

Use the following guidelines to attach a PTO. Be sure to have the correct rpm PTO shaft for the attachment being connected.

Step 1 Set the wheel brakes.

Step 2 Disengage the power to the PTO shaft.

Step 3 Shut off the engine.

WARNING!

Always shut off the engine and disengage the PTO before attempting to connect or disconnect attachments.

Trade Terms

Each module presents a list of Trade Terms that are discussed within the text and defined in the Glossary at the end of the module. These terms are denoted in the text with **bold, blue type** upon their first occurrence. To make searches for key information easier, a comprehensive Glossary of Trade Terms from all modules is located at the back of this book.

Before operating a bulldozer, the operator must understand how to operate the blade and its controls. The blade position can be changed in lift, angle, tilt, and **pitch**. Changing the position of the blade allows the bulldozer to perform different grading operations. Refer to the O&M manual for each dozer for the location and operation of the blade controls.

The lift control lowers or raises the blade. Lowering the blade allows the operator to change the amount of bite or depth to which the blade will dig into the material. Raising the blade permits the operator to travel, shape slopes, or create stockpiles. The lift lever can also be set to **float**. The blade adjusts freely to the contour of the

Review Questions

Review Questions are provided to reinforce the knowledge you have gained. This makes them a useful tool for measuring what you have learned.

Review Questions

1. Caution signs with a yellow background and black letters are used to inform workers of _____.
   a. dangerous conditions
   b. potential hazards or unsafe conditions
   c. hazardous materials in the area
   d. weather conditions that could become hazardous
2. Safety tags can be used in place of safety warning signs.
   a. True
   b. False
3. Barricades and barriers are used _____.
   a. only for traffic diversion on highway construction sites
   b. when there is not enough time to put up a sign
   c. to alert workers to potential hazards
   d. as a last resort to keep vehicles out of the work area
4. In areas where the ground is too soft to permit a machine to operate safely on a grade, it is a good idea to put up a _____.
   a. barrier
   b. flag
   c. caution sign
   d. danger sign
5. The spacing of temporary traffic control signs will depend on the road's _____.
   a. amount of traffic
   b. surface condition
   c. normal posted speed
   d. proximity to pedestrian traffic
6. Channeling devices are used in a temporary traffic control area _____.
   a. when flaggers are not available
   b. to guide traffic into a desired path
   c. to protect workers from speeding traffic
   d. to keep heavy equipment out of the road
7. The area next to the border that is designed to protect workers from closely passing vehicles is called the _____.
   a. buffer zone
   b. transition area
   c. pedestrian zone
   d. advance warning area
8. To use the Stop/Slow paddle, the flagger faces road users and _____.
   a. waves the flag rapidly across the lane of oncoming traffic
   b. steps into the lane of oncoming traffic and raises the flag
   c. repeatedly raises and lowers the flag to shoulder level
   d. extends the arm holding the paddle horizontally away from the body
9. For each hazardous substance in a shipment, OSHA and MSHA require that suppliers of hazardous chemicals provide _____.
   a. a respirator
   b. splash-proof goggles
   c. chemical-resistant gloves
   d. a material safety data sheet (MSDS)
10. The maximum amount of gasoline that OSHA allows to be stored outside an approved storage enclosure is _____.
    a. 5 gallons
    b. 10 gallons
    c. 25 gallons
    d. 50 gallons
11. What are the advanced symptoms of carbon monoxide poisoning?
    a. Dizziness
    b. Headaches
    c. Weakness and confusion
    d. Staggering and falling
12. Which of the following noise sources has the highest dBA level?
    a. Thunder
    b. Stock car races
    c. A jet plane taking off
    d. Shouting near the ear

44 NCCER eavy Equipme Level 22

NCCER Curricula

NCCER's training programs comprise more than 80 construction, maintenance, pipeline, and utility areas and include skills assessments, safety training, and management education.

Boilermaking
Cabinetmaking
Carpentry
Concrete Finishing
Construction Craft Laborer
Construction Technology
Core Curriculum: Introductory Craft Skills
Drywall
Electrical
Electronic Systems Technician
Heating, Ventilating, and Air Conditioning
Heavy Equipment Operations
Highway/Heavy Construction
Hydroblasting
Industrial Coating and Lining Application Specialist
Industrial Maintenance Electrical and Instrumentation Technician
Industrial Maintenance Mechanic
Instrumentation
Insulating
Ironworking
Masonry
Millwright
Mobile Crane Operations
Painting
Painting, Industrial
Pipefitting
Pipelayer
Plumbing
Reinforcing Ironwork
Rigging
Scaffolding
Sheet Metal
Signal Person
Site Layout
Sprinkler Fitting
Tower Crane Operator
Welding

Green/Sustainable Construction

Building Auditor
Fundamentals of Weatherization
Introduction to Weatherization
Sustainable Construction Supervisor
Weatherization Crew Chief
Weatherization Technician
Your Role in the Green Environment

Energy

Alternative Energy
Introduction to the Power Industry
Introduction to Solar Photovoltaics
Introduction to Wind Energy
Power Industry Fundamentals
Power Generation Maintenance Electrician
Power Generation I&C Maintenance Technician
Power Generation Maintenance Mechanic
Power Line Worker
Power Line Worker: Distribution
Power Line Worker: Substation
Power Line Worker: Transmission
Solar Photovoltaic Systems Installer
Wind Turbine Maintenance Technician

Pipeline

Control Center Operations, Liquid
Corrosion Control
Electrical and Instrumentation
Field Operations, Liquid
Field Operations, Gas
Maintenance
Mechanical

Safety

Field Safety
Safety Orientation
Safety Technology

Management

Fundamentals of Crew Leadership
Project Management
Project Supervision

Supplemental Titles

Applied Construction Math
Careers in Construction
Tools for Success

Spanish Translations

Basic Rigging (Principios Básicos de Maniobras)
Carpentry Fundamentals (Introducción a la Carpintería, Nivel Uno)
Carpentry Forms (Formas para Carpintería, Nivel Trés)
Concrete Finishing, Level One (Acabado de Concreto, Nivel Uno)
Core Curriculum: Introductory Craft Skills (Currículo Básico: Habilidades Introductorias del Oficio)
Drywall, Level One (Paneles de Yeso, Nivel Uno)
Electrical, Level One (Electricidad, Nivel Uno)
Field Safety (Seguridad de Campo)
Insulating, Level One (Aislamiento, Nivel Uno)
Ironworking, Level One (Herrería, Nivel Uno)
Masonry, Level One (Albañilería, Nivel Uno)
Pipefitting, Level One (Instalación de Tubería Industrial, Nivel Uno)
Reinforcing Ironwork, Level One (Herreria de Refuerzo, Nivel Uno)
Safety Orientation (Orientación de Seguridad)
Scaffolding (Andamios)
Sprinkler Fitting, Level One (Instalación de Rociadores, Nivel Uno)

Acknowledgments

This curriculum was revised as a result of the farsightedness and leadership of the following sponsors:

Caterpillar, Inc.
KBR, Inc.
Saiia Construction Company, LLC
Skyview Construction and Engineering, Inc.
Southland Safety, LLC

This curriculum would not exist were it not for the dedication and unselfish energy of those volunteers who served on the Authoring Team. A sincere thanks is extended to the following:

Phillip Allen
Roger Arnett
Mark Jones
Larry Proemsey
Joseph Watts

NCCER Partners

American Fire Sprinkler Association
Associated Builders and Contractors, Inc.
Associated General Contractors of America
Association for Career and Technical Education
Association for Skilled and Technical Sciences
Carolinas AGC, Inc.
Carolinas Electrical Contractors Association
Center for the Improvement of Construction Management and Processes
Construction Industry Institute
Construction Users Roundtable
Construction Workforce Development Center
Design Build Institute of America
Merit Contractors Association of Canada
Metal Building Manufacturers Association
NACE International
National Association of Minority Contractors
National Association of Women in Construction
National Insulation Association
National Ready Mixed Concrete Association
National Technical Honor Society
National Utility Contractors Association
NAWIC Education Foundation
North American Technician Excellence
Painting & Decorating Contractors of America
Portland Cement Association
Skills USA
Steel Erectors Association of America
The Manufacturers Institute
U.S. Army Corps of Engineers
University of Florida, M. E. Rinker School of Building Construction
Women Construction Owners & Executives, USA

NCCER Business Partners

Contents

Module Six

Introduction to Earthmoving

This module provides a broad introduction to the process of planning and executing earthmoving activities on various types of construction projects. The use of heavy equipment such as bulldozers, scrapers, excavators, and loaders is explained. (Module ID 22201-12; 12.5 Hours)

Module Seven

Grades

Introduces the concept of preparing graded surfaces using heavy equipment. Covers identification of construction stakes and interpretation of marks on each type of stake. Describes the process for grading slopes. (Module ID 22106-12; 15 Hours)

Glossary

Index

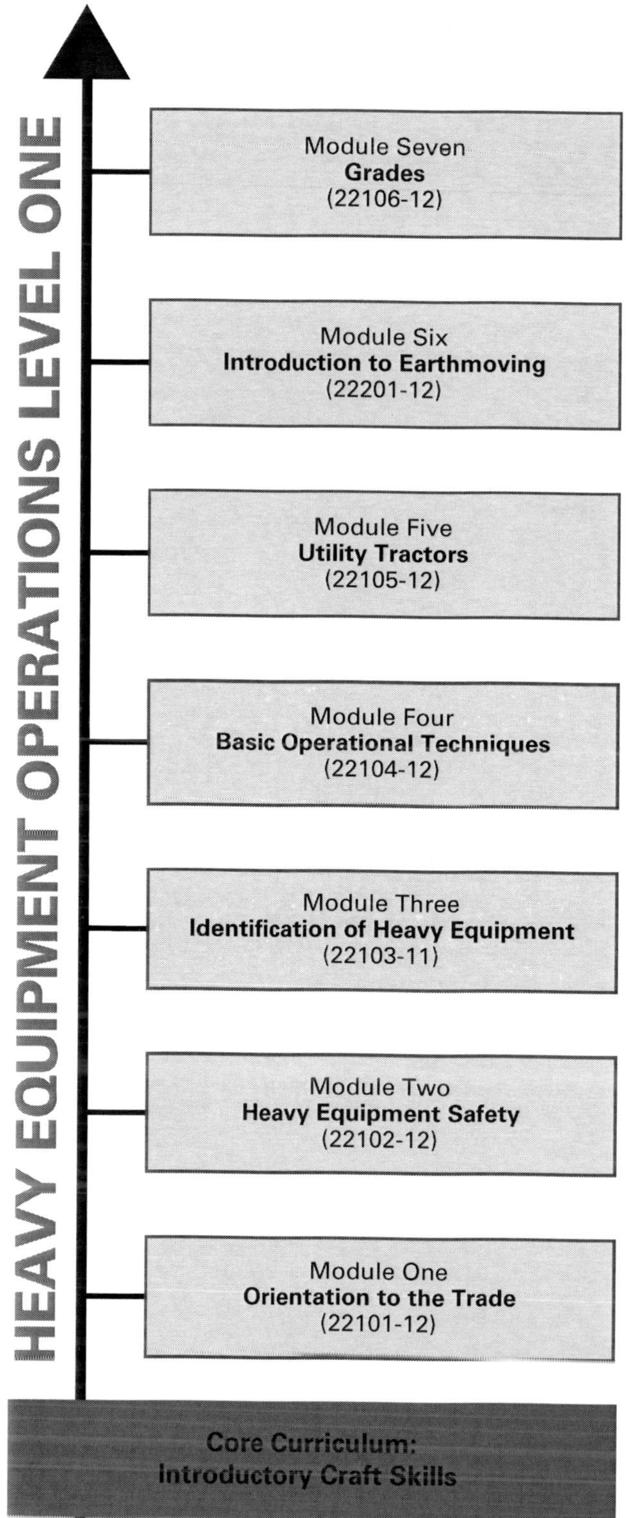

This course map shows all of the modules in *Heavy Equipment Operations Level One.* The suggested training order begins at the bottom and proceeds up. Skill levels increase as you advance on the course map. The local Training Program Sponsor may adjust the training order.

Supplemental Resources

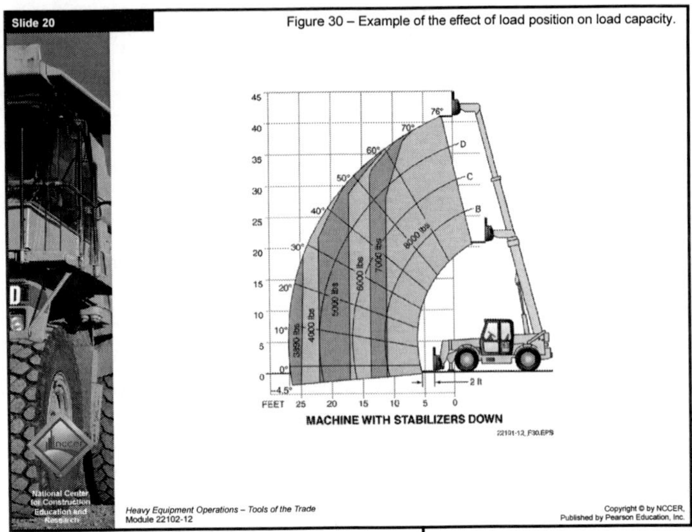

PowerPoint® Presentation Slides

A powerful addition to your classroom discussion, NCCER's Microsoft® PowerPoint® presentation CD features all the drawings and photos from the Trainee Guide.

Heavy Equipment Operations Level One: ISBN 978-0-13-292167-1

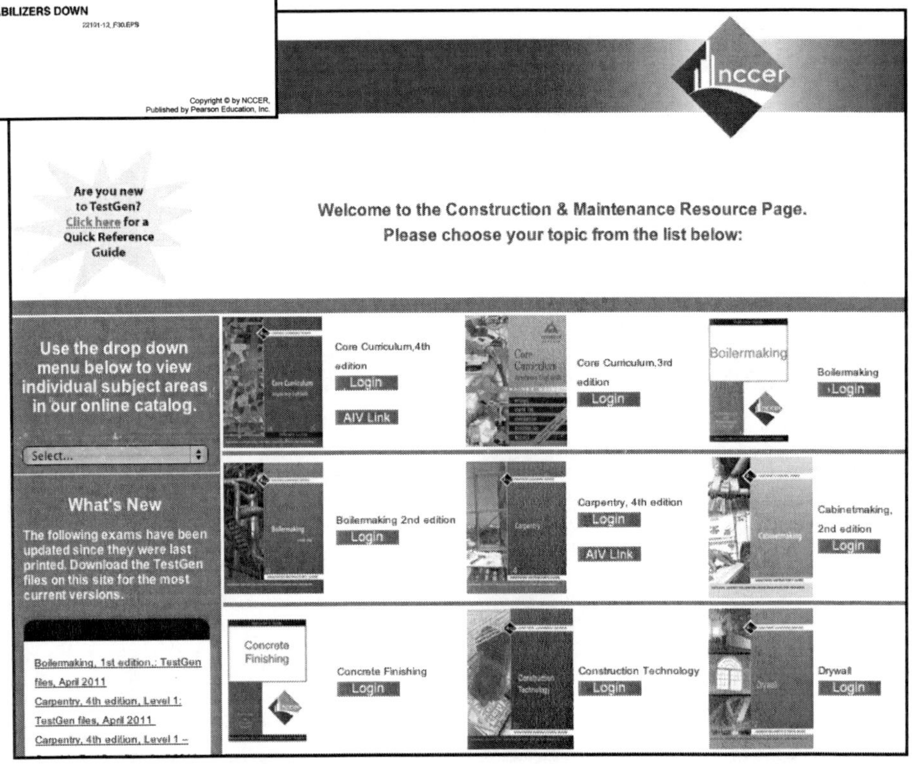

Online Instructor Resource Center

www.NCCERContrenIRC.com

Available for PC and Mac

TestGen software, module exams, and Performance Profile Sheets are now available through a secure website, known as the Instructor Resource Center. Printed module exams no longer come shrinkwrapped with Annotated Instructor's Guides (AIGs). Instead, every AIG will include a craft-specific access code that needs to be redeemed to gain entry to the Instructor Resource Center where the TestGen software, module exams, and Performance Profile Sheets can be downloaded. You will find the access card immediately following the front cover of this book. This change has been made for a couple of reasons:

- *Improved Security* — The website is pass-code protected and meant for instructor use only.
- *Enhanced Flexibility* — Instructors are able to download the TestGen files (previously only available on a separate CD), module exams, and PDFs of the Performance Profile Sheets. The question sequence for exams can be scrambled, and instructors can also add their own questions for quizzing purposes.

Additional Access Codes

Access codes are assigned one per instructor. If other instructors would like to purchase access to the TestGen software without buying an AIG, they may be purchased by referencing **ISBN 978-0-13-292207-4**.

22101-12

Orientation to the Trade

Module One

Materials

reminds you to gather materials needed for classes, labs, and testing.

Homework

suggests that you assign reading for the next session.

Classroom

prompts you to discuss a topic, stress a point, explain a concept, or provide examples.

Audiovisual

indicates an appropriate time to show a slide or transparency.

! Safety

used to emphasize safety issues.

Demonstration

directs you to show trainees how to perform tasks.

Laboratory

used when trainees are to practice performing tasks.

✓ Performance Testing

recommends administering a performance test, or a portion thereof.

PT / Laboratory

indicates that lab practice corresponds with a performance task. If desired, you can note the proficiency of the trainees during the lab to satisfy performance testing requirements.

Teaching Tip

indicates tips for practice exercises, Quick Quizzes, field trips, or other activities designed to encourage class participation.

? Examination

recommends administering the written module exam.

NCCER Standardized Craft Training Program

NCCER provides a standardized national program of accredited craft training. Key features of the program include instructor certification, competency-based training, and performance testing. The program provides trainees, instructors, and companies with a standard form of recognition through the National Registry. The program is described in full in the *Guidelines for Accreditation*, published by NCCER. For more information on standardized craft training, contact NCCER by writing to 13614 Progress Blvd., Alachua FL 32615; calling 1-888-NCCER20; or emailing **info@nccer.org**. More information is available at **www.nccer.org**.

How to Use this Annotated Instructor's Guide

Each page presents two sections of information. The larger section displays each page exactly as it appears in the Trainee Module. The narrow column ties suggested trainee and instructor actions to each page and provides icons (detailed on the left) to call your attention to material, safety, audiovisual, or testing requirements. The bottom of each page includes space for your notes, or when available, teaching tips that suggest activities or class exercises that will help your students learn the material.

Preparation

Before teaching this module, you should review the Objectives, Performance Tasks, Materials and Equipment List, and Module Outline. Be sure to allow ample time to prepare your own training or lesson plan and gather all required materials and equipment.

Instructors purchasing a single module receive the printed exam and performance tasks with the AIG module. Perfect-bound versions of this title no longer include the printed exam booklet and performance task sheets. Log on to the Instructor Resource Center at **www.NCCERIRC.com** to download these materials. Your access code is on the card bound into the front of this AIG. This code will allow you to access exams and performance tasks for all levels of this craft. Be sure to keep all testing materials in a secure place at all times.

Orientation to the Trade
Annotated Instructor's Guide

Module 22101-12

Module Overview

This module introduces the trainee to the basic terminology and equipment used in the heavy equipment trade. The trainee will learn what to expect from an apprenticeship program in heavy equipment, what makes a good operator, and the importance of safety.

Objectives

Upon completion of this module, the trainee will be able to do the following:

1. Explain the basic terminology, types, and uses of equipment.
2. Identify career opportunities available to heavy equipment operators and explain the purpose and objectives of an apprentice training program.
3. Explain the responsibilities and characteristics of a good operator.
4. Explain the importance of heavy equipment safety.

Performance Tasks

This is a knowledge-based module. There are no performance tasks.

Materials and Equipment

Markers/chalk
Pencils and scratch paper
Whiteboard/chalkboard
Heavy Equipment Operations Level One PowerPoint® Presentation Slides (ISBN 978-0-13-292167-1)
Multimedia projector and screen
Computer
Appropriate personal protective equipment
Trade Terms Quiz*
Module Examinations**

* Located at the back of the Trainee Guide module
** Single-module AIG purchases include the printed exam. If you have purchased the perfect-bound version of this title, download these materials from the IRC using your access code.

Safety Considerations

Ensure that the trainees are equipped with appropriate personal protective equipment and know how to use it properly. Review safety guidelines associated with working on or around heavy equipment. Emphasize the importance of proper housekeeping.

Additional Resources

This module presents thorough resources for task training. The following resource material is suggested for further study.

The Earthmover Encyclopedia, Latest Edition. Minneapolis, MN: Quayside Publishing Group.

Teaching Time for This Module

An outline for use in developing your lesson plan is presented below. Note that each Roman numeral in the outline equates to one session of instruction. Each session has a suggested time period of 2½ hours. This includes 10 minutes at the beginning of each session for administrative tasks and one 10-minute break during the session. Approximately 5 hours are suggested to cover *Orientation to the Trade*. You will need to adjust the time required for hands-on activity and testing based on your class size and resources.

Topic	Planned Time
Session I. Introduction; Heavy Equipment Applications; Career Opportunities in Heavy Equipment	
A. Introduction	________
B. Heavy Equipment Applications	________
C. Career Opportunities in Heavy Equipment	________
1. Getting Started	________
2. Other Job-Related Opportunities	________
3. Management Positions	________
4. Self-Employed Contractor	________
Session II. Standardized Training By NCCER; Responsibilities of the Employee; Human Relations; Employer and Employee Safety Obligations; Review and Testing	
A. Standardized Training By NCCER	________
1. Apprenticeship Training	________
2. Youth Training and Apprenticeship Programs	________
B. Responsibilities of the Employee	________
1. Professionalism	________
2. Honesty	________
3. Loyalty	________
4. Willingness to Learn	________
5. Willingness to Take Responsibility	________
6. Willingness to Cooperate	________
7. Rules and Regulations	________
8. Tardiness and Absenteeism	________
C. Human Relations	________
1. Making Human Relations Work	________
2. Human Relations and Productivity	________
3. Attitude	________
4. Maintaining a Positive Attitude	________
D. Employer and Employee Safety Obligations	________
1. OSHA	________
2. MSHA	________
3. Safety in General	________
E. Review	________
F. Module Examination	________
1. Trainees must score 70% or higher to receive recognition from NCCER.	
2. Record the testing results on Training Report Form 200 and submit the results to the Training Program Sponsor.	

Homework

Assign the reading of this module.

22101-12

Orientation to the Trade

Module One

Teaching Tip/Notes

Trainees with successful module completions may be eligible for credentialing through NCCER's National Registry. To learn more, go to **www.nccer.org** or contact us at **1.888.622.3720**. Our website has information on the latest product releases and training, as well as online versions of our *Cornerstone* newsletter and Pearson's product catalog.

Your feedback is welcome. You may email your comments to **curriculum@nccer.org**, send general comments and inquiries to **info@nccer.org**, or fill in the User Update form at the back of this module.

V.1 4/12

Teaching Tip/Notes

22101-12
Orientation to the Trade

Objectives

When you have completed this module, you will be able to do the following:

1. Explain the basic terminology, types, and uses of equipment.
2. Identify career opportunities available to heavy equipment operators and explain the purpose and objectives of an apprentice training program.
3. Explain the responsibilities and characteristics of a good operator.
4. Explain the importance of heavy equipment safety.

Performance Tasks

This is a knowledge-based module; there are no performance tasks.

Trade Terms

Demolition
Infrastructure
Mine Safety and Health Administration (MSHA)
On-the-job learning (OJL)
Occupational Safety and Health Administration (OSHA)

Industry Recognized Credentials

If you're training through an NCCER-accredited sponsor you may be eligible for credentials from NCCER's Registry. The ID number for this module is 22101-12. Note that this module may have been used in other NCCER curricula and may apply to other level completions. Contact NCCER's Registry at 888.622.3720 or go to nccer.org for more information.

Materials

Ensure that you have everything required to teach the course. Check the Materials and Equipment list at the front of this module.

Classroom

Explain that terms shown in bold (blue) are defined in the Glossary at the back of this module.

Audiovisual

Show Slide 1, Objectives. Review the goals of the module, and explain what will be expected of the trainees.

Teaching Tip/Notes

When you call on someone to answer a question, the rest of the class relaxes or even tunes out because they expect that the question and answer will take place only between you and the trainee you called on. Instead, use this technique to involve more trainees in answering questions and to keep them on their toes:

1. Ask trainees to define a term or explain a concept.
2. After one trainee has answered, ask a trainee seated nearby if the answer is right. Then ask whether a trainee in the back of the room agrees.
3. Ask trainees to explain why they think an answer is right or wrong.
4. Use the session to clear up incorrect ideas and encourage trainees to learn from their mistakes.

Contents

Topics to be presented in this module include:

Teaching Tip/Notes

Figures and Tables

Teaching Tip/Notes

1.0.0 Introduction

Heavy construction equipment is used primarily in preparing a site for the construction of buildings and other structures. Ten types of heavy equipment are covered in this module. A heavy equipment operator can specialize in one machine, or can learn to operate several.

At the start of a project, bulldozers, excavators, loaders, and dump trucks are used either to demolish and remove existing structures or to clear a new site of trees, brush, and boulders. At the next phase of the project this same equipment, along with scrapers, is used to add, remove, and relocate earth in order bring the site to the grade established by the site design engineers. It is often necessary to move thousands of tons of material in preparing a site for construction. Excavators and backhoes are used to dig trenches for utilities, such as water, sewer, and gas piping. At the final stage of site development, soil compactors and motor graders are used to prepare roads and parking lots for paving, and to prepare the building site for landscaping.

Heavy equipment operators are highly skilled workers who are needed on every construction site. A skilled operator has many opportunities for advancement with an employer or as an entrepreneur. Many of the large excavating and site-development companies operating today were founded by equipment operators who started with a single machine.

This module introduces the various types of heavy equipment used on construction sites. *Figure 1* shows examples of this equipment, which includes the following:

- Bulldozer
- Dump truck
- Excavator
- Backhoe loader
- Loader
- Motor grader
- Roller/compactor
- Forklift
- Scraper
- Construction tractor

In Level One, you have an opportunity to get some seat time in different heavy equipment. This level's purpose is to familiarize you with the equipment and its controls. The modules also address operator-performed maintenance activities for given machines. Most importantly, the safety aspects of operating heavy equipment are also covered. The proper ways to mount and dismount a machine, how to safely move machinery, and job-site safety requirements in general are all explained.

This level provides detail on the operation of tractors used in construction work to help you become proficient in the operation of this type of vehicle. In higher levels, each remaining types of heavy equipment are covered in detail. By the time you finish this three-level program, you should be proficient in the operation of all 10 heavy equipment machines.

This module provides an overview of heavy equipment used in construction and mining work, including a description of each type of equipment and its uses. The module also explains the structure of an apprentice training program and addresses the values that help a person succeed in the industry. Because safety is such an important concern for anyone in the construction industry, the module also introduces safety issues that are expanded upon in later modules.

2.0.0 Heavy Equipment Applications

Heavy equipment is used in the construction of buildings, bridges, dams, and roads. It is also used extensively in mining operations. In fact, some of the largest equipment is designed for mining work (*Figure 2*).

The primary applications of heavy equipment in construction are in **demolition** work, earth moving, trench digging, road construction, and site grading. An office building complex being developed on wooded land is a good example of when heavy equipments is needed. The first step would be to remove brush, trees, and rocks (*Figure 3*). This is known as clearing and grubbing, and is typically done with a dozer. Then the site must be brought to subgrade, also known as rough grade, by removing or relocating soil and filling depressions (*Figure 4*). Quite often, it is necessary to truck in material from outside sources, or to move excess material from one part of the site to another, in order to achieve designed grades (*Figure 5*).

Excavators are earth-moving machines used to dig foundations or move large quantities of dirt. The soil and rocks that are dug out by an excavator are usually removed from the site by dump trucks (*Figure 6*). The type of dump truck being used depends upon the location of the worksite. There are on-road dump trucks and off-road dump trucks. The trucks may be loaded by excavators, loaders, or backhoes.

Excavators and backhoes are used to dig trenches for utilities, such as water and sewer lines, storm drains, and cables for electricity, telephone systems, and data lines. *Figure 7* shows a backhoe digging a trench.

Classroom

Describe typical uses of heavy equipment.

Audiovisual

Show Slides 2 and 3 (Figure 1). List the ten types of heavy equipment covered in this module.

Classroom

Discuss the different types of heavy equipment and their applications.

Point out differences between mining and construction equipment.

Audiovisual

Show Slide 4 (Figure 2). Discuss uses of heavy equipment in mining.

Show Slide 5 (Figure 3). Explain how heavy equipment is used in various phases of land development.

Show Slide 6 (Figure 4).

Show Slide 7 (Figure 5). Describe excavators and their functions.

Show Slide 8 (Figure 6).

Show Slide 9 (Figure 7). Explain that excavators and backhoes are used for trenching.

Teaching Tip/Notes

22101-12 Orientation to the Trade Module One 1

(A) BULLDOZER (B) ARTICULATED OFF-ROAD DUMP TRUCK

(C) EXCAVATOR (D) BACKHOE LOADER

(E) SKID STEER (F) MOTOR GRADER

22101-12_F01A.EPS

Figure 1 Heavy construction equipment. (1 of 2)

Teaching Tip/Notes

2 NCCER – *Heavy Equipment Operations Level One* 22101-12

(G) VIBRATORY ASPHALT COMPACTOR

(H) STRAIGHT MAST ROUGH TERRAIN FORKLIFT

(I) TANDEM POWERED SCRAPER

(J) UTILITY TRACTOR

(K) OFF-HIGHWAY HEAVY CONSTRUCTION TRUCK

22101-12_F01B.EPS

Figure 1 Heavy construction equipment. (2 of 2)

Teaching Tip/Notes

22101-12 Orientation to the Trade Module One 3

Audiovisual

Show Slide 10 (Figure 8). Explain how a site is brought to a finish grade.

Show Slide 11 (Figure 9). Describe different ways that scrapers are used in construction.

Show Slide 12 (Figure 10). Discuss the uses of heavy equipment in road building.

Show Slides 13 and 14 (Figures 11 and 12).

Classroom

Discuss career opportunities for heavy equipment operators.

Point out the Department of Labor statistics for heavy equipment operators in 2008.

On Site

Dealing with Water Runoff

In most areas of the country, rainwater is absorbed by the soil. When land is developed, however, the soil is covered with buildings, roads, and parking lots, so the water can no longer be absorbed in those areas. If the site is not properly engineered, the runoff will have no place to go except onto neighboring lower property, which may cause flooding. For this reason, many construction sites are required to have detention ponds to receive the excess water. Although storm drainage systems are installed on most properties, a heavy rain can easily overwhelm these systems. The detention pond holds the water and lets it slowly drain.

A retention pond, on the other hand, always contains water. If there are natural water sources such as springs on the property, for example, these sources may be diverted to a retention pond. Retention ponds are often used to create attractive water features for a residential or commercial development.

DETENTION POND 22101-12_SA01.EPS

RETENTION POND 22101-12_SA02.EPS

After the site has been cleared and otherwise prepared, most of the heavy equipment work is finished and the operators go on to the next project. The exception is the equipment used to bring the site and its roads to finish grade (*Figure 8*).

When an existing building or other structure must be removed from the site in order to make way for new construction, heavy equipment operators are involved in the demolition (*Figure 9*). Those operators must be extremely careful in their actions because demolition activities can cause other parts of a building to fall. Excavators, loaders, and dump trucks are used to remove the debris from the site.

Much of the work performed with heavy equipment is associated with road building. The machines that scrape up and move dirt on those road-building projects are generally called scrapers. *Figure 10* shows a few different varieties of scrapers. Top soil, when removed, is stockpiled for reuse.

After the roadways have been roughly graded to the desired levels, additional equipment is brought in to shape, pack, and smooth the dirt and gravel that is used to form the road bed. *Figure 11* shows a soil compactor working with a dozer to form and pack roadway gravel.

After the soil and gravel have been properly shaped and packed for a new road, the paving equipment is brought in. The new road may be poured concrete or asphalt. If it is asphalt, heavy asphalt compactors (*Figure 12*) are used to pack and smooth the freshly laid asphalt.

3.0.0 Career Opportunities in Heavy Equipment

It is difficult to imagine a construction project that does not require heavy equipment operators. Large commercial and industrial projects often need hundreds of them. According to the US Department of Labor's *2010–2011 Occupational Outlook Handbook*, the job opportunities for heavy equipment operators are expected to be good with a projected growth of 12 percent between 2008 and 2018. According to the Department of Labor's statistics, construction equipment operators held

Teaching Tip/Notes

UNDERGROUND MINING LOADER

UNDERGROUND MINING HAULER

22101-12_F02.EPS

Figure 2 Examples of mining equipment.

almost 470,000 jobs in 2008. Over 404,000 of those jobs were as operating engineers and other construction equipment operators. In addition, there were approximately 60,000 paving, surfacing, and tamping equipment-operator positions, and 4,600 pile-driver operators.

Approximately three out of five heavy equipment operators work in the construction industry.

Teaching Tip/Notes

22101-12 Orientation to the Trade Module One 5

Classroom

Discuss job expectations in connection with an aging infrastructure.

Explain typical job progression from lighter to heavier equipment.

Describe different aspects of training, such as equipment simulators.

22101-12_F03.EPS

Figure 3 Opening the site.

22101-12_F04.EPS

Figure 4 Bringing the site to rough grade.

Others work on road, bridge, railroad, and mining projects. One of the new construction areas involves power-generating facilities, including wind farms, solar farms, and biomass facilities. Due to increasing populations, and the deterioration of existing **infrastructure**, such as roads and bridges, jobs for heavy equipment operators are expected to be plentiful for the foreseeable future as those structures are repaired or replaced.

3.1.0 Getting Started

There are many opportunities for heavy equipment operators. Operators typically start out on lighter, simpler equipment, such as forklifts, trucks, skid steer loaders, and roller/compactors.

22101-12_F05.EPS

Figure 5 Fill is loaded onto a truck and transferred from the site.

22101-12_F06.EPS

Figure 6 Excavator loading truck.

To standardize a company's training programs and to document the training each trainee receives, most companies start their trainees out on interactive programs run on a computer. Such programs teach the trainee about the machine to be operated, and ensure that the trainee understands both the company's general safety policies and those associated with each machine being. After a trainee has a proven understanding of the machinery to be operated and the related safety guidelines, the company will move the trainee on to the hands-on portion of the training.

With experience and further training, the operator can advance to more complex equipment, such as bulldozers, excavators, and graders. The operator's pay is generally linked to the type of equipment he or she is operating. The more types

Teaching Tip/Notes

Ask each trainee to explain why they entered a heavy equipment operator course and what they think they would like to do with the knowledge they gain from the program.

22101-12_F07.EPS

Figure 7 Backhoe trenching.

22101-12_F08.EPS

Figure 8 Bringing the site to finish grade.

of machines the operator is qualified to operate, the more likely they are to find steady employment and earn a top income. Some heavy equipment items such as cranes require specialized training and certification. The operator jobs involving the higher-risk activities require the specialized training and certification. The pay for such jobs increases accordingly.

22101-12_F09.EPS

Figure 9 Excavators are often used in demolition work.

3.2.0 Other Job-Related Opportunities

Many companies run their own training programs for equipment operators, so there are opportunities for qualified operators to teach others within their company. In addition to those internal jobs, there are many independent schools that offer equipment operator training programs and need qualified instructors. Construction companies often rent or lease additional equipment as needed. They get that extra equipment from companies that specialize in renting construction equipment. Another employment opportunity for equipment operators is to move into a sales or management role with such a rental company.

3.3.0 Management Positions

Advancement into administrative and management positions is also possible for equipment operators. Many construction companies employ equipment coordinators to make sure that the required types and amounts of equipment, along with qualified operators, are available at each job site when they are needed. The equipment coordinator is typically someone with experience as an equipment operator. Most companies own some heavy equipment and rent or lease the rest as needed. Part of the equipment coordinator's responsibility is to arrange for timely rental of equipment to meet project needs.

Availability of equipment is one problem a construction company faces. Equipment also needs to be serviced and repaired. Large construction companies often establish maintenance facilities to maintain their fleet of equipment. Such a company is likely to have equipment superintendents who are responsible for both the acquisition and

Classroom

Point out teaching opportunities.

Describe management positions, such as equipment coordinator.

Discuss situations an equipment coordinator must handle.

Teaching Tip/Notes

22101-12 Orientation to the Trade Module One 7

Classroom

Point out the opportunities available to a self-employed contractor.

Homework

Have trainees review Sections 4.0.0–7.3.0.

Materials

Ensure that you have everything required for teaching this session.

Classroom

Explain NCCER's role in construction industry training.

Explain the concept of standardized, competency-based training.

Point out the advantages of receiving accredited training through NCCER.

Review basic federal and state requirements for formal apprenticeship programs.

On Site

Mechanical T-Rex

This demolition excavator has a special concrete pulverizing attachment.

22101-12_SA03.EPS

maintenance of the company's equipment. Since experienced heavy equipment operators already have some experience maintaining heavy equipment, they may also work their way into equipment maintenance positions.

3.4.0 Self-Employed Contractor

Being your own boss is part of the American dream. If you look at the history of many excavation and site development companies you will find that they began with an equipment operator who bought a backhoe/loader or some other piece of equipment, along with a trailer and truck, to become an independent contractor. Many construction companies hire independent operators either as a matter of course, or simply to help with peak loads. The way to be successful as an independent operator is to develop a reputation for being on the job when you are needed and for doing good quality work.

4.0.0 Standardized Training By NCCER

NCCER is a not-for-profit education foundation established by the nation's leading construction companies. NCCER was created to provide the industry with standardized construction education materials and a system for tracking and recognizing students' training accomplishments—NCCER's National Registry. Refer to the *Appendix* for examples of NCCER credentials.

NCCER also offers accreditation, instructor certification, and skills assessments. NCCER is committed to developing and maintaining a training process that is internationally recognized, standardized, portable, and competency-based.

Working in partnership with industry and academia, NCCER has developed a system for program accreditation that is similar to those found in institutions of higher learning. NCCER's accreditation process ensures that students receive quality training based on uniform standards and criteria. These standards are outlined in NCCER's Accreditation Guidelines and must be adhered to by NCCER Accredited Training Sponsors.

More than 550 training and assessment centers across the United States and eight other countries are proud to be NCCER Accredited Training Sponsors. Millions of craft professionals and construction managers have received quality construction education through NCCER's network of Accredited Training Sponsors and the thousands of Training Units associated with the Sponsors. Every year the number of NCCER Accredited Training Sponsors increases significantly.

A craft instructor is a journeyman craft professional or career and technical educator trained and certified to teach the NCCER curriculum. This network of certified instructors ensures that NCCER training programs will meet the standards of instruction set by the industry. At the time of this writing, there are more than 4,900 master trainers and 51,402 craft instructors within the NCCER instructor network. More information is available at www.nccer.org.

4.1.0 Apprenticeship Training

Formal apprenticeship programs conform to federal and state requirements under *CFR Titles 29:29* and *29:30*. All approved apprenticeship programs provide **on-the-job learning (OJL)** as well as classroom instruction. The related training requirement is fulfilled by all NCCER craft training

Teaching Tip/Notes

22101-12_F10.EPS

Figure 10 Earth-moving scrapers.

Teaching Tip/Notes

Classroom

Explain the three ways to complete a registered apprenticeship program.

Point out the available programs and advantages for high school students.

Discuss the requirements for students entering the program.

22101-12_F11.EPS

Figure 11 Soil compactor.

programs. The main difference between NCCER training and registered apprenticeship programs is that apprenticeship has specific time limits in which the training must be completed. Apprenticeship standards set guidelines for recruiting and outreach, and a specific time limit for each of a variety of OJL tasks. Additionally, there are reporting requirements and audits to ensure adherence to the apprenticeship standards. Companies and employer associations register their individual apprenticeship programs with the Office of Apprenticeship within the US Department of Labor, and in some instances, with a state apprenticeship council (SAC). The typical heavy equipment operator apprenticeship program is three years.

The traditional apprenticeship program required OJL of 2,000 hours per year and a minimum of 144 hours of classroom-related training. Apprenticeship programs therefore varied in length from 2,000 hours to 10,000 hours. However, on October 29, 2008, the US Department of Labor published new regulations to modernize the National Apprenticeship System. These regulations provide for more flexibility in how related technical instruction (RTI) can be delivered. They also provide registration agencies with the option of issuing interim credentials to offer active apprentices official recognition of their accomplishments and equip them with a portfolio of skills and incentives to complete their programs and continue their career preparation. Finally, these regulations allow program sponsors to offer three different ways for apprentices to complete a registered apprenticeship program:

- The traditional, time-based approach, which requires the apprentice to complete a specific number of OJL and RTI hours.

22101-12_F12.EPS

Figure 12 Asphalt compactors.

- A competency based approach, which requires the apprentice to demonstrate competency in the defined subject areas and requires OJL and RTI.
- A hybrid approach, which requires the apprentice to complete a minimum number of OJL and RTI hours and demonstrate competency in the defined subject areas.

4.2.0 Youth Training and Apprenticeship Programs

Youth apprenticeship programs are available that allow students to begin their apprenticeship or craft training while still in high school. A student entering the program in the eleventh grade may complete as much as one year of the NCCER training program by high school graduation. In addition, programs (in cooperation with local construction industry employers) allow students to work in the craft and earn money while still in school. Upon graduation, students can enter the industry at a higher level and with more pay than someone just starting in a training program.

Students participating in the NCCER or youth apprenticeship training are recognized through official transcripts and can enter the second level or year of the program wherever it is offered. They may also have the option of applying credits at two-year or four-year colleges that offer degree or certificate programs in their selected field of study.

Teaching Tip/Notes

Classroom

List the characteristics needed for success.

Discuss professionalism and the characteristics it includes.

On Site

Equipment Simulators

Equipment simulators, like the one shown here, can be used to train new operators, assess the skills of experienced operators, and screen potential new employees. The simulators duplicate the cab environment and controls. The simulation software that works behind the scenes causes the screen image to respond to the controls in exactly the same way as a real machine would. Trainees using the simulator learn to perform tasks the safest and most efficient way. There are significant savings in terms of machine ownership or rental, cost of fuel, and reduced wear and tear on machines. Simulators are available for most types of heavy equipment.

22101-12_SA04.EPS

Did You Know?

Apprenticeships in the United States

In 2008, more than 500,000 apprentices received registered apprenticeship training in the United States.

5.0.0 Responsibilities of the Employee

In order to be successful, professionals must be able to use current trade materials, tools, and equipment to finish tasks quickly and efficiently. An equipment operator must be adept at adjusting methods to meet each situation. The successful equipment operator must keep abreast of technical advancements and continually gain the skills to use them. A professional never takes chances with regard to personal safety or the safety of others.

5.1.0 Professionalism

The word *professionalism* is a broad term that describes the desired overall behavior and attitude expected in the workplace. Professionalism is too often absent from the construction site and the various trades. Most people would argue that professionalism must start at the top in order to

Teaching Tip/Notes

Ask each trainee to identify personal characteristics that would make them good employees. Also ask them to identify one personality trait that might cause them problems on the job. Ask questions like: Do you have trouble getting up in the morning? Do you have a problem taking direction from a supervisor? How would you react if a supervisor instructed you to do something you disagreed with or thought was wrong?

Classroom

Discuss why honesty and integrity are so important on the job.

Explain how an employee can demonstrate loyalty.

On Site

Child Labor Laws

Federal law establishes the minimum standards for workers under the age of 18. Some municipal jurisdictions may enforce stricter regulations. Employers are required to abide by the laws that apply to them.

The Child Labor Provisions of the Fair Labor Standards Act forbid employers from using illegal child labor, and also forbid companies from doing business with any other business that does. DOL investigates alleged abuses of the law. In such cases, employers have to provide proof of age for their employees.

In addition to the Child Labor Provisions, employers in the construction trades are required to follow DOL's Child Labor Bulletin No. 101, Child Labor Requirements in Nonagricultural Occupations Under the Fair Labor Standards Act. Bulletin No. 101 does the following:

- Explains the coverage of the Child Labor Provisions
- Identifies minimum age standards
- Lists the exemptions from the Child Labor Provisions
- Sets out employment standards for 14- and 15-year-old workers
- Defines the work that can be performed in hazardous occupations
- Provides penalties for violations of the Child Labor Provisions
- Recommends the use of age certificates for employees

be successful. It is true that management support of professionalism is important to its success in the workplace, but it is more important that individuals recognize their own responsibility for professionalism.

Professionalism includes honesty, productivity, safety, civility, cooperation, teamwork, clear and concise communication, being on time and prepared for work, and regard for one's impact on one's co-workers. It can be demonstrated in a variety of ways every minute you are in the workplace. Most importantly, do not tolerate unprofessional behavior from co-workers. This is not to say that you shun the unprofessional worker; instead, you should demonstrate the benefits of professional behavior.

Professionalism is both a benefit to the employer and the employee. It is a personal responsibility. The industry is what each individual chooses to make of it; choose professionalism and the industry image will follow.

5.2.0 Honesty

Honesty and personal integrity are important traits of the successful professional. Professionals pride themselves in performing a job well, and in being punctual and dependable. Each job is completed in a professional way, never by cutting corners or reducing materials. A valued professional maintains work attitudes and ethics that protect property, such as tools and materials belonging to employers, customers, and other trades from damage or theft at the shop or job site.

Honesty and success go hand-in-hand. It is not simply a choice between good and bad, but a choice between success and failure. Dishonesty always catches up with you. Whether you are stealing materials, tools, or equipment from the job site or simply lying about your work, it will not take long for your employer to find out. Of course, you can always go and find another employer, but this option will ultimately run out on you.

If you plan to be successful and enjoy continuous employment, consistency of earnings, and being sought after, as opposed to seeking employment, then start out with the basic understanding of honesty in the workplace and you will reap the benefits.

Honesty means more, however, than just not taking things that do not belong to you. It means giving a fair day's work for a fair day's pay. It means that your words convey true meanings and actual happenings. Thoughts as well as actions should be honest. Employers place a high value on an employee who is strictly honest.

5.3.0 Loyalty

Employees expect employers to look out for their interests, to provide them with steady employment, and to promote them to better jobs as openings occur. Employers feel that they, too, have a right to expect their employees to be loyal to them—to keep company interests in mind, to speak well of it to others, to keep any minor troubles strictly within the plant or office, and to keep absolutely confidential all matters that pertain

Teaching Tip/Notes

to the business. Both employers and employees should keep in mind that loyalty is not something to be demanded; rather, it is something to be earned.

5.4.0 Willingness to Learn

Every office and plant has its own way of doing things. Employers expect their workers to be willing to learn these ways. Adapting to change and being willing to learn new methods and procedures as quickly as possible is key. Sometimes, a change in safety regulations or the purchase of new equipment makes it necessary for even experienced employees to learn new methods and operations. Employees often resent having to accept improvements because of the retraining that is involved. However, employers no doubt think they have a right to expect employees to put forth the necessary effort. Methods must be kept up to date in order to meet competition and show a profit. It is this profit that enables the owner to continue in business and provide jobs for the employees.

5.5.0 Willingness to Take Responsibility

Most employers expect their employees to see what needs to be done, then go ahead and do it. It is very tiresome to have to ask again and again that a certain job be done. It is obvious that having been asked once, an employee should assume the responsibility from then on. Once the responsibility has been delegated, the employee should continue to perform the duties without further direction. Every employee has the responsibility for working safely.

5.6.0 Willingness to Cooperate

To cooperate means to work together. In the modern business world, cooperation is the key to getting things done. Learn to work as a member of a team with your employer, supervisor, and fellow workers in a common effort to get the work done efficiently, safely, and on time.

5.7.0 Rules and Regulations

People can work well together only if there is some understanding about what work is to be done, when and how it will be done, and who will do it. Rules and regulations are a necessity in any work situation and should be so considered by all employees.

5.8.0 Tardiness and Absenteeism

Tardiness means being late for work, and absenteeism means being off the job for one reason or another. Consistent tardiness and frequent absences are an indication of poor work habits, unprofessional conduct, and a lack of commitment.

Humans are all creatures of habit. The habit of always being late may have begun during school days when it was hard to get up in the morning. This habit can get us into trouble at school, and it can continue getting us into trouble when we are through with school and go to work.

Work life is governed by the clock. People are required to be at work at a definite time. Failure to get to work on time results in confusion, lost time, and resentment on the part of those who do come on time. In addition, it may lead to penalties, including dismissal. Although it may be true that a few minutes out of a day are not very important, you must remember that a principle is involved. It is your obligation to be at work at the time indicated.

The terms of work are agreed upon when a job is accepted. Perhaps it will help you to see things more clearly if you try to look at the matter from the point of view of the boss. Supervisors cannot keep track of people if they come in any time they please. It is not fair to others to ignore tardiness. Failure to be on time may hold up the work of fellow workers. Better planning of your morning routine will often keep you from being delayed and so prevent a breathless, late arrival. In fact, arriving a little early indicates your interest and enthusiasm for your work, which is appreciated by employers. The habit of being late is another one of those things that can stand in the way of promotion.

It is sometimes necessary to take time off from work. No one should be expected to work when sick or when there is serious trouble at home. However, it is possible to get into the habit of letting unimportant and unnecessary matters keep us from the job. This results in lost production and hardship on those who try to carry on the work with less help. Again, there is a principle involved. Employers have a right to expect you to be on the job unless there is some very good reason for staying away. Certainly, some trivial reason should not keep you at home. You should not stay up nights until you are too tired to go to work the next day. If you are ill, you should use the time at home to do all you can to recover quickly. This, after all, is no more than most employers would expect of a person depended upon to do a certain job.

Classroom

Explain what being willing to learn says about an employee.

Explain why willingness to take responsibility is important.

Discuss why being cooperative about the job and as a team member matters.

Discuss the repercussions of tardiness and failing to notify the employer of a problem.

Teaching Tip/Notes

Classroom

Discuss different aspects of positive human relations.

On Site

Ethical Principles for Members of the Construction Trades

- *Honesty* – Be honest and truthful in all dealings. Conduct business according to the highest professional standards. Faithfully fulfill all contracts and commitments. Do not deliberately mislead or deceive others.
- *Commitment to safety* – Focus on working safely to protect yourself and your co-workers from harm. Make sure the work environment is as safe as you can make it.
- *Integrity* – Demonstrate personal integrity and the courage of your convictions by doing what is right even where there is pressure to do otherwise. Do not sacrifice your principles because it seems easier.
- *Loyalty* – Be worthy of trust. Demonstrate fidelity and loyalty to companies, employers and sponsors, co-workers, and trade institutions and organizations.
- *Fairness* – Be fair and just in all dealings. Do not take undue advantage of another's mistakes or difficulties. Fair people are open-minded and committed to justice, equal treatment of individuals, and tolerance for and acceptance of diversity.
- *Respect for others* – Be courteous and treat all people with equal respect and dignity.
- *Obedience* : Abide by laws, rules, and regulations relating to all personal and business activities.
- *Commitment to excellence* – Pursue excellence in performing your duties, be well-informed and prepared, and constantly try to increase your proficiency by gaining new skills and knowledge.
- *Leadership* – By your own conduct, seek to be a positive role model for others.

If it is necessary to stay home, then at least phone the office early in the morning so that the boss can find another worker for the day. Time and again, employees have remained home without sending any word to the employer. This is the worst possible way to handle the matter. It leaves those at work uncertain about what to expect. They have no way of knowing whether you have merely been held up and will be in later, or whether immediate steps should be taken to assign your work to someone else. Courtesy alone demands that you let the boss know if you cannot come to work.

The most frequent causes of absenteeism are illness, death in the family, accidents, personal business, and dissatisfaction with the job. Some of the causes are legitimate and unavoidable, while others can be controlled. One can usually plan to carry on most personal business affairs after working hours. Frequent absences reflect unfavorably on a worker when promotions are being considered.

Employers sometimes resort to docking pay, demotion, and even dismissal in an effort to control tardiness and absenteeism. No employer likes to impose restrictions of this kind. However, in fairness to those workers who do come on time and who do not stay away from the job, an employer is sometimes forced to discipline those who do not follow the rules.

6.0.0 HUMAN RELATIONS

Most people underestimate the importance of working well with others. There is a tendency to pass off human relations as nothing more than common sense. What exactly is involved in human relations? Part of human relations is being friendly, pleasant, courteous, cooperative, adaptable, and sociable.

6.1.0 Making Human Relations Work

As important as the previously noted characteristics are for personal success, they are not enough. Human relations is much more than just getting people to like you. It is also knowing how to handle difficult situations as they arise.

Human relations is knowing how to work with supervisors who are often demanding and sometimes seem unfair. It involves understanding your own personality traits and those of your co-workers. Building sound working relationships in various situations is important. If working relationships have deteriorated for one reason or another, restoring them is essential. Human relations is learning how to handle frustration without hurting others.

Teaching Tip/Notes

6.2.0 Human Relations and Productivity

Effective human relations is directly related to productivity. Productivity is the key to business success. Every employee is expected to produce at a certain level. Employers quickly lose interest in an employee who has a great attitude but is able to produce very little. There are work schedules to be met and jobs that must be completed.

All employees, both new and experienced, are measured by the amount of quality work they can safely turn out. The employer expects every employee to do their share of the workload.

However, doing one's share in itself is not enough. If you are to be productive, you must do your share (or more than your share) without antagonizing your fellow workers. You must perform your duties in a manner that encourages others to follow your example. It makes little difference how ambitious you are or how capably you perform. You cannot become the kind of employee you want to be or the type of worker management wants you to be without learning how to work with your peers.

Employees must do everything they can to build strong, professional working relationships with fellow employees, supervisors, and clients.

6.3.0 Attitude

A positive attitude is essential to a successful career. First, being positive means being energetic, highly motivated, attentive, and alert. A positive attitude is essential to safety on the job. Second, a positive employee contributes to the productivity of others. Both negative and positive attitudes are transmitted to others on the job. A persistent negative attitude can spoil the positive attitudes of others. It is very difficult to maintain a high level of productivity while working next to a person with a negative attitude. Third, people favor a person who is positive. Being positive makes a person's job more interesting and exciting. Fourth, the kind of attitude transmitted to management has a great deal to do with an employee's future success in the company. Supervisors can determine a subordinate's attitude by their approach to the job, reactions to directives, and the way they handle problems.

6.4.0 Maintaining a Positive Attitude

A positive attitude is far more than a smile, which is only one example of an inner positive attitude. As a matter of fact, some people transmit a positive attitude even though they seldom smile. They do this by the way they treat others, the way they look at their responsibilities, and the approach they take when faced with problems. Here are a few suggestions to help you to maintain a positive attitude:

- Remember that your attitude follows you wherever you go. If someone makes a greater effort to be a more positive person in their social and personal lives, it automatically helps them on the job. The reverse is also true. One effort will complement the other.
- Negative comments are seldom welcomed by fellow workers on the job. Neither are they welcome on the social scene. The solution is to talk about positive things and be complimentary. Constant complainers do not build healthy and fulfilling relationships.
- Look for the good things in people on the job, especially your supervisor. Nobody is perfect, and almost everyone has a few worthwhile qualities. If you dwell on people's good features, it will be easier to work with them.
- Look for the good things where you work. What are the factors that make it a good place to work? Is it the hours, the physical environment, the people, or the actual work being done? Or is it the atmosphere? Keep in mind that you cannot expect to like everything. No work assignment is perfect. But if you concentrate on the good things, the negative factors will seem less important and bothersome.
- Look for the good things in the company. Just as there are no perfect assignments, there are no perfect companies. Nevertheless, almost all organizations have good features. Is the company progressive? What about promotional opportunities? Are there chances for self-improvement? What about the wage and benefit package? Is there a good training program? You cannot expect to have everything you would like, but there should be enough to keep you positive. In fact, if you decide to stick with a company for a long period of time, it is wise to look at the good features and think about them. If you think positively, you will act the same way.
- You may not be able to change the negative attitude of another employee, but you can protect your own attitude from becoming negative.

7.0.0 Employer and Employee Safety Obligations

An obligation is like a promise or a contract. In exchange for the benefits of your employment and your own well-being, you agree to work

Classroom

Describe how productivity is connected to human relations.

Ask the trainees about their reactions to a workmate with a positive attitude.

Ask them how they feel around someone who has a negative view of things.

Discuss the suggestions for attitude improvement.

Explain employer's and employees' obligations regarding safety.

Teaching Tip/Notes

Classroom

Point out OSHA's purpose and role in workplace safety.

Explain what an employee is entitled to under OSHA.

List OSHA's most important general requirements.

On Site

Tips for a Positive Attitude

Here is a short checklist of things that you should keep in mind to help you develop and maintain a positive attitude:

- Remember that your attitude follows you wherever you go.
- Helpful suggestions and compliments are much more effective than negative ones.
- Look for the positive characteristics of your teammates and supervisors.

safely. In other words, you are obligated to work safely. You are also obligated to make sure anyone you happen to supervise or work with is working safely. Your employer is obligated to maintain a safe workplace for all employees. Safety is everyone's responsibility.

Some employers have safety committees. If you work for such an employer, you are then obligated to that committee to maintain a safe working environment. This means that you should follow the safety committee's rules for proper working procedures and practices, and report any unsafe equipment and conditions directly to the committee or your supervisor.

On the job, if you see something that is not safe, report it! Do not ignore it. It will not correct itself. You have an obligation to report it.

In the long run, even if you do not think an unsafe condition affects you, it does. Do not mess around; always report unsafe conditions. Do not think your employer will be angry because your productivity suffers while the condition is corrected. On the contrary, your employer will be more likely to criticize you for not reporting a problem.

Your employer knows that the short time lost in making conditions safe again is nothing compared with shutting down the whole job because of a major disaster. If disaster happens, you are out of work anyway. So do not ignore an unsafe condition.

7.1.0 OSHA

In 1970, the US Congress passed the Occupational Safety and Health Act. This act also created the **Occupational Safety and Health Administration (OSHA)**. OSHA is part of the US Department of Labor. The job of OSHA is to set occupational safety and health standards for all places of employment, enforce these standards, ensure that employers provide and maintain a safe workplace for all employees, and provide research and educational programs to support safe working practices. This applies to every part of the construction industry. Whether you work for a large contractor or a small subcontractor, you are obligated to report unsafe conditions. The easiest way to do this is to tell your supervisor. If that person ignores the unsafe condition, report it to the next highest supervisor. If it is the owner who is being unsafe, let that person know your concerns. If nothing is done about it, report it to OSHA. If you are worried about your job being on the line, think about it in terms of your life, or someone else's, being at risk.

OSHA was adopted with the stated purpose "to assure as far as possible every working man and woman in the nation safe and healthful working conditions and to preserve our human resources." OSHA requires each employer to provide a safe and hazard-free working environment. OSHA also requires that employees comply with OSHA rules and regulations that relate to their conduct on the job. To gain compliance, OSHA can perform spot inspections of job sites, impose fines for violations, and even stop any more work from proceeding until the job site is safe.

According to OSHA standards, you are entitled to on-the-job safety training. As a new employee, you are entitled to the following:

- Being shown how to do your job safely
- Being provided with the required personal protective equipment
- Being warned about specific hazards
- Being supervised for safety while performing the work

The enforcement of this act of Congress is provided by the federal and state safety inspectors, who have the legal authority to make employers pay fines for safety violations. The law allows states to have their own safety regulations and agencies to enforce them, but they must first be approved by the US Secretary of Labor. For states that do not develop such regulations and agencies, federal OSHA standards must be obeyed.

These standards are listed in *OSHA Safety and Health Standards for the Construction Industry (29 CFR, Part 1926)*, sometimes called *OSHA Standards*

Teaching Tip/Notes

1926. Other safety standards that apply to construction are published in *OSHA Safety and Health Standards for General Industry (29 CFR, Parts 1900 to 1910)*.

The most important general requirements that OSHA places on employers in the construction industry are as follows:

- The employer must perform frequent and regular job site inspections of equipment.
- The employer must instruct all employees to recognize and avoid unsafe conditions, and to know the regulations that pertain to the job so they may control or eliminate any hazards.
- No one may use any tools, equipment, machines, or materials that do not comply with *OSHA Standards 1926*.
- The employer must ensure that only qualified individuals operate tools, equipment, and machines.

7.2.0 MSHA

As mentioned earlier, heavy equipment operators may be operating machinery in mining environments, both above ground and under Earth's surface. In addition to the OSHA safety guidelines, there are safety guidelines set forth by the **Mine Safety and Health Administration (MSHA),** which also falls under the US Department of Labor. In 1977, a new Mine Safety and Health Act was created. The following year (1978), MSHA was put in place. During 1977, 242 miners died in mining accidents. As a result of the stricter MSHA safety guidelines, mining accidents resulting in death dropped by more than 50 percent over the next 10 years. The mining deaths continued to drop over the next 20-plus years to a point where there were a record low 34 fatalities reported in 2009.

MSHA breaks mining into two major groups: coal mining and mining of metal and nonmetal resources. Anyone expecting to operate heavy equipment in mining operations can expect to receive additional training related only to mining safety. This training must be site-specific and task-oriented, and must be delivered by a certified mine instructor. The MSHA web site (www.msha.gov) has numerous links related to mining safety. This information should be reviewed by heavy equipment operators going into any mining operations.

7.3.0 Safety in General

The equipment you will be operating is expensive, so one of your major responsibilities as an operator will be to take good care of the equipment. Never overload it, or perform tasks for which it was not intended. All heavy equipment manuals have load charts that explain both how much the machine can carry, and how it must be operated when loaded. Always perform the required checks and inspections prior to, during, and after operations. Above all, keep in mind that the equipment, because of its size, represents a potential safety hazard to you and other workers on the site. Always operate the equipment in a safe manner and stay within the operating guidelines provided by the manufacturer.

Summary

A variety of heavy equipment is used on a construction site, especially in the site development phase of a project. In many instances, demolition of existing structures must be done using heavy equipment before new construction can begin. In addition to construction work, heavy equipment is used extensively in mining and road construction work.

Heavy equipment operators have many opportunities for career advancement and personal growth. If they choose to do so, they can advance into supervisory and management positions within a company and even start their own business using the skills they develop as heavy equipment operators.

Construction work is a team effort in which every person must do their job on time and in a professional manner. This means that all members of a construction crew must learn to work together and must develop interpersonal skills and good work habits.

By its very nature, heavy equipment is potentially hazardous to the operator and others working around the equipment. This means that everyone must be particularly safety conscious when heavy equipment is being used on a job site.

Classroom

Explain what brought MSHA into being.

Define the two groups under MSHA.

Stress the need to review mining safety before beginning any mining operations.

Discuss the operator's responsibility for safely operating equipment and performing periodic maintenance.

Summarize the major concepts presented in this module.

Teaching Tip/Notes

Classroom

Have the trainees complete the Review Questions, and go over the answers prior to administering the Module Examination.

Review Questions

1. Which of the following items is likely to be used for clearing land?
   a. Bulldozer
   b. Forklift
   c. Motor grader
   d. Compactor

22101-12_RQ01.EPS

Figure 1

2. The equipment shown in *Figure 1* is a(n) _____.
   a. backhoe loader
   b. skid steer
   c. motor grader
   d. excavator

22101-12_RQ02.EPS

Figure 2

3. The heavy equipment item shown in *Figure 2* is a(n) _____.
   a. excavator
   b. motor grader
   c. off-road truck
   d. underground mining hauler

4. Demolition of a building is commonly done with a(n) _____.
   a. motor grader
   b. compactor
   c. excavator
   d. forklift

18 NCCER – *Heavy Equipment Operations Level One* 22101-12

Teaching Tip/Notes

22101-12_RQ03.EPS

Figure 3

5. The heavy equipment item shown in *Figure 3* is a(n) _____.
   a. off-road truck
   b. motor grader
   c. excavator
   d. scraper

6. Of almost 470,000 construction equipment operator jobs in 2008, approximately how many were operating engineers and other construction equipment operators?
   a. 1,600
   b. 6,500
   c. 352,000
   d. 404,000

7. Among heavy equipment operators, approximately how many work in construction?
   a. One out of three
   b. Two out of five
   c. Three out of five
   d. Five out of ten

8. To standardize a company's training programs and to document the training each trainee receives, most modern companies start their trainees out _____.
   a. on interactive programs run on a computer
   b. as an assistant to a senior operator
   c. as part-time operators
   d. as maintenance helpers

9. An experienced qualified heavy equipment operator may have the opportunity to train others within their own company.
   a. True
   b. False

10. If one of your co-workers does not act in a professional manner, the best approach is to _____.
    a. refuse to cooperate with the person
    b. report the person to your supervisor
    c. advise other workers to shun the person
    d. demonstrate the benefits of professional behavior

Teaching Tip/Notes

22101-12 Orientation to the Trade Module One 19

Classroom

Have the trainees complete the Trade Terms Quiz, and go over the answers prior to administering the Module Examination.

? Examination

Administer the Module Examination. Record the results on Training Report Form 200, and submit the results to the Training Program Sponsor.

Trade Terms Quiz

Fill in the blank with the correct term that you learned from your study of this module.

1. ________ is the federal government agency established to ensure a safe and healthy environment in the workplace.
2. ________ is the destruction and removal of a structure.
3. Job-related learning acquired while working is known as ________.
4. is the federal government agency established to ensure the safety of mining operations.
5. A system of public works, such as highways, bridges, and dams, is called ________.

Trade Terms

Demolition
Infrastructure
On-the-job learning (OJL)
Occupational Safety and Health Administration (OSHA)
Mine Safety and Health Administration (MSHA)

Teaching Tip/Notes

Cornerstone of Craftsmanship

Roger Arnett

Machine Systems and Operation Training Consultant
Caterpillar Tractor Inc.

Roger Arnett tried a lot of different things before winding up as a training consultant for Caterpillar. His desire to learn and his willingness to try different things, combined with a good foundation of education and training, led him to a rewarding career.

How did you choose a career in the heavy equipment field?
I think it chose me. I started out in farming, but then moved into surface mining. I also received a degree in Education and Technology. After 10 years of mining experience, including machine repair and operation, I was fortunate to be hired by Caterpillar.

Who inspired you to enter the industry?
My inspiration came from family and friends, combined with a love of working outdoors and working with equipment.

What types of training have you been through?
I have been through and received certifications in service training for all machine systems. I have also had operator training in mining and at Caterpillar, along with classroom certifications in instruction, material development, and manufacturing technology for both wood and metals.

How important is education and training in construction?
The quality of work heavy equipment operators can supply to employers is directly proportional to the amount of good training they have received. The knowledge they have gained through experience and running equipment certainly helps, but formal training can greatly speed up this process. Operating experience is valuable if, and only if, the correct skills have been learned and practiced and the operator has been trained to operate equipment safely. Today, new operators who have been professionally trained with NCCER and Caterpillar formal training can work for their pick of great companies. The other good operators without formal training have a harder challenge today finding employment with a great company.

How important are NCCER credentials to a career?
NCCER credentials will greatly enhance any operator's opportunities to first land a good job and build a great career and then to move upwards towards a supervisory position or a higher position in management within a company.

How has training/construction impacted your life?
It has allowed me to build my career with a top company and future personal growth is still ahead. Without professional training I would still be struggling to learn things "the hard way."

What kinds of work have you done in your career?
I have been a professional homebuilder, union cement mason and laborer, and farmer. I've taught both academic and vocational trades on the secondary and junior college levels. I worked in surface coal mining for 10 years, with both operating and machine repair experience on all machines and their systems. I worked as a service training instructor for eight machine families and their systems. I've conducted operator training and been a curriculum developer, as well as a custom operator training consultant for global machine training. I recently worked in Costa Rica training 180 new machine operators how to operate equipment using simulators and real machines.

Teaching Tip/Notes

Tell us about your present job.
Currently I work for Caterpillar as a machine training consultant to assist customers in setting up their own training programs for operator or service training. The work involves assessing customers' current training programs and materials, assisting them with building their own programs to a higher standard, and standardizing their written curricula so it can assist their operators to work safely and productively, and get the most out of their equipment. I work with the companies to help them decrease their machine downtime by training operators to run their machines correctly. I also help them change the way they are using their equipment at their different sites to improve their production and efficiency, whether it is at a quarry, mine site, pipeline application, or general construction site.

What do you enjoy most about your job?
I enjoy witnessing the learning experience taking place as I instruct students everywhere on heavy equipment. I take pleasure from assisting in making their job safer, easier, and more productive, which in turn increases their personal value to their companies. It is gratifying to see industries increase their fleet health and productivity, while working safely and smartly and improving the overall health of the equipment the operators are running as they learn to maintain and service their equipment.

What factors have contributed most to your success?
I have always been willing to receive training with an open mind and listen to those with experience. I'm not afraid of working hard and learning to work smart at the same time -- in short, a strong work ethic. I have always believed that no matter how great a system is, it can usually be improved in some way, because nothing is perfect.

Would you suggest construction as a career to others? Why?
Construction is a very interesting business, as you get to experience the development of projects from the ground up, starting with nothing but plans and ending with finished work. To see something appear from nothing and to work on large scale construction projects is truly exciting.

What advice would you give to those new to the heavy equipment field?
Get some good foundational training with NCCER materials and at least a two-year associates degree in your field. Have patience and work hard; the world of opportunity is really out there for those who want to work.

Can you share an interesting career-related fact or accomplishment?
Working with the Carlindo dealership and INADEH training center in Panama City, Panama, we developed a training program for 730 new heavy equipment operators who had never even run a lawnmower. Using simulators and real machines, we screened 1,200 applicants to train the 730 new operators in two weeks. Within 5½ months, they were competent enough on HEX, articulated trucks, and track-type tractors to replace the operators in the city so they could move out to start the widening of the Panama Canal. NCCER and Caterpillar materials were both used in this huge project that had never been attempted before.

How do you define craftsmanship?
Craftmanship is combining the basic ability of an individual with their personal values of knowledge, patience, and empathy to produce a product of high quality. Craftsmen must care about their work, their skills, and have pride in their finished product so they produce work of consistent high quality. Most true craftsman have always had good training and work experience as their foundation. They combine this with a great work ethic and caring about their craft. I used to tell my high school shop students to not turn their projects into me for a grade until they were proud enough to put their name on it!

Teaching Tip/Notes

Trade Terms Introduced in This Module

Demolition: The destruction and removal of a structure.

Infrastructure: A system of public works such as highways, bridges, and dams.

Mine Safety and Health Administration (MSHA): A federal government agency established to ensure the safety of mining operations.

On-the-job learning (OJL): Job-related learning acquired while working.

Occupational Safety and Health Administration (OSHA): The federal government agency established to ensure a safe and healthy environment in the workplace.

Teaching Tip/Notes

Appendix

SAMPLES OF NCCER TRAINING CREDENTIALS

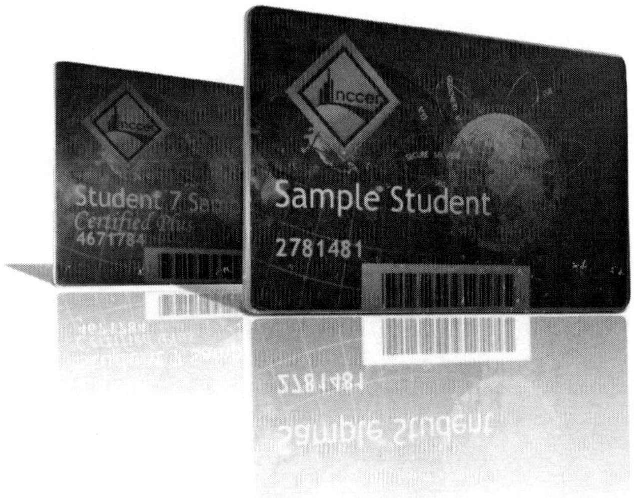

Teaching Tip/Notes

24 NCCER – *Heavy Equipment Operations Level One* 22101-12

NCCER
The Standard for Developing Craft Professionals

This is to certify that

Steven Whitaker

has fulfilled the requirements for

Heavy Equipment Operations Level One

in NCCER's standardized training curriculum
on this Sixteenth day of September, 2012

The Standard for Developing Craft Professionals
Founded 1996

Donald E. Whyte
President, NCCER

Teaching Tip/Notes

22101-12 Orientation to the Trade Module One 25

THE STANDARD FOR DEVELOPING CRAFT PROFESSIONALS

13614 Progress Boulevard, Alachua, Florida 32615 • p. 888.622.3720 f. 386.518.6255 • www.nccer.org

Official Transcript

January 17, 2012

NCCER Card #: 1720726
Trainee Name: John Q Smith
Sponsor: Austin Industrial Incorporated
Address: 2801 E 13th St
La Porte, TX 77571

Current Employer/School:
Solomon Plumbing Company

Module	Description	Instructor	Training Location	Date Completed
00101-04	Basic Safety	Kevin Jenkins	Solomon Plumbing Company	2/20/2008
00102-04	Introduction to Construction Math	Dave Buck	Building Trades Institute, LLC	8/8/2008
00103-04	Introduction to Hand Tools	Kevin Jenkins	Solomon Plumbing Company	1/1/2008
00104-04	Introduction to Power Tools	Dave Buck	Building Trades Institute, LLC	8/8/2008
00105-04	Introduction to Blueprints	Kevin Jenkins	Solomon Plumbing Company	3/20/2008
00106-04	Basic Rigging	Dave Buck	Building Trades Institute, LLC	8/8/2008
00108-04	Basic Employability Skills	Rod Blackburn	Utility Contractors, Inc.	3/15/2009
02101-05	Introduction to the Plumbing Profession	Kevin Jenkins	Solomon Plumbing Company	3/22/2008
26101-02	Electrical Safety	Don Whyte	National Center for Construction Education &	7/29/2002
26102-02	Hand Bending	Don Whyte	National Center for Construction Education &	7/29/2002
26103-02	Fasteners and Anchors	Don Whyte	National Center for Construction Education &	7/29/2002
26104-02	Electrical Theory One	Don Whyte	National Center for Construction Education &	7/29/2002
26105-02	Electrical Theory Two	Don Whyte	National Center for Construction Education &	7/29/2002
26106-02	Electrical Test Equipment	Don Whyte	National Center for Construction Education &	7/29/2002
26107-02	Introduction to the National Electrical Code	Don Whyte	National Center for Construction Education &	7/29/2002
26108-02	Raceways, Boxes, and Fittings	Don Whyte	National Center for Construction Education &	7/29/2002
26109-02	Conductors	Don Whyte	National Center for Construction Education &	7/29/2002

Page 1

Donald E. Whyte
President, NCCER

26 NCCER – *Heavy Equipment Operations Level One* 22101-12

Teaching Tip/Notes

Additional Resources

This module presents thorough resources for task training. The following resource material is suggested for further study.

The Earthmover Encyclopedia, Latest Edition. Minneapolis, MN: Quayside Publishing Group.

Figure Credits

Reprinted courtesy of Caterpillar Inc., Module opener, Figures 1 (B, I, K), 2, 3, 5–7, 9, 10 (top and middle photos), 11, 12, and SA04

Deere & Company, Figure 1 (A, C–F), 4, and 8

BOMAG Americas, Inc., Figure 1 (G)

Sellick Equipment Limited, Figure 1 (H)

CNH America LLC, Figure 1 (J)

Topaz Publications, Inc., SA01 and SA02

Stanley LaBounty, SA03

Mark Jones, Figure 10 (bottom photo)

Teaching Tip/Notes

ANSWERS TO REVIEW QUESTIONS

Answer		Section Reference
1.	a	1.0.0
2.	b	1.0.0; Figure 1
3.	d	2.0.0; Figure 2
4.	c	2.0.0; Figure 9
5.	d	2.0.0; Figure 10
6.	d	3.0.0
7.	c	3.0.0
8.	a	3.1.0
9.	a	3.2.0
10.	d	5.1.0

ANSWERS TO TRADE TERMS QUIZ

1. Occupational Safety and Health Administration (OSHA)
2. Demolition
3. On-the-job learning (OJL)
4. Mine Safety and Health Administration (MSHA)
5. Infrastructure

NCCER CURRICULA — USER UPDATE

NCCER makes every effort to keep its textbooks up-to-date and free of technical errors. We appreciate your help in this process. If you find an error, a typographical mistake, or an inaccuracy in NCCER's curricula, please fill out this form (or a photocopy), or complete the online form at **www.nccer.org/olf**. Be sure to include the exact module ID number, page number, a detailed description, and your recommended correction. Your input will be brought to the attention of the Authoring Team. Thank you for your assistance.

Instructors – If you have an idea for improving this textbook, or have found that additional materials were necessary to teach this module effectively, please let us know so that we may present your suggestions to the Authoring Team.

NCCER Product Development and Revision

13614 Progress Blvd., Alachua, FL 32615

Email: curriculum@nccer.org

Online: www.nccer.org/olf

❑ Trainee Guide ❑ AIG ❑ Exam ❑ PowerPoints Other ____________

Craft / Level: Copyright Date:

Module ID Number / Title:

Section Number(s):

Description:

Recommended Correction:

Your Name:

Address:

Email: Phone:

22102-12

Heavy Equipment Safety

Module Two

Materials

reminds you to gather materials needed for classes, labs, and testing.

Homework

suggests that you assign reading for the next session.

Classroom

prompts you to discuss a topic, stress a point, explain a concept, or provide examples.

Audiovisual

indicates an appropriate time to show a slide or transparency.

! Safety

used to emphasize safety issues.

Demonstration

directs you to show trainees how to perform tasks.

Laboratory

used when trainees are to practice performing tasks.

✓ Performance Testing

recommends administering a performance test, or a portion thereof.

PT / Laboratory

indicates that lab practice corresponds with a performance task. If desired, you can note the proficiency of the trainees during the lab to satisfy performance testing requirements.

Teaching Tip

indicates tips for practice exercises, Quick Quizzes, field trips, or other activities designed to encourage class participation.

? Examination

recommends administering the written module exam.

NCCER Standardized Craft Training Program

NCCER provides a standardized national program of accredited craft training. Key features of the program include instructor certification, competency-based training, and performance testing. The program provides trainees, instructors, and companies with a standard form of recognition through the National Registry. The program is described in full in the *Guidelines for Accreditation*, published by NCCER. For more information on standardized craft training, contact NCCER by writing to 13614 Progress Blvd., Alachua FL 32615; calling 1-888-NCCER20; or emailing **info@nccer.org**. More information is available at **www.nccer.org**.

How to Use this Annotated Instructor's Guide

Each page presents two sections of information. The larger section displays each page exactly as it appears in the Trainee Module. The narrow column ties suggested trainee and instructor actions to each page and provides icons (detailed on the left) to call your attention to material, safety, audiovisual, or testing requirements. The bottom of each page includes space for your notes, or when available, teaching tips that suggest activities or class exercises that will help your students learn the material.

Preparation

Before teaching this module, you should review the Objectives, Performance Tasks, Materials and Equipment List, and Module Outline. Be sure to allow ample time to prepare your own training or lesson plan and gather all required materials and equipment.

Instructors purchasing a single module receive the printed exam and performance tasks with the AIG module. Perfect-bound versions of this title no longer include the printed exam booklet and performance task sheets. Log on to the Instructor Resource Center at **www.NCCERIRC.com** to download these materials. Your access code is on the card bound into the front of this AIG. This code will allow you to access exams and performance tasks for all levels of this craft. Be sure to keep all testing materials in a secure place at all times.

Heavy Equipment Safety
Annotated Instructor's Guide

Module 22102-12

Module Overview

This module introduces the trainee to working around heavy equipment in a safe and responsible manner. The trainee will learn how to use personal protective equipment, set up barricades and barriers, and use flags and paddles to control traffic. This module also covers trenching and excavation safety precautions.

Objectives

Upon completion of this module, the trainee will be able to do the following:

1. Explain the importance of safety when working with heavy equipment.
2. State the purposes of signs, tags, barricades, and lockout/tagout devices used on construction sites.
3. Describe the long- and short-term health effects, first-aid measures, handling and storage, and/or required personal protective equipment (PPE) for a chemical using a material data safety sheet (MSDS).
4. Identify safeguards used in a highway construction work zone.
5. State general guidelines for safe operation, maintenance, and transportation of heavy equipment.
6. Explain the dangers of working around an excavation area with heavy equipment.

Performance Tasks

Under the supervision of the instructor, the trainee should be able to do the following:

1. Demonstrate how to use various types of personal protective equipment (PPE):
   - Hard hat
   - Safety glasses
   - Ear protection
   - Gloves
   - Safety harness
   - Reflective safety vest
2. Place barricades and temporary traffic control devices for a highway construction zone.
3. Demonstrate how to use flags or paddles to control traffic.

Materials and Equipment

Markers/chalk
Pencils and scratch paper
Whiteboard/chalkboard
Heavy Equipment Operations Level One PowerPoint® Presentation Slides (ISBN 978-0-13-292167-1)
Multimedia projector and screen
Computer
Appropriate personal protective equipment
Copy of the MUTCD
Several different respirator cartridges
Barricades
Temporary traffic control devices
Flags and paddles
Personal protective equipment (PPE), including:
- Hard hat
- Safety glasses
- Ear protection
- Gloves
- Safety harness

Samples of various soil types
Trade Terms Quiz*
Module Examinations**
Performance Profile Sheets**

* Located at the back of the Trainee Guide module

** Single-module AIG purchases include the printed exam and performance task sheet. If you have purchased the perfect-bound version of this title, download these materials from the IRC using your access code.

Safety Considerations

Ensure that the trainees are equipped with appropriate personal protective equipment and know how to use it properly. Review safety guidelines associated with working on or around heavy equipment. Emphasize the importance of proper housekeeping.

Additional Resources

This module presents thorough resources for task training. The following resource material is suggested for further study.

Construction Safety. 1996. Jimmie Hinze. Englewood Cliffs, NJ: Prentice Hall.

Construction Safety Council, www.buildsafe.org.

Field Safety Participant Guide, 2003. NCCER. Upper Saddle River, NJ: Prentice Hall.

Handbook of OSHA Construction Safety and Health, Second Edition. Charles D. Reece and James V. Eidson: CRC Press.

HazCom for Construction. DVD. DuPont Sustainable Solutions – Training Solutions. Virginia Beach, VA.

Manual on Uniform Traffic Control Devices for Streets and Highways. Latest edition. Washington, DC: U.S. Department of Transportation, Federal Highway Administration

MSHA excavation guidelines, www.usbr.gov.

NAHB-OSHA Jobsite Safety Handbook, 1999. Washington, DC: Home Builder Press. Available online at www.osha.gov.

OSHA Standards 29 CFR, Part 1926 Safety and Health Regulations for the Construction Industry. Washington, DC: Occupational Safety and Health Administration (OSHA), US Department of Labor, US Government Printing Office.

Safety Orientation Pocket Guide, 2004. NCCER. Upper Saddle River, NJ: Prentice Hall.

Safety Technology Participant Guide, 2003. NCCER. Upper Saddle River, NJ: Prentice Hall.

The National Institute for Occupational Safety and Health (NOISH) Centers for Disease Control and Prevention. Atlanta, GA. Available online at www.cdc.gov.

United States Department of Labor, Occupational Safety and Health Administration Home Page, www.osha.gov.

Teaching Time for This Module

An outline for use in developing your lesson plan is presented below. Note that each Roman numeral in the outline equates to one session of instruction. Each session has a suggested time period of 2½ hours. This includes 10 minutes at the beginning of each session for administrative tasks and one 10-minute break during the session. Approximately 10 hours are suggested to cover *Heavy Equipment Safety*. You will need to adjust the time required for hands-on activity and testing based on your class size and resources. Because laboratories often correspond to Performance Tasks, the proficiency of the trainees may be noted during these exercises for Performance Testing purposes.

Topic	Planned Time
Session I. Introduction; Work Zone Safety; Hazard Communication	
A. Introduction	____________
B. Work Zone Safety	____________
1. Tools for Safety	____________
2. Traffic Control	____________
C. PT/Laboratory	____________
1. Have the trainees place barricades and temporary traffic control devices for a highway construction zone. This laboratory corresponds to Performance Task 2.	
2. Have the trainees demonstrate how to use flags or paddles to control traffic. This laboratory corresponds to Performance Task 3.	
D. Hazard Communication	____________
1. Material Safety Data Sheets	____________
2. Noise	____________

Session II. Heavy Equipment Safety

A. Heavy Equipment Safety
   1. Job-Site Safety
   2. Equipment Safety
   3. Personal Safety
   4. PT/Laboratory

      Have the trainees demonstrate how to use various types of personal protective equipment (PPE). This laboratory corresponds to Performance Task 1.
   5. Weather Hazards
   6. Moving Equipment Safely
   7. Heavy Equipment Maintenance

Session III. Trenching and Excavation Safety

A. Trenching and Excavation Safety
   1. Excavation Planning
   2. Trenching Hazards
   3. Trench Failure
   4. Soil Hazards
   5. Guidelines for Working Near a Trench
   6. Making the Trench Safe
   7. Personal Safety

Session IV. Review and Testing

A. Review

B. Module Examination
   1. Trainees must score 70% or higher to receive recognition from NCCER.
   2. Record the testing results on Training Report Form 200 and submit the results to the Training Program Sponsor.

C. Performance Testing
   1. Trainees must perform each task to the satisfaction of the instructor to receive recognition from NCCER. If applicable, proficiency noted during laboratory exercises can be used to satisfy the Performance Testing requirements.
   2. Record the testing results on Training Report Form 200 and submit the results to the Training Program Sponsor.

Homework

Assign the reading of this module.

Teaching Tip/Notes

Trainees with successful module completions may be eligible for credentialing through NCCER's National Registry. To learn more, go to **www.nccer.org** or contact us at **1.888.622.3720**. Our website has information on the latest product releases and training, as well as online versions of our *Cornerstone* newsletter and Pearson's product catalog.

Your feedback is welcome. You may email your comments to **curriculum@nccer.org**, send general comments and inquiries to **info@nccer.org**, or fill in the User Update form at the back of this module.

V.1 4/12

Teaching Tip/Notes

22102-12

Heavy Equipment Safety

Objectives

When you have completed this module, you will be able to do the following:

1. Explain the importance of safety when working with heavy equipment.
2. State the purposes of signs, tags, barricades, and lockout/tagout devices used on construction sites.
3. Describe the long- and short-term health effects, first-aid measures, handling and storage, and/or required personal protective equipment (PPE) for a chemical using a material data safety sheet (MSDS).
4. Identify safeguards used in a highway construction work zone.
5. State general guidelines for safe operation, maintenance, and transportation of heavy equipment.
6. Explain the dangers of working around an excavation area with heavy equipment.

Performance Tasks

Under the supervision of your instructor, you should be able to do the following:

1. Demonstrate how to use various types of personal protections equipment (PPE):
   - Hard hat
   - Safety glasses
   - Ear protection
   - Gloves
   - Safety harness
   - Reflective safety vest
2. Place barricades and temporary traffic control devices for a highway construction zone.
3. Demonstrate how to use flags or paddles to control traffic.

Trade Terms

Center of gravity
Channeling device
Flagger
Pinch points
Temporary traffic control (TTC)

Industry Recognized Credentials

If you're training through an NCCER-accredited sponsor you may be eligible for credentials from NCCER's Registry. The ID number for this module is 22102-12. Note that this module may have been used in other NCCER curricula and may apply to other level completions. Contact NCCER's Registry at 888.622.3720 or go to nccer.org for more information.

Materials

Ensure that you have everything required to teach the course. Check the Materials and Equipment list at the front of this module.

Classroom

Explain that terms shown in bold (blue) are defined in the Glossary at the back of this module.

Audiovisual

Show Slide 1, Objectives, and Slide 2, Performance Tasks. Review the goals of the module, and explain what will be expected of the trainees.

Teaching Tip/Notes

When you call on someone to answer a question, the rest of the class relaxes or even tunes out because they expect that the question and answer will take place only between you and the trainee you called on. Instead, use this technique to involve more trainees in answering questions and to keep them on their toes:

1. Ask trainees to define a term or explain a concept.
2. After one trainee has answered, ask a trainee seated nearby if the answer is right. Then ask whether a trainee in the back of the room agrees.
3. Ask trainees to explain why they think an answer is right or wrong.
4. Use the session to clear up incorrect ideas and encourage trainees to learn from their mistakes.

Contents

Topics to be presented in this module include:

Teaching Tip/Notes

Teaching Tip/Notes

Figures and Tables

Teaching Tip/Notes

1.0.0 Introduction

Working in and around heavy equipment is hazardous. Heavy equipment operators and other workers need to work together to keep the job site safe. The heavy equipment operator's job is to operate the equipment in a manner that protects both the operator and other workers. When working around heavy equipment, it is your job to act in a way that does not place you or your co-workers in danger of an accident.

This module covers some of the dangers about working around heavy equipment. Because heavy equipment work is hazardous, there are a number of specific rules and procedures put in place to decrease the chances for an accident. This module provides basic information about how to stay safe around heavy equipment.

> **Did You Know?**
>
> **Safety Training**
>
> The general principle is that unsafe and unhealthful working conditions on mine sites fall under the Mine Safety and Health Administration (MSHA), and all else comes under the authority of the Occupational Safety and Health Administration (OSHA). Except for a few industry-specific requirements, the regulations are the same. Safety training is required for all activities. Never operate tools, machinery, or equipment without prior training.

2.0.0 Work Zone Safety

One of the ways accidents can be avoided is the use of various safety devices, such as signs, safety tags, barricades and barriers, and other devices. When used properly, these devices can not only prevent accidents but can also assist in the smooth flow of work and permit the job to be completed on schedule.

Construction work zones are often located near public areas and sometimes even on public roads (*Figure 1*). This creates the additional danger of pedestrian and motor traffic in the surrounding area. Work on public roads is so dangerous that every state in the United States has enacted laws that allow added fines and penalties for motorists who speed and commit other traffic violations in roadway construction zones. Specific standard procedures must be used without fail in these types of work zones to ensure everyone's safety.

Safety devices, policies, and procedures are just tools. They cannot prevent accidents unless used correctly. You are the best tool for preventing accidents. Stay alert for information and conditions that can affect everyone's safety (*Figure 2*). Do not rely on someone else to keep you safe. Think a job through before starting it. Look around for potential hazards and stay alert for changes in the work area. Remember the following:

- Accidents can kill and disable.
- Accidents cost you and your company money.
- Accidents can be prevented.

22102-12_F01.EPS

Figure 1 Traffic control.

Classroom

Point out that working with heavy equipment is hazardous, and it is the operator's job to operate in a safe and responsible manner.

Explain how signs, tags, and other safety devices can help reduce accidents.

Audiovisual

Show Slides 3 and 4 (Figures 1 and 2). Stress the need to be aware of conditions and situations that may affect safety.

Teaching Tip/Notes

22102-12 Heavy Equipment Safety Module Two 1

Audiovisual

Show Slide 5 (Figure 3). Discuss typical uses of barricades.

Show Slide 6 (Figure 4). Point out the colors used to indicate danger.

Show Slide 7 (Figure 5). Describe typical caution signs.

Show Slide 8 (Figure 6). Describe various traffic control signs.

2.1.0 Tools for Safety

To ensure everyone's safety, keep unauthorized people and vehicles away from the work area. Creating a clear work zone is an important part of working safely (*Figure 3*). Signs, tags, lockout/tagout devices, and barricades and barriers are some of the ways to mark the job site and alert workers and the public of hazardous conditions. Operators should always be aware of state and local regulations before beginning a job.

2.1.1 Signs

All work sites have specific markings and signs to identify hazards and provide important information. It is essential to recognize all types of signs related to each job site and make sure they are properly placed and working correctly. Doing this can save a life.

One type of sign is the danger sign (*Figure 4*). Danger signs are usually red, black, and white. They are used to inform workers that an immediate hazard exists.

Another type of sign is the caution sign (*Figure 5*). Caution signs are used to inform workers of potential hazards or unsafe conditions. They are usually yellow with black letters. When you see a caution sign, take action to protect yourself.

Another type of sign seen at work sites located on or near roads is a **temporary traffic control (TTC)** sign (*Figure 6*). These signs are used to redirect traffic and provide information to motorists and pedestrians during construction and maintenance of roadways and at other work sites. These signs are orange with black letters.

22102-12_F02.EPS

Figure 2 Stay alert for important safety information.

22102-12_F03.EPS

Figure 3 Typical use of barricades at a job site.

Case History

A 23-year-old apprentice power line worker died of injuries he received after being run over by the tandem dual rear tires of a digger derrick truck. He was part of a five-person crew that had been working on the job for three days. At the time of the incident, the victim and two other crew members had just finished setting and back-filling around a utility pole. They proceeded to the next pole requiring framing and setting, walking 30 feet ahead of the digger derrick in one lane of the road. The digger derrick moved slowly in reverse to the same pole. At a point approximately midway between the two poles, the victim knelt with his back to the truck to apparently inscribe a word or initials into some seal coating on the roadway. He was hit and run over by the backing truck's passenger-side tandem dual rear tires. The digger derrick truck did not have a back-up alarm system.

The Bottom Line: Keep your mind on the job. Communication with other workers can save your life. Check equipment for warning signals before you use it. If it does not have warning signals, establish hand or radio signals. Always check the surrounding area for potential hazards.

Source: The National Institute for Occupational Safety and Health (NIOSH)

2 NCCER – *Heavy Equipment Operations Level One* 22102-12

Teaching Tip/Notes

22102-12_F04.EPS

Figure 4 Typical danger sign.

22102-12_F05.EPS

Figure 5 Typical caution sign.

Safety alert symbols are sometimes seen on signs. The safety alert symbol is an exclamation point inside of a triangle. It is an internationally recognized symbol that means: Attention! Become Alert! These signs are seen in manuals, on machines, and on safety signs. When you see a safety alert symbol, pay attention to the message. It could save your life. If unsure of the meaning, check with a supervisor.

Safety alert symbols are organized into three safety levels:

- *Danger* – Danger symbols indicate immediate hazards that, if not avoided, will result in death or serious injury (see *Figure 7*).
- *Warning* – Warning symbols indicate potential hazards that, if not avoided, could result in death or serious injury (see *Figure 8*).
- *Caution* – Caution symbols indicate hazards that, if not avoided, may result in minor or moderate injury (see *Figure 9*).

22102-12_F06.EPS

Figure 6 Temporary traffic control signs.

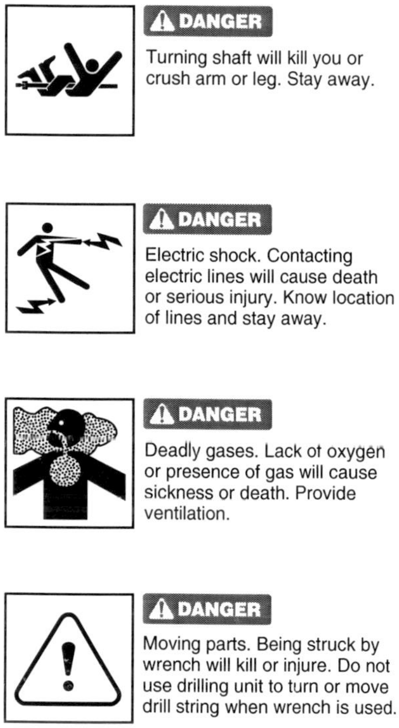

22102-12_F07.EPS

Figure 7 Danger symbols.

On-site traffic signs help with the safe movement of vehicles and pedestrians. Just as traffic signs must be obeyed on public highways and streets, they must also be observed in the workplace.

A slow-moving-vehicle emblem is used on vehicles that move at speeds of 25 miles per hour (mph) or less. The emblem is a fluorescent yellow-orange triangle with dark red reflective border (*Figure 10*). It is frequently seen on construction and farm equipment.

Audiovisual

Show Slides 9–11 (Figures 7–9). Explain the meaning of danger, warning, and caution symbols.

Show Slide 12 (Figure 10). Point out that a slow-moving vehicle travels at or below 25 mph.

Teaching Tip/Notes

22102-12 Heavy Equipment Safety Module Two 3

Audiovisual

Show Slide 13 (Figure 11). Describe safety tags and give examples of their uses.

Safety

Warn the trainees never to remove a safety tag they did not personally place.

Audiovisual

Show Slide 14 (Figure 12). Stress the importance of using and respecting lockout devices.

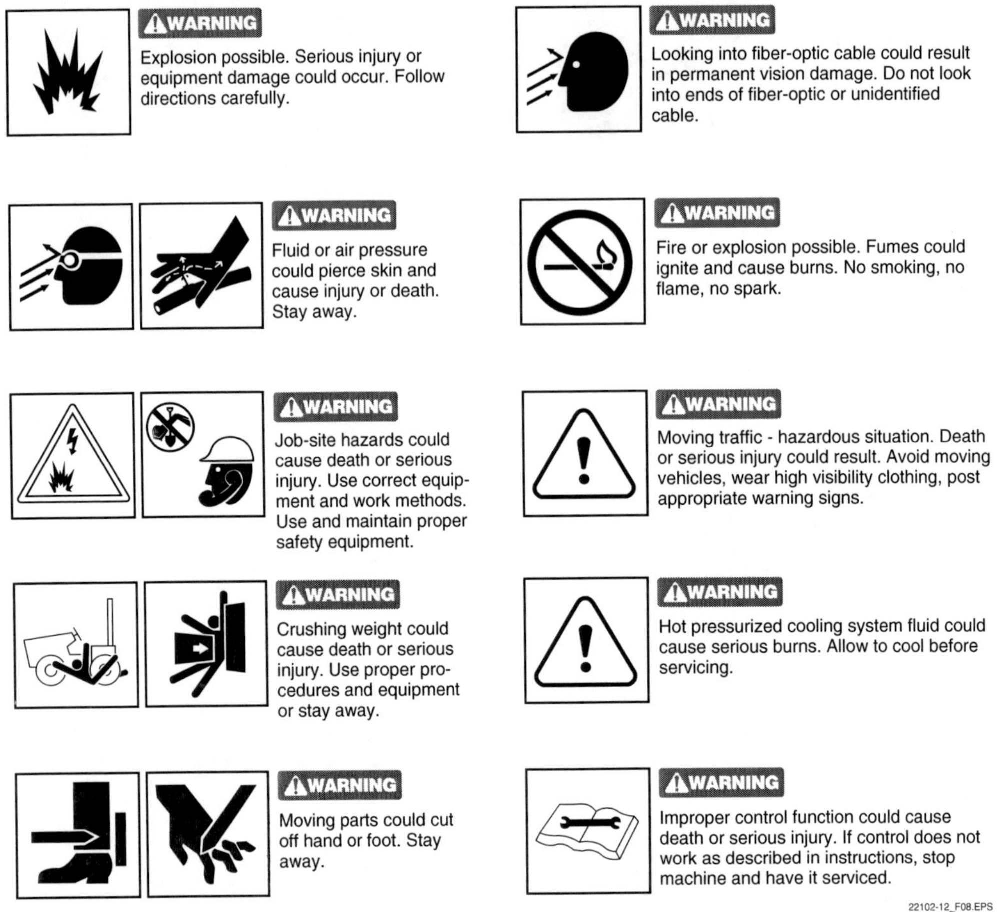

Figure 8 Warning symbols.

2.1.2 Safety Tags

Accident prevention tags are used as a temporary warning for workers of immediate and potential hazards (*Figure 11*). Tags are similar to signs, but they are not meant to be used in place of signs or as a permanent means of protection. For example, a tag that states DO NOT START may be used on a forklift while it is being serviced.

Tags can be an effective way of protecting workers and property when used correctly. Use tags on the equipment whenever it is out of order, when it is being serviced, or when other workers need special information about the equipment.

> **WARNING!**
> Never remove a safety tag that you did not place. You could place others in danger.

2.1.3 Lockout Devices

Another safety tool that you may be required to use is a lockout/tagout device. It is used to be certain that equipment is not energized. Some of these devices are very secure, because they permit a part or switch to be locked with a keyed padlock. Others are less secure than the padlock because they may simply be tags that have the words LOCKED

Teaching Tip/Notes

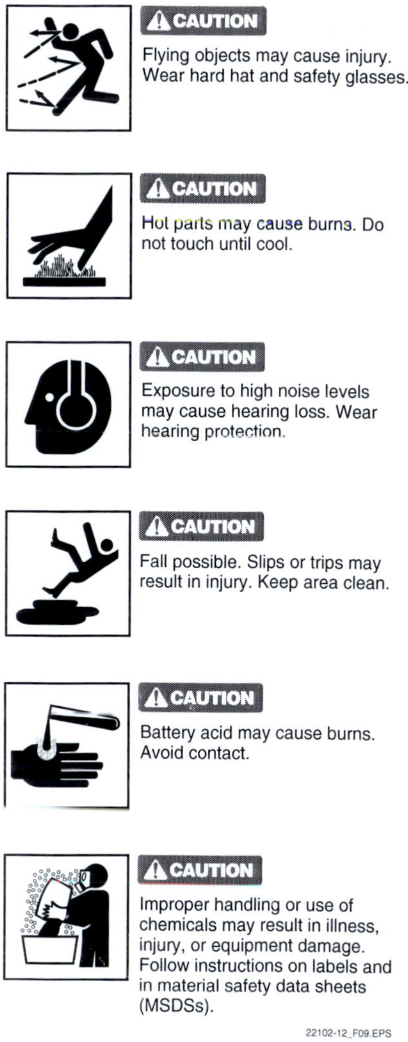

Figure 9 Caution symbols.

OUT on them (*Figure 12*). Both devices mean the same thing and should be removed only by the person who attached the device.

2.1.4 Barricades and Barriers

Barricades and barriers are used to alert workers to potential hazards and to help prevent accidents. Each work zone may have different policies and procedures for how and when to use them, so learn and follow the rules at the job site. Sometimes barriers are placed in an area to prevent in-

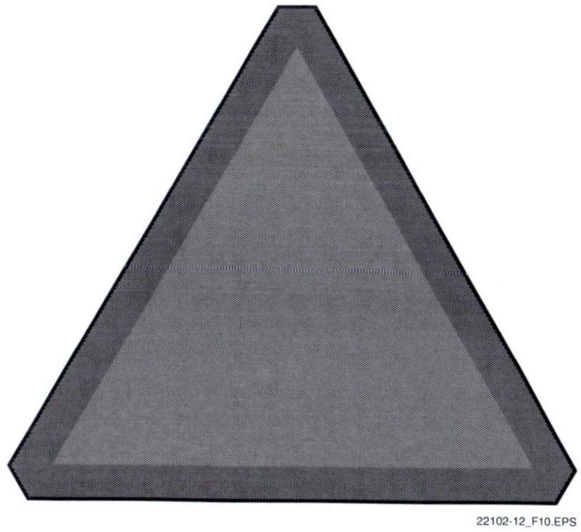

Figure 10 Slow-moving-vehicle sign.

jury to workers, and other times they are placed to alert heavy equipment operators of dangers. For example, a barrier may be erected to keep a bulldozer off steeply graded ground. *Figure 13* shows the various uses for barricades and barriers.

Barricades and barriers help to prevent accidents and injuries, but they are not foolproof. When you are assigned to a job site, carefully examine the area where you are working. Before moving heavy equipment, walk the area to ensure that there are no hidden hazards. This is especially important when there are obstructions, such as snow, mud, leaves, and other debris, in the area. Never move a barricade, even in an emergency, unless certain that it is safe to enter the barricaded area.

2.1.5 Audible and Hand Signals

Audible signals such as alarms, bells, buzzers, whistles, and horns can be used to communicate hazards to workers. For example, back-up alarms are used on forklifts, trucks, and other heavy equipment.

Case History

Many sidewalks and roadways use grates as barriers to cover vaults or pits. During a snowstorm in 2003, a skid-steer loader that was being used to clear snow fell into a vault when the grate system failed under the weight of the loader. The operator died.

The Bottom Line: Obstacles can make an unsafe area appear safe. Do not take chances. Examine the work area before moving any equipment.

Audiovisual

Show Slide 15 (Figure 13). Explain the purpose of barricades and barriers.

Teaching Tip/Notes

22102-12 Heavy Equipment Safety Module Two 5

Audiovisual

Show Slide 16 (Figure 14). Go over typical hand signals.

Show Slide 17 (Table 1). Emphasize the importance of persons using hand signals being in eye contact with each other.

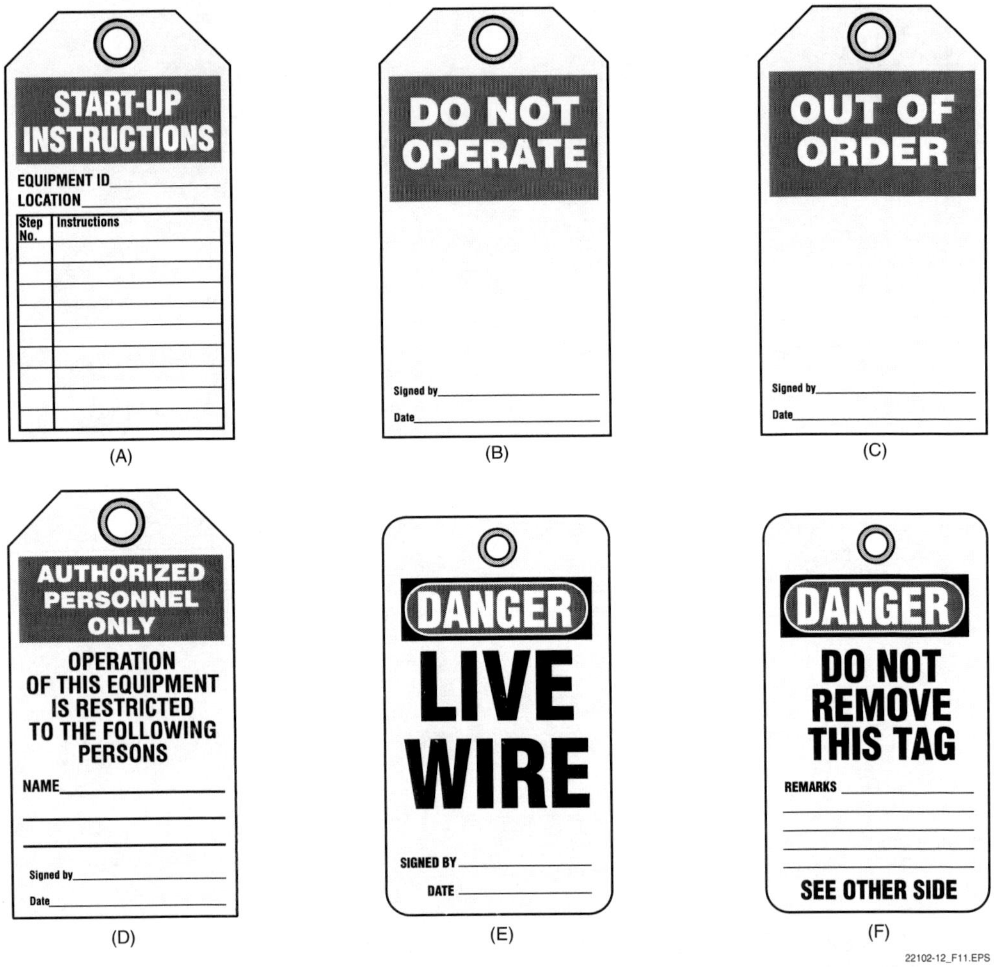

Figure 11 Examples of safety tags.

Fire alarms are used to clear work areas. Conveyer belt lines have buzzers, bells, and/or whistles to let workers know that they are about to start. However, when heavy equipment is present on a work site, there may be so much background noise that these types of alarms cannot be heard. When audible signals and voice communications are not practical because of noise or distance, it may be useful to communicate with hand signals (*Figure 14*).

There can be various meanings for the same hand signal. Always establish with your co-workers the hand signals that will be used and their meanings before starting the task. Some of the common hand signals and their meanings are listed in *Table 1*.

> **NOTE**
> Standards may vary from state to state. Be sure to use hand signals for your local jurisdiction.

Hand signals can be safely used over long distances, but those using hand signals must never be out of each other's sight. Always use slow and exaggerated gestures to be sure that you are understood. Heavy equipment operators who lose sight of their co-worker must immediately bring their vehicle to a safe stop. Always make eye contact with nearby personnel to be sure they are aware of your presence.

Teaching Tip/Notes

2.2.0 Traffic Control

When construction sites are near or in public roads, moving traffic presents an additional hazard to construction workers. Dangers to construction personnel working near moving traffic are so great that all states have laws that provide for additional fines for drivers who speed or commit other traffic violations in a traffic work zone. When the work site is located near traffic, supervisors study the area and make plans to ensure that the job can be completed safely and efficiently, but safety is your responsibility. Stay alert.

When the normal use of a road is disrupted, temporary traffic control (TTC) measures must be used. TTC measures help to ensure that motorists, pedestrians, and bicyclists can use the road safely while construction is going on. Planning traffic flow is key to keeping traffic moving smoothly and managing traffic incidents. Specially trained workers usually create traffic flow plans. They also decide on the need for and position of TTC devices such as barricades, cones, and signs, but you may be called on to help place some of these devices. In addition, it may be necessary to temporarily change the TTC during an accident or other emergency.

The US Department of Transportation Federal Highway Administration publishes a manual that states the basic principles for changing the flow

22102-12_F12.EPS

Figure 12 Lockout/tagout devices.

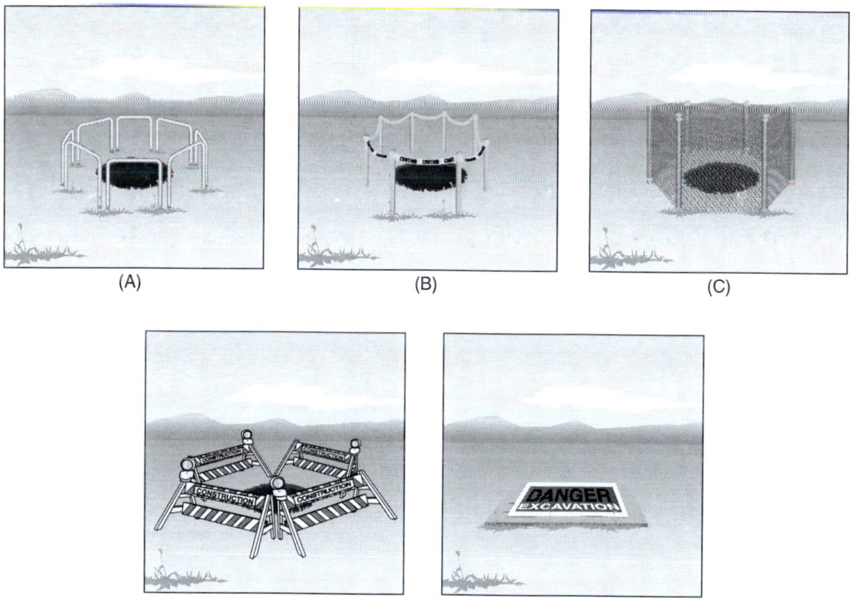

(A) (B) (C) (D) (E)

22102-12_F13.EPS

Figure 13 Common types of barricades and barriers.

Classroom

Discuss the additional hazards of working on or near public roads.

Explain when to use temporary traffic control (TTC) measures.

Teaching Tip/Notes

22102-12 Heavy Equipment Safety Module Two 7

Classroom

Discuss the basic requirements for an effective traffic control device.

Stress the need to be alert for changing conditions that could affect traffic.

22102-12_F14.EPS

Figure 14 Example of a hand signal (come here).

of traffic at a work site. It also explains how to design and use traffic control devices. This manual is called the *Manual on Uniform Traffic Control Devices (MUTCD) for Streets and Highways.* It is used for streets and highways open to the public, regardless of type or class or the public agency having control of the road.

For a traffic control device to be useful, it must meet five basic requirements:

- Fulfill a need
- Attract attention
- Have a clear and simple meaning
- Command respect from road users
- Give enough time for proper response

TTC measures and devices are used to provide for safe and orderly movement of traffic through or around work zones and to protect workers, responders to traffic accidents, and equipment. At the same time, the TTC zone must permit workers to complete the project quickly and well.

Conditions in TTC zones are always changing. These conditions include lane closures and merges, speed changes, **flaggers,** trucks and other

Table 1 Common Hand Signals

Signal	Meaning
Point index finger toward self	I/me
Point index finger toward object	It/them
Point index finger toward person	You/them
Circle index finger at group	We/us/all of us
Beckon by moving arm toward self	Come here
Extend arm and point hand to the right	Move right
Extend arm and point hand to the left	Move left
Point with thumb in a particular direction	Move this way/go this way
Bring index finger across throat	Quit
Slowly ease palm face down	Relax/slow down
Put palm over brow	Scout it out/check it out
Hold index finger up near head	Wait
Hands on top of head	I'm OK
Thumb up	Good/OK
Thumb down	Bad/not OK
Slap forehead	Bad idea
Palm down and rotated from side to side	Unsure/can't decide
Wave goodbye	Goodbye
Form a circle with thumb and index finger	OK/I understand/agree
Military salute	I understand and will comply
Shake head from side to side	No/disagree

22102-12_T01.EPS

Teaching Tip/Notes

equipment entering and leaving the roadway, and uneven pavement. This can be very confusing to motorists and dangerous for construction workers and heavy equipment operators.

Various methods and devices are used to mark the TTC zone (*Figure 15*). It is important to understand the how the TTC zone is set up and how TTC devices are used to keep workers safe. It is also important to know the purpose and use of flaggers in the TTC zone, and to be aware of the signals that flaggers use, just in case you need to control traffic in an emergency.

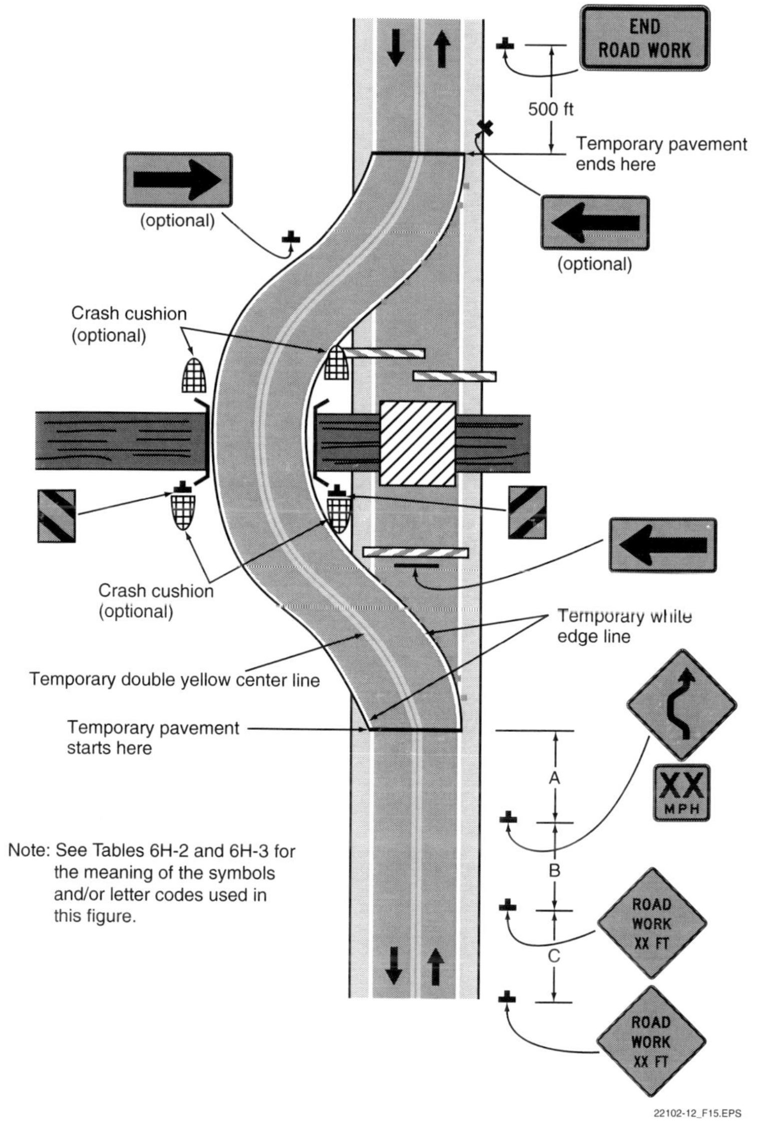

Figure 15 TTC zones can be confusing.

Audiovisual

Show Slide 18 (Figure 15).

Teaching Tip/Notes

22102-12 Heavy Equipment Safety Module Two 9

Audiovisual

Show Slide 19 (Figure 16). Describe the areas in a TTC zone.

Show Slide 20 (Figure 17). Explain how the posted speed limit affects the placement of signs.

Show Slides 21–23 (Figures 18–20). Point out different messages on TTC signs.

! Safety

Warn the trainees to always stay inside the workspace borders.

Stress the need for heavy equipment operators to be aware of the location of the person monitoring the barriers and channeling devices.

Classroom

Describe the duties and responsibilities of a flagger.

2.2.1 Temporary Traffic Control Zone

Because the TTC zone is always changing, a single set of TTC devices cannot satisfy all conditions for a project. At the same time, making detailed plans that cover all situations is not practical. This section covers some common uses for TTC devices. The exact devices used on each job site depend on the type of road, road conditions, length of the job, physical environment, and how close the workspace is to vehicular and pedestrian traffic. Although you will not be planning TTC zones, you need to understand the layout and purpose of the zone in order to work safely (*Figure 16*).

The first part of the TTC zone is the advance warning area. This area is where warning signs (*Figure 17*) appear to let motorists know what to expect ahead. The actual signs depend on the type of work being done. The spacing of the signs depends on the normal posted speed of the road.

The second part of the TTC zone is the transition area. This area is used to gradually move traffic out of its normal flow and into the desired path. Sometimes this area has flashing, lighted signs (*Figure 18*). Orange and white barrels, cones, and other types of markers, which are called **channeling devices** (*Figure 19*), are used to guide traffic into the desired path. The length of this area depends on the amount of traffic normal for the road, as well as other road conditions.

The next area in the TTC zone is a longitudinal buffer zone. If there is no room for it, truck-mounted attenuators are required. The buffer zone has some type of barrier device on the traffic side to mark its border. This is the work area. It is divided into two areas: the buffer and the workspace (*Figure 20*).

The lateral buffer is next to the border and is designed to protect workers from closely passing vehicles. Buffer zones must be clear areas. It is there to give reaction time in case a vehicle breaks through the barriers.

The workspace is for workers, equipment, and material. Stay in this area. Do not leave this area on foot or move the equipment from this area unless authorized to do so.

WARNING!

Be aware of the workspace borders at all times. Leaving the workspace can cause an accident.

The final area in the TTC zone is the termination area. This is the end of the construction area. It uses channeling devices and is marked with signs to let motorists know that they are leaving the construction zone and to merge back into their lane.

WARNING!

While working in the TTC zone, someone may be assigned to monitor the barriers, channeling devices, and signs to ensure that they stay in position. Heavy equipment operators need to be aware of the movements of this worker to prevent injuries or death.

2.2.2 Flagger Control

Flaggers are part of the construction team. They are responsible for public safety, so they must be trained in safe traffic control practices. In some states, flaggers must be certified. Flaggers must keep themselves and the construction team safe, as well as direct traffic using signs, paddles, or flags. On most jobs, it is unlikely that an equipment operator needs to be a flagger, but in the case of an emergency, the operator may need to perform this task for a short period. A flagger must be able to do the following:

- Clearly, firmly, and politely give specific instructions.
- Move quickly to avoid danger.

On Site

Safety Gear for Flaggers Standard

Flaggers must wear high-visibility safety apparel. Class 2 material is required for daytime work and Class 3 is needed at night. Class 3 can be used for both day and night. The apparel background (outer) material color must be fluorescent orange-red, fluorescent yellow-green, or a combination of the two as defined in the American National Standards Institute (ANSI) standard. The retroreflective material must be orange, yellow, white, silver, yellow-green, or a fluorescent version of these colors. It must be visible at a minimum distance of 1,000 feet. The retroreflective safety apparel must clearly identify the wearer as a person.

Source: Department of Transportation (DOT), *Manual on Uniform Traffic Control Devices 2009.* Section 6E.02 High-Visibility Clothing

Teaching Tip/Notes

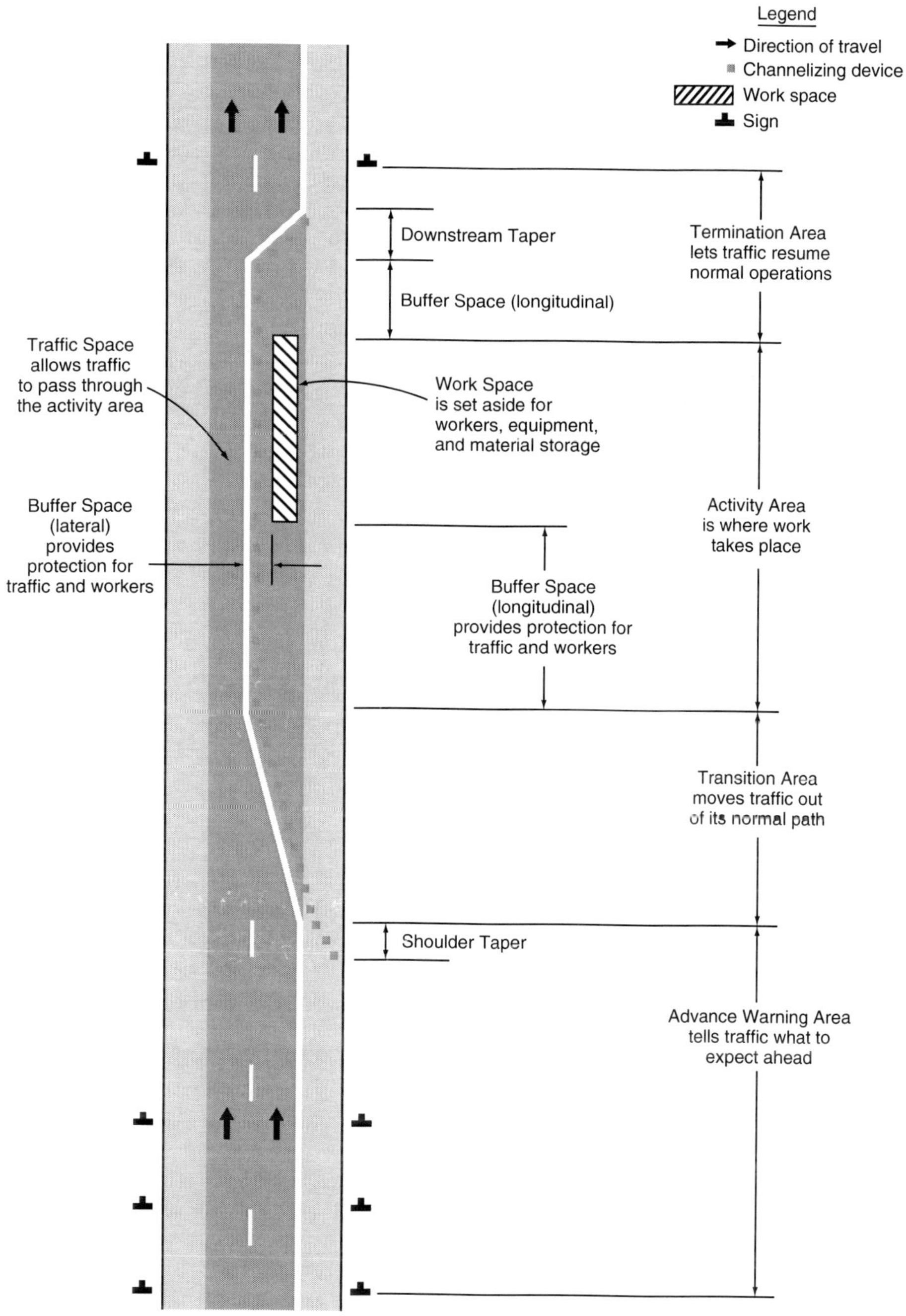

Figure 16 A typical TTC zone.

Teaching Tip/Notes

Classroom

Point out the highly visible clothing that a flagger is required to wear.

Figure 17 Warning signs.

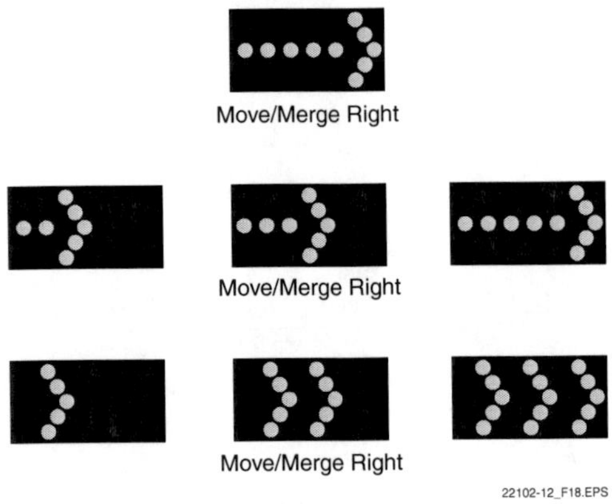

Figure 18 Merge right arrow signs.

- Use signaling devices, such as paddles and flags, to provide clear direction to drivers.
- Apply safe traffic control practices, sometimes in stressful or emergency conditions.
- Identify unsafe traffic situations and warn workers in time to avoid injury.

One of the ways that flaggers keep themselves safe is by wearing highly visible safety apparel, usually a vest that is either fluorescent orange-red or fluorescent yellow-green. The vest must be ANSI-approved and needs to be visible at a minimum of 1,000 feet.

Hand-signaling devices, such as Stop/Slow paddles, and lights are used to direct motorists through TTC zones. The Stop/Slow paddle is the best device to use because it is easy for drivers to understand. Red flags may be used in emergencies.

The Stop/Slow paddle is octagonal in shape, and the Stop side looks like the Stop signs you see at intersections. The Slow side has an orange background with black letters and border. When used at night, the paddle must be made of material that reflects light. The Stop/Slow paddle can also have either white or red flashing lights on the Stop side and either white or yellow flashing lights on the Slow side.

When flags are used they must be at least 24 inches square and made of a good grade of red material. The free edge of a flag must be weighted so the flag hangs vertically, even in heavy winds. When used at night, flags must be made of red retro reflective material.

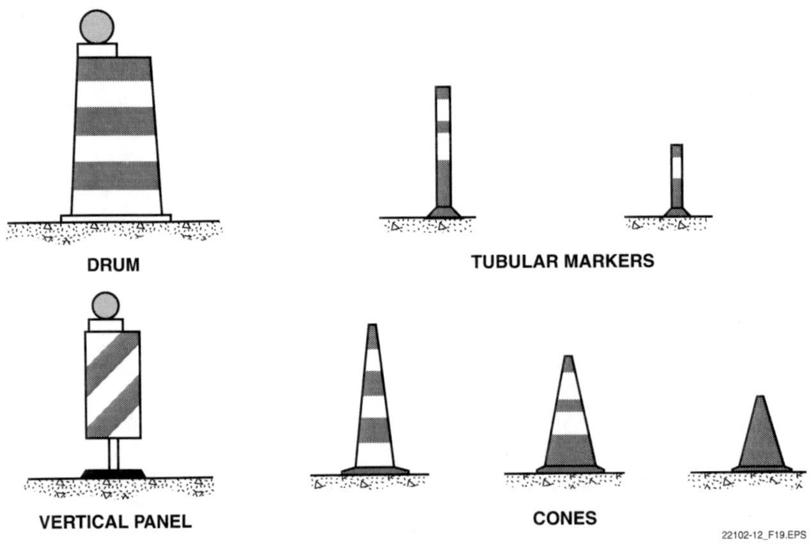

Figure 19 Channeling devices.

Teaching Tip/Notes

22102-12_F20.EPS

Figure 20 Typical barricades used in road construction.

> **WARNING!**
>
> Never step in front of a vehicle in order to stop it.

The paddle must be held still, with the flagger's arm extended horizontally away from the body (*Figure 21*). The other arm is used to give motorists additional directions. To stop traffic, the flagger's free hand is held above shoulder level with the palm facing oncoming traffic. To direct stopped traffic to continue, the Slow side of the paddle must face oncoming traffic, and the flagger's free arm is used to motion drivers to continue. To slow moving traffic, the flagger can slowly motion up and down with the free hand, palm down.

When a red flag is used in an emergency situation, the flagger must stand facing oncoming traffic with the red flag in the hand closest to traffic. To stop traffic, the flagger extends the flag horizontally across the road users' lane so that the full area of the flag is visibly hanging below the staff. The free arm is used in the same way as described when using the Stop/Slow paddle. To release stopped traffic, the flag is hung down at the flagger's side and the free arm is used to wave traffic on. To slow traffic, the flagger must stand with the free hand down and slowly wave the flag up and down without raising the arm above shoulder level. Motorists often mistake this gesture as a stop signal.

Case History

A seven-person crew was paving two lanes of a four-lane highway. One worker was walking back and forth along the highway to ensure that the traffic cones in the center of the highway and the construction-zone warning signs remained in position. The foreman operated the paving machine, and a ride-on roller/compactor followed behind compacting the newly laid asphalt. The roller operator made a pass with the roller, stopped, and then reversed the machine. The machine traveled approximately 10 feet when the operator sensed that something was wrong and stopped the roller. The worker assigned to monitor the cones and signs was discovered lying face down with his head crushed. Emergency rescue personnel arrived at the site within 15 minutes, but the worker was pronounced dead at the site.

The Bottom Line: The TTC work zone is always changing. Heavy equipment operators and other workers need to be constantly aware of their surroundings.

Source: The National Institute for Occupational Safety and Health (NIOSH)

Teaching Tip/Notes

Provide a copy of the MUTCD for the trainees to examine or refer them to the Internet. Access to the MUTCD can be obtained at http://mutcd.fhwa.dot.gov. As a homework or extra-credit assignment, have the trainees use the MUTCD to draw an appropriate traffic control plan for a given situation.

Safety !

Stress that one should never step in front of a vehicle in order to signal it to stop.

Audiovisual

Show Slide 24 (Figure 21). Describe paddle signals and their meaning.

Demonstration

Take the trainees to an area where they can set up safety equipment. Show them how to position barricades and barriers.

PT/Laboratory

Under your supervision, have the trainees place barricades and temporary traffic control devices for a highway construction zone. Note the proficiency of each trainee. This laboratory corresponds to Performance Task 2.

Demonstration

Take the trainees to an area where they can demonstrate flags and paddles. Show them how to use the flags and paddles to control traffic.

PT/Laboratory

Under your supervision, have the trainees demonstrate how to use flags or paddles to control traffic. Note the proficiency of each trainee. This laboratory corresponds to Performance Task 3.

Classroom

Point out conditions that can lull operators into a trance-like state and the hazards that may result.

Describe methods to help operators stay alert.

Figure 21 Flagger signals.

2.3.0 Personal Safety

Working in and around heavy equipment at any job site adds another hazardous element to any work. It is easy to become so engrossed in a task that you forget what is going on around you. When you are operating heavy equipment, you may find yourself lulled into a trance by the sound and movement of the equipment. If working in a vehicle with a temperature-controlled cab, you may forget about outside conditions. Responsible operators remain alert and stay aware of the surroundings at all times.

Heavy equipment operators need to stay alert by taking every opportunity to get out of the operator's seat. Walking increases blood circulation, thereby decreasing drowsiness, and reducing muscle discomfort from sitting in the same posi-

Teaching Tip/Notes

tion for long periods. In newer models of heavy equipment, the cabs are designed to reduce vibration and noise and to increase driver visibility. But heavy equipment is built to last, and an older model of equipment may be used. You can still help to relieve stress and muscle fatigue even when unable to leave the operator's seat. First, sit up straight while working. Good posture goes a long way in not only reducing fatigue, but also reducing muscle strain. Next, every time the vehicle stops, set the brake and stretch your arms and legs while sitting in the operator's seat, then do a few neck and shoulder rolls. Once in the habit of stretching, it becomes second nature.

When heavy equipment is in use, it is best to minimize the number of workers on foot around the work site. Sometimes this is not practical; so make eye contact with the driver when in the vicinity of the equipment. Drivers must make eye contact and acknowledge the presence of other workers, and take care not to lose sight of them. Workers on foot need to stay in the driver's view and when leaving the area must do so in plain sight of the driver. If conditions make it impossible for the driver and other personnel to maintain eye contact, a third person should be positioned so that he or she can maintain eye contact with both the driver and other personnel at all times.

3.0.0 Hazard Communication

Construction workers are exposed to a wide range of health hazards, especially when working around heavy equipment. In addition to the obvious hazards of roll-over injuries and collisions, heavy equipment introduces less obvious hazards, some of which may not be apparent until working in construction for many years.

As a heavy equipment operator, you will be responsible for the operation, and perhaps the maintenance, of the equipment. While performing these tasks, you will be exposed to many types of chemicals used in products, such as diesel fuel, hydraulic fluid, and cleaning solvents. In addition, you will be exposed to diesel exhaust and noise from the equipment. Take steps to protect yourself from these hazards.

3.1.0 Material Safety Data Sheets

OSHA and MSHA require that all employers inform and educate their employees about hazardous chemicals at the job site as part of a Hazard Communication (HazCom) program. As part of this program, OSHA and MSHA require that all hazardous chemicals at a workplace be clearly identified. Further, they require that suppliers of hazardous chemicals provide a material safety data sheet (MSDS) for each hazardous substance in a shipment. This information must be available at the work site, but it is up to you to use the information. A file containing the MSDSs for any potentially harmful materials on the site, or associated with the job, should be located near the work area. For construction jobs, such files may

Case History

A 32-year-old construction surveyor, who had been on the job for only nine days, was doing a grade check on a roadway when the grader operator spotted a manhole cover that he could not go over without damaging it. The grader operator stopped as he approached the manhole cover, then checked his side mirrors for anything that might be behind him. Seeing nothing, he started to back up. The surveyor was bending down with his back to the grader to tie a ribbon on a stake. A co-worker saw the surveyor and heard the grader's back-up alarm. He yelled for the victim to move but it was too late. The surveyor died from multiple injuries.

The Bottom Line: Stay alert. Heavy equipment is dangerous to work around.

Source: The National Institute for Occupational Safety and Health (NIOSH)

Case History

An 18-year-old flagger, outfitted in full reflective vest, pants, and hard hat, was directing traffic at one end of a bridge approach during a night milling operation. The work zone was correctly marked with cones and signs, and the entire bridge was lit with streetlights. The flagger was standing under portable floodlights in the opposing lane close to the center line, facing oncoming traffic. A pickup truck traveling in the wrong lane at an estimated speed of 55 to 60 miles per hour struck the flagger head on and carried him approximately 200 feet. He died of multiple traumatic injuries at the scene.

The Bottom Line: Highway construction is very dangerous. Even when following all of the safety rules, you are still at risk. Stay alert.

Source: The National Institute for Occupational Safety and Health (NIOSH)

Classroom

Describe some of the less-visible hazards of working around heavy equipment, such as chemicals and diesel fumes.

Explain the purpose of MSDSs.

Stress the importance of reading MSDSs appropriate to the work area.

Teaching Tip/Notes

Classroom

Describe how MSDS sections are arranged.

Describe the product and company identification section.

Explain what is found in the hazardous ingredients section.

Point out that the hazards identification section lists typical effects on the body but does not include allergic reactions.

! Safety

Emphasize that severe allergic reactions to hazardous chemicals should be treated by a doctor as soon as possible.

Classroom

Give examples of the information found in the first-aid section, such as treating for inhalation or splatters.

be kept in the supervisor's truck or in the project manager's office.

MSDSs are available so that workers can protect themselves. Employers are required to supply you with safety equipment that you may need in order to work around hazardous chemicals. Although the employer provides the MSDSs and the safety equipment, it is still up to employees to use the MSDS information and the protective equipment to protect themselves. It is part of your job to check the MSDS and evaluate the work environment to determine whether protective equipment is needed.

> **NOTE**
> A complete MSDS for diesel fuel is available in the *Appendix* of this module.

In 1985, OSHA established a voluntary format for MSDSs. In more recent years, a 16-section format was created and has been accepted on the international basis. MSDSs do not all look alike, but they all should have the 16 sections containing specific bits of information that may be needed. It will take a little practice to quickly find the information on the MSDS. The 16 sections are as follows:

- Identification
- Composition/information on ingredients
- Hazard(s) identification
- First-aid measures
- Fire-fighting measures
- Accidental-release measures
- Handling and storage
- Exposure controls/personal protection
- Physical and chemical properties
- Stability and reactivity
- Toxicological information
- Ecological information
- Disposal considerations
- Transport information
- Regulatory information
- Other information

3.1.1 Section 1 – Product and Company Identification

The product and company identification section of a MSDS provides the proper name(s) of the product; the chemical family that the product is associated with; the name of the manufacturing company; and the phone numbers for contacting the manufacturer.

3.1.2 Section 2 – Composition

Section 2 identifies the hazardous ingredients that make up the product. This section also lists the maximum exposure allowed in order to prevent skin damage.

3.1.3 Section 3 – Hazards Identification

The Hazards Identification section gives an emergency overview of the product. The product's appearance is described, and its combustibility is addressed. Warnings are given about actions to avoid, such as inadvertent suction of the liquid into the lungs. This section also explains what reactions you may experience if the product is inhaled or ingested, or if it comes in contact with skin or eyes. If the product is a carcinogen, which can cause some form of cancer, that information is also listed in this section.

> **WARNING!**
> The MSDS lists typical immediate effects of the hazardous chemical on the body. A reaction that is not listed may be an allergic reaction. Allergic reactions can be very mild, like redness on the skin, or they can be so severe that they are life threatening. Severe reactions need to be treated by a doctor as soon as possible. Symptoms of a severe allergic reaction include shortness of breath.

3.1.4 Section 4 – First Aid

Anyone working with heavy equipment will sooner or later be involved in refueling the machines. Even when spills do not occur, vapors from the diesel fuel will still fill the fueling area. Section 4 of an MSDS lists the first-aid actions to take if the product is inhaled, ingested, or splattered onto skin or into eyes. Such splashes may aggravate existing skin conditions.

Understand that these measures are the minimum actions to be taken after exposure to diesel fuel. Each work site may have stricter policies that must be followed. Know where to find first-aid equipment before you need it. *OSHA CFR 1926.50* requires that employers provide first-aid kits at every job site. These kits typically contain gauze pads, rolled gauze bandage, wound cleaning agent, scissors, adhesive tape, bandages, Band-Aids, a splint, tweezers, resuscitation equipment, and a blanket. Because of the potential for workers to be exposed to blood-borne pathogens, there should also be a bodily fluid pickup kit or personal protection kit. These kits usually contain

Teaching Tip/Notes

Provide an MSDS for the trainees to examine and follow throughout this section of the module.

eye and hand protection, fluid absorbents, sanitizing and disposal products, and a CPR mask.

3.1.5 Section 5 – Fire-Fighting Measures

When heavy equipment is being used to clear land, the potential for fire exists as the equipment moves through underbrush that may be dry enough to easily catch fire. A diesel fuel leak could lead to a fire. Heavy equipment being used in mining operations also are operating in a potentially dangerous environment where fires can create explosions. Section 5 lists specific hazards associated with the product, and the products recommended for extinguishing any fire involving the product. This section also lists any special PPE for fighting a fire fueled by this product.

> **WARNING!**
> In some situations, there are certain substances that should not be used on fires. Make sure to review such warnings before attempting to extinguish any fire.

3.1.6 Section 6 – Accidental Release Measures

Some products simply need to be isolated from public access. The Accidental Release Section of the MSDS addresses what to do if a product is accidentally released by way of spills, leaks, or unintentional releases. If applicable, this section may list specific products recommended for absorbing the spilled or leaked product.

3.1.7 Section 7 – Handling and Storage

Section 7 explains how a product should be handled and stored. With products that may ignite, special grounding procedures may be needed to keep any static electricity from igniting the product. Products such as fuels often need to be kept cool and ventilated. Never cut, drill, grind, or weld on an empty container that has held potentially flammable materials.

> **GOING GREEN**
> **Containing Runoff**
> If water is used to cool exposed surfaces close to a fire, make sure to contain any run-off water. Keep it out of sewers and streams.

Proper storage of diesel fuel is needed to ensure that it does not represent a hazard to the workplace. There are ways of handling and storing flammable and combustible materials such as diesel fuel that decrease explosion and fire risks. Only approved containers and approved portable tanks may be used for flammable and combustible liquid storage and handling. Approved storage and handling containers are UL-Listed items, such as safety cans and safety cabinets. It is your responsibility to know what the handling and storage procedures are on the job site.

Safety cans are special vessels used for safe transportation and storage of flammable materials (*Figure 22*). These essential pieces of equipment are available in a variety of styles and colors. Many job sites use color-coded safety cans to designate different types of liquids, such as gasoline, kerosene, diesel fuel, and oils. Color coding not only helps segregate liquids that should not be used together; it also helps ensure that the right liquids are used for each job. Safety cans have many built-in safety features such as vapor control, flame arresters, and pressure-relief valves. The features and usage requirements vary by manufacturer, so make sure you are familiar with the specifics for the safety cans on your site.

> **WARNING!**
> If you do not know how to use safety cans properly, ask your supervisor for instruction. Improper use is dangerous!

Safety cabinets (*Figure 23*) are important for the proper storage of flammables and combustibles. These cabinets are common on job sites, including construction, maintenance, and renovation locations. Just like with safety cans, it is important

22102-12_F22.EPS

Figure 22 Safety can.

Classroom

Explain the potential for fires around heavy equipment.

Point out the need for specific instructions on fire fighting for particular chemicals.

Safety !

Stress the need to be prepared by reviewing the fire fighting section before a problem occurs.

Classroom

Discuss some of the handling and storage requirements for particular chemicals.

Point out that some chemicals require special containers.

Audiovisual

Show Slides 25 and 26 (Figures 22 and 23).

Safety !

Point out the need for instruction on using safety cans and cabinets.

Teaching Tip/Notes

Ask the trainees for examples of chemicals they use frequently.

Classroom

Explain the section on PPE and exposure controls, such as ventilation or splashing.

Discuss the reasons why knowing the proper PPE for each chemical is so important.

Audiovisual

Show Slides 27 and 28 (Figures 24 and 25).

Safety

Warn the trainees to be sure their safety glasses are always in good condition with clean lenses.

Case History

At approximately midnight, the operator of a paving machine checked the level of fuel in his machine. Since it was night, he had difficulty seeing, so he used his cigarette lighter to see the fuel level. The fuel vapors ignited, and the operator was burned badly enough to need hospitalization.

The Bottom Line: Use the information in the MSDS. Stay Alert. Think Ahead.

Source: Occupational Safety and Health Administration (OSHA)

that these cabinets be used properly. Appropriate safety cabinets are designed specifically to reduce the risk of explosion and fire. Features include vents with flame arresters, special lips on the shelves to catch spills, and grounding connectors to prevent buildup of static electricity. They are made of steel designed to withstand intense heat and are tailored to meet OSHA, MSHA, and National Fire Protection Association (NFPA) guidelines. Some manufacturers produce self-closing models that automatically close in response to fire-level heat. Because characteristics of these cabinets vary by model and manufacturer, familiarize yourself with the specific cabinets and procedures used on your site.

WARNING

If you do not know how to use safety cabinets properly, ask your supervisor for instruction. Improper use is dangerous!

22102-12_F23.EPS

Figure 23 Safety cabinet.

NOTE

OSHA 29 CFR 1910.106 regulates the amount of flammable or combustible liquids that may be located near or inside of a gas storage cabinet or in any one fire area of a building. The standard prohibits more than 60 gallons of Class I or Class II liquids or more than 120 gallons of Class III liquids in one storage cabinet. No more than 25 gallons may be stored outside a protected storage area or cabinet.

Gas storage cabinets must be designed to limit the internal temperature not to exceed 325°F. Metal cabinets should be constructed of 18-gauge sheet metal and should be double-walled. For added safety, gas storage cabinets should be self-closing. Cabinets must be labeled in conspicuous lettering with the statement Flammable—Keep Fire Away. No more than 25 gallons of flammable or combustible liquids may be stored in a room outside of an approved storage cabinet.

Source: Occupational Safety and Health Administration web site. Regulations (Standards – 29 CFR), Flammable and Combustible Liquids – 1926.152.

3.1.8 Section 8 – Exposure Controls/ PPE

Section 8 addresses both the personal protection equipment (PPE) needed to prevent injuries that could occur if exposed to the product, and any engineering measures that should be taken to exhaust any vapors or fumes generated by the product. This section lists specific PPE, such as specific respiratory filters (*Figure 24*) or specific types of gloves, that should be used for protection. If splashing is possible, this section may indicate that safety glasses/goggles (*Figure 25*), or even face shields be used.

CAUTION

Handle safety glasses and goggles with care. If they get scratched, replace them. The scratches will interfere with your vision. Clean the lenses regularly with lens tissues or a soft cloth.

Section 8 is very important. Refer to this section often for all chemicals that are handled. The MSDS for diesel fuel (shown in the *Appendix*) recommends wearing safety glasses with side shields when there is a danger of splashing. It recommends protective gloves when there is a danger of skin contact. Several types of protective gloves are listed. Always check the label on the glove package to ensure that the proper gloves are used. The MSDS also recommends the use of air-purifying respirators when there is danger from fumes. When using an air-purifying respirator,

Teaching Tip/Notes

18 NCCER – *Heavy Equipment Operations Level One* 22102-12

be aware that this type of respirator needs a filter cartridge to work correctly. *Table 2* lists the color codes for cartridges. Always use the correct filter. For example, a black or yellow filter (organic vapor) is the correct filter to select for diesel fumes.

Diesel exhaust, also called diesel fumes, is made up of a mixture of gases and particles. One of these gases is carbon monoxide, which is a deadly gas that is produced when any fuel is burned. Soot is a major part of diesel exhaust (60 to 80 percent). This is the black smoke seen coming out of the exhaust pipe.

In the short term, breathing diesel exhaust can cause coughing, itchy or burning eyes, wheezing, and difficulty breathing. These effects disappear when the worker is away from the source of the exhaust.

Longer exposure to diesel exhausts can cause carbon monoxide poisoning. Carbon monoxide gas cannot be seen or smelled, and it can build up quickly, even in areas where there appears to be adequate air circulation. It can overcome workers without warning, producing weakness and confusion.

Recognizing carbon monoxide poisoning can be difficult, because early symptoms, such as headache, dizziness, and nausea, can be easily mistaken for a minor illness. Advanced symptoms of carbon monoxide poisoning are confusion and weakness. Victims of carbon monoxide poisoning must be immediately removed from the site and given 100-percent oxygen. Some cases are so bad that the victim needs to be placed in a special device called a hyperbaric chamber, which provides oxygen under pressure.

The soot in diesel exhaust can get into a worker's lungs and cause a number of lung problems, maybe even cancer. There is also evidence that the fine particles in soot can worsen heart problems and respiratory illnesses such as asthma.

There are safety measures that can be taken to reduce exposure to diesel exhaust. The first is to use low-sulfur diesel fuel, which burns cleaner than high-sulfur fuel. Another is to direct the equipment exhaust pipe away from the driver and nearby workers. This can make a big difference in the amount of exhaust to which workers are exposed.

Another form of protection is to wear a respirator, but workers must have a medical evaluation, be fit-tested by a competent person, and undergo training before wearing them. Beards and even stubble may prevent a proper seal. The respirator must be approved by the National Institute of Occupational Safety and Health (NIOSH) and carry an approval number. Respirators are only a temporary measure. The primary goal is to eliminate or control the diesel emissions.

Finally, turn off equipment when it is not needed. Idling engines burn fuel and put out poisonous gases. Taking some of these steps and using protective equipment and can mean a healthier life for everyone.

3.1.9 Section 9 – Physical and Chemical Properties

Section 9 lists the physical and chemical properties of the product. Those properties include

Did You Know?

Latex

Latex is a common material for protective gloves, but latex gloves are not recommended for use around diesel fuel because diesel fuel dissolves latex.

On Site

Static Electricity Can Kill

In recent incidents reported to the National Institute for Occupational Safety and Health (NIOSH), fires spontaneously ignited when workers or others attempted to fill portable gasoline containers (gas cans) in the backs of pickup trucks equipped with plastic bed liners or in cars with carpeted surfaces. Serious skin burns and other injuries resulted. Similar incidents in the last few years have resulted in warning bulletins from several private and government organizations.

These fires result from the buildup of static electricity. The insulating effect of the bed liner or carpet prevents the static charge generated by gasoline flowing into the container or other sources from grounding. The discharge of this buildup to the grounded gasoline dispenser nozzle may cause a spark and ignite the gasoline. Both ungrounded metal (most hazardous) and plastic gas containers have been involved in these incidents.

Source: The National Institute for Occupational Safety and Health (NIOSH)

Audiovisual

Show Slide 29 (Table 2).

Classroom

Describe the symptoms of carbon monoxide poisoning.

Discuss the hazards of diesel fumes and soot.

Describe preventive measures for working around diesel engines.

Describe some of the physical and chemical properties of the product in the MSDS.

Safety !

Discuss the potential danger of static electricity.

Teaching Tip/Notes

Provide several different respirator cartridges for the trainees to examine. Have them identify the type of cartridge by its color.

22102-12 Heavy Equipment Safety Module Two 19

Classroom

Define *flash point* and discuss conditions where it might occur.

Define *stability* and *reactivity.*

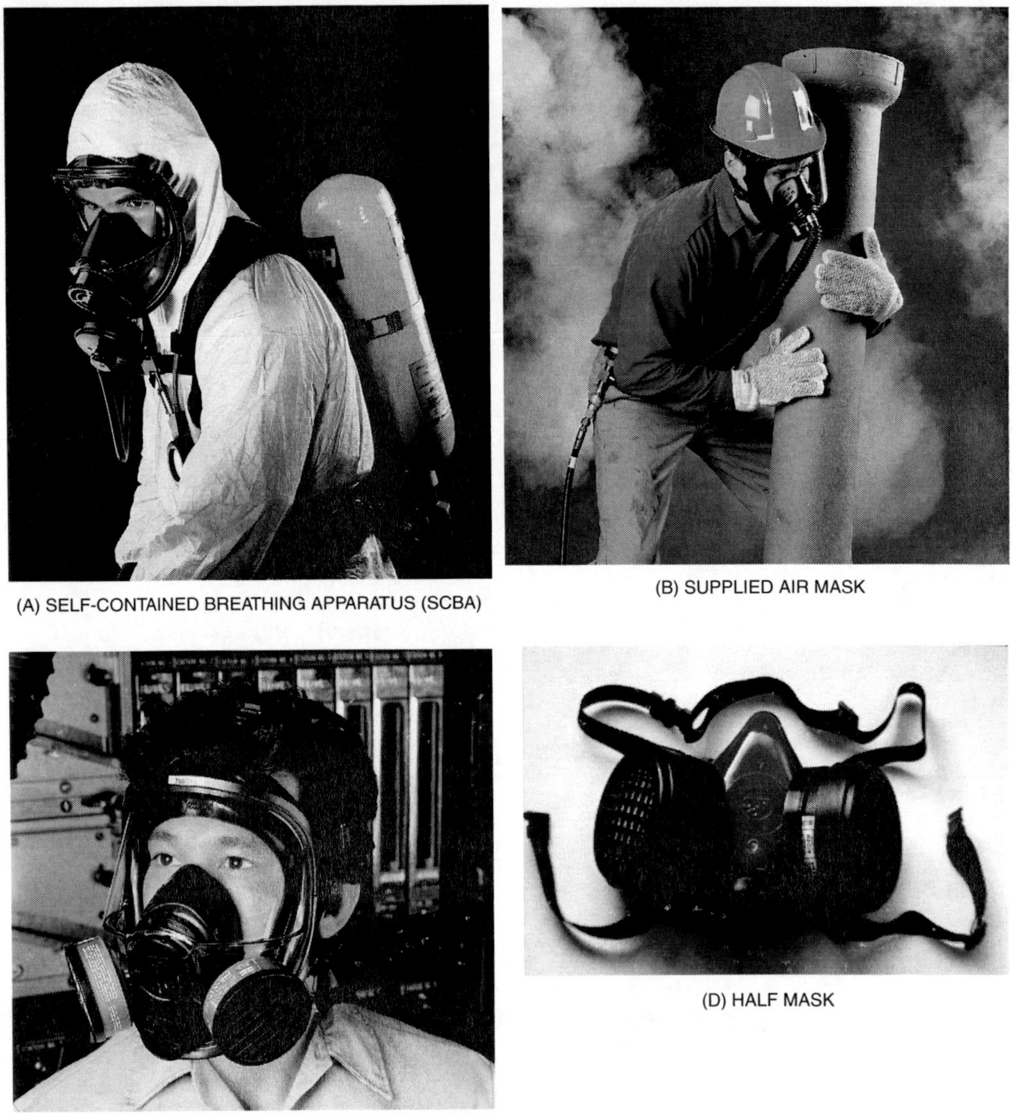

Figure 24 Respirators.

things like its appearance and whether it is a solid, liquid, or a gas. The section also lists other bits of information, such as the product's boiling point, melting point, and flash point, that are related to how the product reacts to different temperatures and pressures.

3.1.10 Section 10 – Stability and Reactivity

Section 10 lists information about how stable the product is at given temperatures and pressures. It also lists materials to keep separated from the product, and conditions to avoid that could otherwise cause the product to ignite or explode.

Teaching Tip/Notes

20 NCCER – *Heavy Equipment Operations Level One* 22102-12

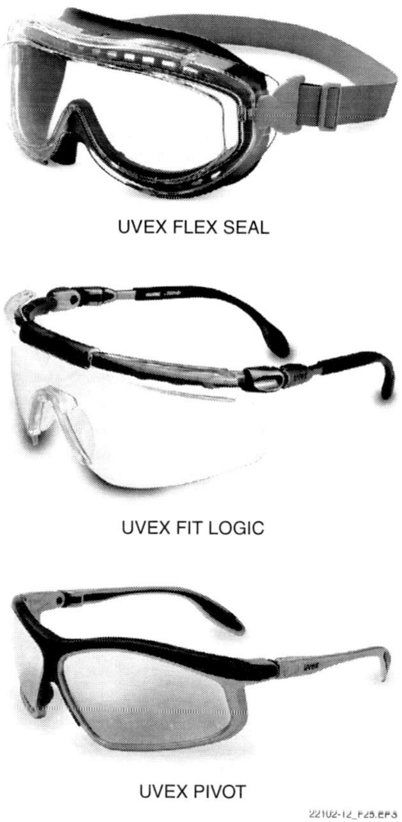

Figure 25 Safety glasses and goggles.

Case History

Two employees were attempting to adjust the brakes on a backhoe. The victim told the backhoe operator to raise the wheels off the ground with the front bucket and the outriggers so that he could get to the brakes. The victim then crawled under the machine and began to adjust the brakes. He did this without considering that there was only a 36-inch space from the ground to the drive shaft. While adjusting the brakes, the hood of his rain jacket wrapped around the drive shaft and broke his neck. He died instantly.

The Bottom Line: Loose clothing can be caught in moving parts of machinery. You must consider all possible dangers when working on equipment.

Source: The Occupational Safety and Health Administration (OSHA)

Table 2 Respirator Cartridge Color Code

Respirator Filter Cartridge Color-Codes	
Organic Vapor	Black
Acid Gas	White
Organic Vapor/Acid Gas	Yellow
Ammonia	Green
Particle (HEPA)	Magenta
Multi-Gas/Vapor	Olive

22102-12_T02.EPS

3.1.11 Section 11 – Toxicological Information

The word *toxic* is usually a red flag indicating a potential health hazard. Toxicity is the degree to which a substance can harm humans or animals. Section 11 lists the toxicity of the product. The section may go into detail about the chemical reactions the product has had on both humans and test animals. Be sure to read and understand the summary statements associated with the product.

3.1.12 Section 12 – Ecological Information

Everyone must be careful not to harm the environment. Section 12 lists the environmentally damaging effects of the product. Diesel fuel, for example, can be toxic to aquatic life.

3.1.13 Section 13 – Disposal Considerations

Section 13 may be a little more important to heavy equipment operators than some of the other sections. This section explains how to clean up and dispose of a given product. According to federal regulations, No. 2 diesel fuel, as produced, is not specifically listed as a hazardous waste. If discarded or disposed of and allowed to mix with water or some other substances, it may meet the criteria of hazardous waste. Check with federal, state, and local guidelines to determine the proper disposal methods.

3.1.14 Section 14 – Transport Information

Different countries, and different states within the United States, have different guidelines on how to ship or transport given products. Section 14 of an MSDS shows the product's proper shipping name, its hazard class, and how it should be packed for transport. When products such as diesel fuel are transported, the shipping vehicle must be labeled with the applicable UN/Identification number. That number allows anyone seeing the vehicle to know the hazard level of the product being transported.

Classroom

Have the trainees discuss the toxicity of the product in the MSDS.

Explain how ecological information is helpful.

Describe methods of disposing of a product such as diesel fuel.

Explain the purpose of transportation information for a product.

Teaching Tip/Notes

22102-12 Heavy Equipment Safety Module Two 21

Classroom

Point out the information relevant to regulatory guidelines.

Explain the type of information that may be found in Section 16.

Describe some of the problems caused by excessive noise on a job site.

Audiovisual

Show Slide 30 (Figure 26).

Classroom

Discuss the decibel levels for various activities.

3.1.15 Section 15 – Regulatory Information

Section 15 lists the various regulatory guidelines that may be associated with the product. This section may list the different Right-to-Know regulations of different states and communities. If a product such as No. 2 diesel is being transported across state lines, such information may be important to the shipper transporting the product.

3.1.16 Section 16 – Other Information

Section 16 is a catch all section. It may contain additional information that the product supplier wants to add.

3.2.0 Noise

Noise on construction sites and in closed spaces such as mines makes it hard to hear voice commands and warning signals. When hearing protection devices (*Figure 26*) are used, care must be taken to ensure that the device does not mute sounds to the point that workers cannot hear safety signals, such as backup alarms and evacuation horns. In any area where hearing is difficult, workers must make eye contact with operators.

Long-term exposure to noise can cause noise-induced hearing loss or a permanent ringing in the ears, a condition called tinnitus. Employees are responsible for wearing hearing protection devices (HPD) and ensuring that they fit properly and offer adequate protection. Workers are required to wear hearing protection anytime there is a likelihood of exposure to loud noise. Even removing the device for a few minutes can seriously reduce its protective ability. If there are problems with hearing or with the device, employees should seek help.

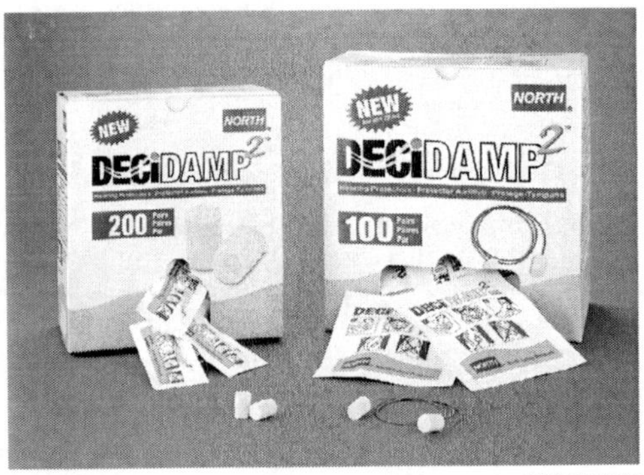

22102-12_F26.EPS

Figure 26 Hearing protection devices.

Management must provide employees with the proper devices and train employees in their use and care. It is also management's responsibility to monitor noise levels and try to lessen excessive levels.

A personal HPD or protector is anything that can be worn to reduce the level of sound entering the ear. Earmuffs, ear canal caps, and earplugs are the three principal types of devices. Each employee reacts individually to the use of these devices, and a successful hearing loss prevention program should respond to the needs of each employee. Making sure these devices protect hearing effectively requires the coordinated effort of management, the hearing loss prevention program operators, and the affected employees.

Noise is measured in units called decibels (acoustic) or dBA. Here are the dBA levels for some common everyday sounds:

- 0 dBA – Softest sound a person with normal hearing can hear
- 10 dBA – Normal breathing
- 30 dBA – Soft whisper
- 60 dBA – Normal conversation
- 110 dBA – Shouting in ear
- 120 dBA – Thunder
- 130 dBA – Stock car races
- 140 dBA – Jet plane (taking off)

The noise level on a typical construction job varies and is often irregular. Since most construction takes place outside, the loudness of the noise decreases rapidly as it travels away from the source. When heavy equipment is present, the sound tends to be more constant. Heavy equipment operators have no choice but to stay close to the source of the noise, which can exceed 100 dBA.

Both OSHA and MSHA have hearing protection and noise level requirements. If the exposure equals or exceeds 105 dBA time-weighted average (TWA), or an 800-percent dosage, in addition to enrollment into the hearing conservation program, MSHA requires miners to use dual protection, including both earplugs and earmuffs.

To protect hearing, avoid sources of loud noise as much as possible. Stay away from the area if not working there. This helps save your hearing. Limiting the number of workers around heavy equipment helps to protect hearing and also to prevent other accidents caused by contact with heavy equipment. Wear appropriate hearing protection devices, but be sure to stay alert.

Teaching Tip/Notes

Heavy equipment operators should be familiar with construction hazard communication standards. Trainees will need pencils and paper. You will need to arrange for a safety professional to give a presentation on hazard communication standards, to include topics such as MSDSs, chemical hazards, and personal protective equipment. Allow 40 to 50 minutes for this exercise.

1. Introduce the speaker who will give a presentation on hazard communication standards for heavy equipment operations.
2. Have the trainees take notes and write down questions during the presentation.
3. Ask the presenter to spend some time after the presentation to answer questions. Have the trainees ask their questions.

4.0.0 HEAVY EQUIPMENT SAFETY

Those who work around heavy equipment need a basic knowledge of all types of heavy equipment, as well as the hazards and safeguards of this equipment. The types of jobs performed by heavy equipment operators vary. The following are just a few of the jobs that use heavy equipment:

- Civil, residential, commercial, and industrial construction work
- Snow and ice control operations
- Maintaining road surfaces
- Loading, lashing, securing, and unloading equipment and materials
- Excavation and maintenance of public utilities

Construction workers may be expected to operate many different types of heavy equipment (*Figure 27*). This equipment can vary greatly in size and weight.

It is important to know how to use this equipment safely and to know the hazards and safeguards on the job site.

The first rule for heavy equipment operation is to never use a piece of equipment that you are not trained to use. Doing so not only puts you at risk of injury, but it also puts your co-workers at risk. Current OSHA and MSHA regulations state that at the time of their initial assignment and then at least annually, employers must instruct every employee in the safe operation and servicing of all equipment that the employee will use.

Some general points must be remembered when operating any heavy equipment follow:

- Before starting any equipment, study the manufacturer's operator manual.
- Learn the equipment's capacities and limitations.
- Learn the location and function of all controls, indicators, and warning devices.
- Be familiar with the safety devices on the equipment.
- Learn to recognize the equipment's warning and safety signals, such as high-temperature or low-pressure indicators.
- Securely fasten the seat belt.
- Do not permit others to ride in the equipment unless the equipment is specified for a rider.
- Operate the equipment smoothly—no jerky turns, starts, or stops.
- When towing something with the equipment, hitch only to points recommended by the manufacturer.
- When stopped, set brakes securely and use park lock if available.
- Maintain three-point contact when mounting or dismounting the equipment. Do not jump from the equipment.
- If there is contact with electrical lines, stay on the equipment unless there is a fire.

> **NOTE**
> OSHA and MSHA require that approved seat belts and rollover protection structures (ROPS) be installed on virtually all heavy equipment (old equipment must be retrofitted). Do not use any heavy equipment that is not equipped with these safety devices.

4.1.0 Job-Site Safety

Safety on the job is everyone's responsibility. The equipment operator has control over equipment operation, inspection, lubrication, and maintenance. Therefore, the operator must use good safety practices in these areas. It's important to understand, however, that even though the operator has much of this responsibility, everyone on the site must work safely. They are also responsible for being aware of their co-workers and their actions.

Follow these general job-safety rules to make sure everyone on the site is safe:

- Be alert. Watch out for moving equipment. If eye contact cannot be made with an equipment operator, assume the operator cannot see you.
- Usually one person at a site is appointed signal person. Find out who this is on your job and then watch for and obey his or her signals.
- Use retaining guards and safety devices on all equipment.
- Report all defective tools, machines, or other equipment, as well as all accidents and near-accidents to your supervisor.
- Make sure all others on the site are at a safe distance away from the equipment.
- Use traffic-control devices where required.
- Follow the manufacturer's safety rules and limitations for all equipment.

4.1.1 Power Lines

Equipment with tall booms, such as excavators and telescoping boom fork lifts, run the risk of coming into contact with live power lines. These lines carry high voltages that can kill anyone who comes into contact with them. For that reason, special care must be taken when it is necessary to work around power lines. A spotter should be always be used in such cases. *Table 3* shows the minimum distance that must be maintained from power lines based on the voltage level.

Homework

Have trainees review Sections 4.0.0–4.6.9.

Materials

Ensure that you have everything required for teaching this session.

Classroom

List some of the jobs that use heavy equipment.

Audiovisual

Show Slide 31 (Figure 27).

Classroom

Review the general safety guidelines for heavy equipment operations.

Discuss the job-site safety rules.

Point out that one of the greatest dangers for heavy equipment operators is contact with power lines.

Teaching Tip/Notes

Ask the trainees if they ever wear hearing protection outside the jobsite. Have them name activities in which the noise level exceeds safe limits.

Audiovisual

Show Slide 32 (Table 3). Discuss the guidelines and need for maintaining clearances and safe distances.

Figure 27 Examples of heavy equipment in use.

Teaching Tip/Notes

> **WARNING!** Equipment does not have to come in direct contact with wires to act as a conductor. If operating near power lines, power to these lines should be shut off. Verify that power is off with your supervisor.

If the vehicle does come into contact with a power line, follow these guidelines:

- Stay inside the vehicle and do not touch any metal in the vehicle.
- If forced to leave the vehicle because of fire or to get help, keep feet together and jump clear without touching any metal on the vehicle. To minimize the possibility of shock, hop to a safe area. Maintaining a single point of contact with the ground prevents current from finding a pathway through the body.

Downed power lines may still be live. When near downed power lines, follow these guidelines:

- Call emergency services for trained help.
- Do not allow other vehicles or people to enter the area.

Table 3 Power Line Clearances

Crane in Operation	
Power Line (kV)	Boom or Mast Minumum Clearances (Feet)
0 to 50	10
50 to 200	15
200 to 350	20
350 to 500	25
500 to 750	35
750 to 1000	45

22102-12_T03.EPS

- Never touch an overhead line if it has been brought down by machinery or has fallen.
- Never assume lines are dead; only the power company or its on-site representative can confirm that the system is de-energized.
- Treat anything in contact with a power line as energized. Anything the downed line touches and the ground surrounding it will also be energized.
- Never touch a person who is in contact with a live power line.
- If someone is trapped in a vehicle, stay well clear, at least 33 feet (more if high voltage), and try to keep the trapped person calm until help arrives.
- Do not attempt to help the victim; you are more likely to be electrocuted and may further endanger the victim.
- Wait for trained help to arrive.

4.2.0 Equipment Safety

The identification of safety hazards begins with a daily machine-safety inspection. Before using any equipment, the operator should inspect around and under it for visible evidence of potential safety hazards or operational problems. These unsafe conditions must either be corrected immediately or reported. Heavy equipment operations should not proceed until the conditions are safe.

The following is a list of basic items that should be checked daily:

- Loose, worn, or leaky components
- Engine performance and gauges
- Oil pressure and levels
- Housekeeping tasks
- Fire extinguishers
- Audible warning devices
- Hydraulic lines and fluid

Case History

Two employees were spreading concrete as it was being delivered by a concrete pumper truck boom. The truck was parked across the street from the work site. Overhead power lines ran above the boom on the pumper truck.

One employee was moving the hose to pour the concrete when the boom of the pumper truck came in contact with the overhead power lines. These lines carry 7,200 volts. The employee was electrocuted and died immediately. He then fell on the employee who was assisting him. The second employee received a massive electrical shock and burns.

No one on the site had received the proper safety training. Otherwise, they would have known how dangerous it is to work under power lines.

The Bottom Line: Always check the job site for hazards before beginning work. Overhead and underground power lines can be deadly.

Source: The Occupational Safety and Health Administration (OSHA)

Safety !

Warn the trainees that operating near power lines is dangerous. Lines should be verified as off and a safe distance maintained.

Classroom

Explain what to do when encountering downed power lines.

Stress the importance of a daily machine inspection for safety.

Review the list of items to check daily.

Teaching Tip/Notes

Classroom

Discuss the guidelines for personal safety.

Audiovisual

Show Slide 33 (Figure 28). Describe the purpose and use of these items of PPE.

Demonstration

Demonstrate the use of personal protective equipment. Show the trainees how to wear PPE.

PT/Laboratory

Under your supervision, have the trainees demonstrate how to use various types of personal protective equipment (PPE). Note the proficiency of each trainee. This laboratory corresponds to Performance Task 1.

Classroom

Discuss examples of how weather conditions affect safety.

- Windshield, wipers, mirrors, and lights
- Batteries and charging system
- Visible hand signal and load charts
- Guards and clutches
- Outriggers, tracks, wheels, and tires
- Brakes, locks, and safety devices
- Drive chains, steering, and all rollers
- Sheaves, drums, and all cables
- Blocks and hooks
- Boom, gantry, jib, and extension
- Carrier assembly
- All controls, including hoisting, swing, travel, and boom

4.3.0 Personal Safety

Personal safety is as important as site safety. Your actions affect everyone on the site. Follow these guidelines to prevent accidents and injury:

- Wear close-fitting clothing that is appropriate for the activity being performed.
- Tie back long hair before operating equipment.
- Wear safety glasses, a suitable hard hat, goggles, hearing protection, and safety shoes where required (*Figure 28*).
- Wear an approved reflective vest.
- Fasten loose sleeves when working around machine tools or rotating equipment.
- Remove rings and other jewelry when working.
- Be alert.
- Do not jump on or off moving equipment.
- Know the location of first-aid equipment, fire extinguishers, and emergency telephone numbers.
- Learn where to get assistance and how to use emergency communication equipment before you need it.

Never operate heavy equipment without a rollover protective structure (ROPS) or falling object protective structure (FOPS). The ROPS/FOPS should only be removed for maintenance and must be reinstalled before the equipment is operated again. Repairing or modifying ROPS/FOPS could weaken the structures and reduce your protection. See manufacturer's manuals for the equipment for complete instructions and inspection requirements.

4.4.0 Weather Hazards

Heavy equipment operators usually work outdoors. Under certain environmental conditions, such as extremely hot or cold weather, work can become uncomfortable and possibly dangerous. The following sections address specific things to be aware of when working under these adverse conditions.

22102-12_F28.EPS

Figure 28 Personal protective equipment.

Teaching Tip/Notes

Ask the trainees to provide examples of situations that could be hazardous with the wrong clothing or personal items.

4.4.1 Cold Weather

The amount of injury caused by exposure to abnormally cold temperatures depends on wind speed, length of exposure, temperature, and humidity. Freezing is increased by wind and humidity or a combination of the two factors. Follow these guidelines to prevent injuries such as frostbite during extremely cold weather:

- Always wear the proper clothing.
- Limit exposure as much as possible.
- Take frequent, short rest periods.
- Keep moving. Exercise fingers and toes if necessary.
- Do not drink alcohol before exposure to cold. Alcohol can dull your sensitivity to cold and make you less aware of over-exposure.
- Avoid icy surfaces.
- Do not expose yourself to extremely cold weather if any part of your clothing or body is wet.
- Do not smoke before exposure to cold. Breathing can be difficult in extremely cold air, and smoking can worsen the effect.
- Learn how to recognize the symptoms of over-exposure and frostbite.
- Place cold hands under dry clothing against the body, such as in the armpits.
- Be careful of ice on ladders and walkway surfaces.

4.4.2 Cold Exposure Symptoms and Treatment

Spending long periods of time in the cold can be dangerous. It's important to know the symptoms of cold weather exposure and how to treat them. Symptoms of cold exposure include the following:

- Shivering
- Numbness
- Low body temperature
- Drowsiness
- Weak muscles

Follow these steps to treat cold exposure:

Step 1 Notify supervision.

Step 2 Get to a warm inside area as quickly as possible.

Step 3 Remove wet or frozen clothing and anything that is binding, such as necklaces, watches, rings, and belts.

Step 4 Rewarm by adding dry clothing or wrapping in a blanket.

Step 5 Drink warm, not hot, liquids, and do not drink alcohol.

Step 6 Check for frostbite. If frostbite is found, seek medical help immediately.

4.4.3 Symptoms and Treatment of Frostbite

Frostbite is an injury resulting from exposure to cold elements. It happens when crystals form in the fluids and underlying soft tissues of the skin. The frozen area is generally small. The nose, cheeks, ears, fingers, and toes are usually affected. Affected skin may be slightly flushed just before frostbite sets in. Symptoms of frostbite include the following:

- Skin that becomes white, gray, or waxy yellow. Color indicates deep tissue damage. Victims are often not aware of frostbite until someone else recognizes the pale, glossy skin.
- Skin tingles and then becomes numb.
- Pain in the affected area starts and stops.
- Blisters show up on the area.
- The area of frostbite swells and feels hard.

Case History

A crane operator operating a 65-ton rubber tire hydraulic crane had completed a pick and was demobilizing the crane from the pick area. He backed the crane from a concrete slab pick-up area using an exit ramp. The boom of the crane was retracted and a lifting beam and load block secured. He intended to lower the lifting beam to a storage area on the ground located to his right rear. Upon exiting the slab, the operator noticed that the crew bus was parked parallel to the slab to his right and a control point was barricaded directly behind the operator's right rear. He stopped the crane, boomed out approximately 55 feet, and swung the boom over the crew bus to his right so the load (lifting beam) would go around the back of the bus and not over it. Once the boom was perpendicular to the bus, the crane tipped over because the operator had failed to extend the outriggers. Fortunately, the crane did not strike the bus and the worker was unhurt. The operator was at fault for the accident because he failed to use the correct procedures.

Source: U.S. Department of Energy

Classroom

Describe the injuries that can be caused by extreme cold.

Review the cold weather safety guidelines.

Describe the symptoms of cold weather exposure.

Explain the steps to take in case of exposure to cold.

Discuss symptoms of frostbite and how to treat it.

Teaching Tip/Notes

Classroom

Describe the stages of overexposure to heat: exhaustion, cramps, and heat stroke.

Discuss ways to prevent adverse reactions to excessive heat.

On Site

Cold Weather Clothing Tips

Use the following tips to prevent injury due to cold weather:

- Dress in layers.
- Wear thermal-type woolen underwear.
- Wear outer clothing that will repel wind and moisture.
- Wear face protection and head and ear coverings.
- Carry an extra pair of dry socks when working in snowy or wet conditions.
- Wear warm boots and make sure that they are not so tight that circulation becomes restricted.
- Wear wool-lined mittens or gloves covered with wind- and water-repellent material.

NOTE

In advanced cases of cold weather exposure or hypothermia, mental confusion and poor judgment occur, the victim staggers, eyesight fails, and the victim falls and may pass out. Shock is evident, and breathing may cease. Death, if it occurs, is usually due to heart failure.

Use the following steps to treat frostbite:

Step 1 Protect the frozen area from refreezing.

Step 2 Do not massage or apply heat to the frostbitten area.

Step 3 Get medical attention immediately.

4.4.4 Hot Weather

Hot weather can be as dangerous as cold weather. When someone is exposed to excessive amounts of heat, they run the risk of overheating. Some conditions associated with overheating include the following:

- Heat exhaustion
- Heat cramps
- Heat stroke

Heat exhaustion is characterized by pale, clammy skin; heavy sweating with nausea and possible vomiting; a fast, weak pulse; and possible fainting.

Heat cramps can occur after an attack of heat exhaustion. Cramps are characterized by abdominal pain, nausea, and dizziness. The skin becomes pale with heavy sweating, muscular twitching, and severe muscle cramps.

Heat stroke is an immediate, life-threatening emergency that requires urgent medical attention. It is characterized by headache, nausea, and visual problems. Body temperature can reach as high as 106°F. This is accompanied by hot, flushed, dry skin; slow, deep breathing; possible convulsions; and loss of consciousness.

Follow these guidelines when working in hot weather in order to prevent heat exhaustion, cramps, or heat stroke:

- Drink plenty of water.
- Do not overexert yourself.
- Wear lightweight clothing.
- Keep your head covered and face shaded.
- Take frequent, short work breaks.
- Rest in the shade whenever possible.

Case History

A miner was adding water to the batteries on a scissor lift being used for drilling rock bolts. He was on both knees, leaning inside the battery compartment when a forklift backed into and struck the compartment door. The forklift moved backwards approximately three feet, closing the door and causing the miner's left and right upper arms and outer back areas to be squeezed inside the compartment. When the miner yelled out, the forklift operator stopped backing up. He then moved forward so the miner could stand. The miner was severely bruised on his left and right upper arms and upper back. The driver of the forklift had not looked around the work site to see if anyone else was there. Fortunately, the miner recovered from his injuries.

The Bottom Line: Always check your work site before operating equipment. Know where your co-workers are at all times.

Source: U.S. Department of Energy

Teaching Tip/Notes

4.5.0 Moving Equipment Safely

All heavy equipment is moved during the course of a job. It is moved to the site, around the site, and away from the site. Many injuries and deaths happen during the movement of heavy equipment. It is important to be especially safety conscious whenever equipment is moving. When operating heavy equipment, be responsible. A responsible operator must be a qualified and authorized operator. To be qualified, you must understand the instruction manual supplied by the manufacturer of the equipment, have training in the actual operation of the equipment, and know the safety rules and regulations for the work site. Never operate any equipment under the influence of alcohol or drugs. If you are taking prescription or over-the-counter drugs, seek medical advice about whether you can safely operate machinery. Never knowingly allow anyone to operate heavy equipment when he or she is impaired.

Always observe the following guidelines when driving equipment on public roads:

- Drive slowly and never speed.
- Know the distance it takes to stop safely.
- Allow extra time to enter traffic.
- Know and follow the traffic control pattern.
- Travel with lights on.
- Use proper warning signs and flags.
- Secure all attachments and loose gear.
- Turn cautiously; allow for extensions or attachments and for structural clearances. Some equipment is top-heavy and will tip over if a turn is made too fast.
- Know the turning radius of the equipment.
- Know the swing radius of the equipment and make sure the swing path is clear of people and objects.
- Know and obey all state and local laws.

Always follow these guidelines for driving equipment on the job site:

- Never drive a machine in a congested area, or around people, without a spotter or flagger. The spotter or flagger is responsible for determining and controlling the driver's speed.
- Be sure everyone is in the clear while backing up, hooking up, or moving attachments.
- Never move buckets or shovels over the heads of other workers.
- If you cannot see the surrounding area clearly from the operator's seat, get a spotter or do not operate the equipment.
- Wait for an all-clear signal before moving.
- Signal a forward move with two blasts of the horn.
- Signal a reverse move with three blasts of the horn.
- Yield the right-of-way to loaded equipment on construction sites.
- Maintain a safe distance from all other vehicles.
- When moving, keep the equipment in gear at all times; never coast.
- Maintain a ground speed consistent with ground conditions and posted speed limits.
- Know where overhead electrical power lines are located.
- Ensure that buried pipes and power and gas lines have been located and marked.

Each piece of heavy equipment has its own unique set of hazards. Be aware of its **pinch points** (*Figure 29*). Pinch points occur when there is motion between equipment parts. Serious injury or death can result from getting part of your body or your clothing caught in a pinch point.

On Site

Ergonomics

Ergonomics is the science that deals with identifying and reducing stress on equipment users. In the past, jarring movements, uncomfortable seats, tight compartments, deafening noise, and hard-to-use controls were common. In the long term, operators were faced with a variety of physical problems, including bad backs and hearing loss, because of the harshness of their working environment.

In recent years, equipment designers have focused on eliminating these problems. Their efforts have resulted in such improvements as:

- Adjustable suspension seats
- Roomier cabs with better visibility
- More leg and foot room
- Easy-to-operate joystick controls
- Quieter operation
- Easier-to-reach service components, such as oil dipsticks and fuel tank nozzles

Classroom

Discuss the guidelines for safely moving heavy equipment on public roads.

Review the rules for driving heavy equipment on the job site.

Audiovisual

Show Slide 34 (Figure 29). Discuss examples of pinch points.

Teaching Tip/Notes

22102-12 Heavy Equipment Safety Module Two 29

Audiovisual

Show Slide 35 (Figure 30). Describe tipping hazards.

Classroom

Explain how maintenance problems can lead to breakdowns and cause safety hazards.

Figure 29 Pinch points.

Another danger that heavy equipment operators must face is tipping over. Tipping can be caused by improper loading, unsafe driving, or unstable terrain. These factors affect the **center of gravity,** which is the point at which the machine and load are in balance. To understand how to prevent tipping, you must first understand center of gravity. The center of gravity is the point where all of an object's weight is evenly distributed. The addition of a load causes a machine's center of gravity to move toward the load. This effect is very noticeable on a machine like a telescoping boom fork lift. *Figure 30* is a load chart for such a forklift. Notice that moving the load up and down, forward and back affects the load capacity of the forklift. These changes are, in effect, changing the machine's center of gravity. If the forks are raised to their maximum height and extended the maximum distance forward, the load capacity is reduced by more than 50 percent (form 11,000 pounds to 4,000 pounds). Presumably, if a 5,000-pound load were extended to the maximum height and distance, the forklift might tip over because its center of gravity shifted too far forward.

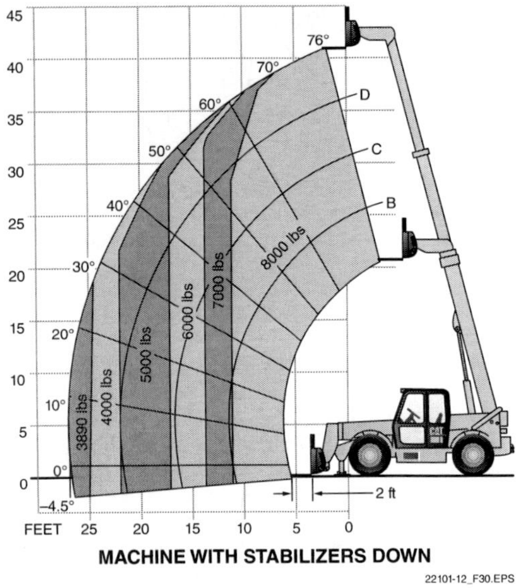

Figure 30 Example of the effect of load position on load capacity.

4.6.0 Heavy Equipment Maintenance

Heavy equipment breakdowns can often be traced to improper maintenance or lack of regular maintenance. These malfunctions can cause unexpected failure of the equipment, resulting in scheduling delays and accidents. An operator's minimum responsibility is to ensure that the equipment is in safe working order before using it. However, operators sometimes perform periodic maintenance.

> **NOTE**
> Only persons qualified to do so and authorized by the employer may perform maintenance on a vehicle or equipment.

Manufacturers supply detailed operating and maintenance manuals for equipment. Workers qualified and authorized to perform maintenance on the equipment should study the manufacturer's manuals, including maintenance schedules, lubrication charts, and parts lists, before starting the equipment. Maintenance can be dangerous unless performed properly. Workers should have the necessary skills, information, tools, and equipment to do the job correctly. Use parts, lubricants, and service techniques recommended by the manufacturer. If you do not know what you are doing—stop!

Teaching Tip/Notes

Case History

A 49-year old bulldozer operator was crushed to death while compacting earthen fill during the construction of an oil exploration island. The victim was operating the bulldozer on level ground, alternately in forward and reverse at half throttle. With the blade completely down, the victim shifted the transmission control lever toward neutral. The transmission was partially disengaged, but not fully in neutral.

He assumed, without checking, that the bulldozer transmission was in a stable neutral position. He exited the cab on the right side. When he did this, the transmission slipped back into the first gear of reverse, causing the bulldozer to suddenly move. As a result, the victim was pulled between the underside of the fender and the top of the track cleats. As the bulldozer continued in reverse, the victim was fatally crushed beneath the track cleats.

The Bottom Line: Always use parking brakes, and double-check settings before exiting the equipment. Never leave equipment when it is still running.

Source: The National Institute for Occupational Safety and Health (NIOSH)

While performing maintenance on equipment, keep the following in mind:

- Disconnect power, then lock and tagout machines before performing maintenance.
- Use properly fitting wrenches on nuts and bolts.
- Keep the work area clear of scraps and litter.
- Dispose of combustible materials properly.
- Clean up any spilled liquids immediately.
- Store oily rags in self-closing metal containers.
- Never use compressed air to clean yourself or your clothing.

4.6.1 Pre-Maintenance Safety

Before servicing equipment, prepare the service area. Begin by emptying your pockets of any items that could fall into machinery. Then collect any PPE that may be needed, such as aprons and gloves; safety glasses, goggles, or shields; safety shoes; welding protection equipment; and a filter mask or respirator. Wear all the protective clothing that the job requires.

If the work site does not have a designated service area, select a clean, level area out of the flow of traffic. Make sure there is enough room to work, as well as adequate light and ventilation. Check overhead clearances if necessary. Before starting, clean the working surface, removing oil, grease, and water to eliminate slippery areas. Put sand or other absorbent material on these areas.

After moving the equipment to the service area, follow the necessary shutdown procedures. After the equipment is shut down, always attach a DO NOT OPERATE tag or similar warning tag to the starter switch or steering lever before performing maintenance. If the engine must not be started during maintenance, remove the ignition key and, if possible, use a lockout device. Many machines have ignition lockout switches in or near the engine compartment for this purpose.

When performing maintenance, always use the correct tools. If guards or covers need to be removed during servicing, remove only those necessary to gain access to other parts and replace them as soon as possible. Never leave guards off while the equipment is unattended. Make sure that any devices that are used during maintenance, like lifting and supporting devices, are strong enough to perform their jobs.

4.6.2 Engine Maintenance

Working on a running engine is dangerous. It must only be done by a qualified technician. If adjustments must be made with the engine running, always work as a two-person team with one person sitting in the operator's seat while the other works on the equipment. Never attempt to get around any safety devices or circuitry to start the equipment. Bypassing normal starting procedures may cause the machine to start and move suddenly, resulting in serious injury or death to anyone in its path.

If the engine needs to be running during maintenance, ensure that there is proper ventilation in the service area because exhaust fumes can kill. Use an exhaust pipe extension if the engine is running in an enclosed area. If there is no exhaust pipe extension, make sure that the doors are open to allow outside air into the area.

WARNING!

While performing any maintenance on any equipment, keep clear of all rotating components. Wrapping or entanglement may result in serious injury or death.

4.6.3 Hydraulic System Maintenance

The hydraulic system of the equipment is under pressure whenever the engine is running, and it can hold pressure even after being shut down.

Classroom

Explain how to prepare for servicing equipment, including preparing the area and collecting tools and PPE.

Point out the dangers of working on a running engine.

Safety !

Warn the trainees to keep clear of all rotating components when performing maintenance.

Classroom

Discuss potential hazards when working on hydraulic systems, such as high pressures, tension, and flammable materials.

Teaching Tip/Notes

Classroom

Describe PPE required for working on hydraulic systems.

Describe maintenance activities for a cooling system.

Emphasize the importance of following the manufacturer's instructions.

Stress the hazards of filling the fuel tank when the engine is hot.

Point out safe practices for fueling.

! Safety

Explain the dangers of grinding, flame-cutting, brazing, and welding on a fuel tank or line.

Classroom

Describe the potential hazards of working with batteries.

Cycle hydraulic controls such as steering after shutdown to relieve trapped pressure in the lines. Whenever possible, remove tension from any attachments or devices. For example, lower the forks of a forklift to the ground. If this is not possible, install manufacturer-approved support devices or block the cylinders and equipment securely before working on the hydraulic system.

When venting or filling the hydraulic system, loosen the filler cap slowly and remove it gradually. If the system is equipped with an accumulator, see the manufacturer's manual for maintenance instructions. Even after releasing the pressure on lines, remember, hot hydraulic fluid can cause severe burns. Wait for fluid to cool before disconnecting lines. Clean up spilled hydraulic fluid immediately.

Do not permit open flame around the hydraulic system. Recharge hydraulic accumulators with dry nitrogen gas only. Do not use oxygen or combustible gases. Use extreme caution when handling pressurized cylinders. Never reset any relief valve in a hydraulic system to a pressure higher than that recommended by the manufacturer. Do not bypass relief devices or close off overflow devices.

Hydraulic fluid under pressure can penetrate skin or eyes and can cause injury, blindness, or death. Fluid leaks under pressure may not be visible. Do not use your hand or any part of your body to check for a leak. Use proper PPE, including a face shield or safety goggles for eye protection. If any fluid is injected under the skin, it must be removed within a few hours by a doctor familiar with this type of injury.

4.6.4 Cooling System Maintenance

Liquid cooling systems build pressure as the engine gets hot. Before removing the radiator cap, stop the engine and let the system cool. Wear protective clothing and safety glasses, goggles, or a shield. Remove the radiator cap only after the coolant is cold. Slowly turn the radiator cap to the first position to allow any residual pressure to escape before removing the cap entirely. For cooling systems with an overflow tank, the coolant can usually be checked at the tank without removing the radiator cap. See the manufacturer's instructions.

4.6.5 Fueling

Stop the engine and shut off electrical equipment while filling the fuel tank. Use extra caution when fueling a hot engine. Always ground the fuel nozzle or container spout against the filler neck to avoid sparks. Never overfill the fuel tank. If any fuel is spilled, clean it up immediately. Keep sparks and open flames away from fuel. Never smoke while handling fuel or working on the fuel system. The fumes in an empty container are explosive. Never cut or weld on fuel lines, tanks, or containers. Do not walk away when fueling equipment. If the fill spout is elevated, climb up and then have someone hand up the fuel hose.

WARNING!

Do not grind, flame-cut, braze, or weld any fuel tanks or lines. Doing so can cause residual vapors to explode, causing serious injuries or even death.

4.6.6 Battery Maintenance

Explosive gases are always present when working around batteries. This is especially true when the battery is being charged. To avoid explosion, never smoke around batteries. Keep arcs, sparks, and open flames away from batteries, and always work on batteries in a well-ventilated area.

On Site

Don't Get Burned

The following are some facts about gasoline:

- One gallon of gasoline contains the same explosive force as 14 sticks of dynamite.
- Gasoline vapors are heavier than air, can travel a number of feet to an ignition source, and can ignite at temperatures of 45°F. To be safe, keep open gasoline containers well removed from all potential ignition sources.
- Gasoline has a low electrical conductivity. As a result, a charge of static electricity builds up on gasoline as it flows through a pipe or hose. Getting into and out of a vehicle during refueling can build up a static charge, especially during dry weather. That charge can cause a spark that can ignite gasoline vapors if it occurs near the fuel nozzle.

Source: U.S. Department of Energy

Teaching Tip/Notes

Case History

An employee was repairing and testing a hydraulic pump. He had his hand over the open valve to the pump on the outflow side when a high-pressure stream of hydraulic fluid struck it. The stream punctured his hand and forced hydraulic fluid under the skin, resulting in injuries that required hospitalization. The cause of this accident was an open hydraulic line capable of developing a pressure of 2,000 psi.

The Bottom Line: Pressurized fluids can cause injuries. Study the service manual before servicing equipment, and follow all instruction and warnings.

Source: Occupational Safety and Health Administration (OSHA)

The liquid in batteries is commonly called battery acid, but it is technically an electrolyte because it permits electrons to flow through it. This electrolyte contains sulfuric acid, which is very poisonous and can cause severe chemical burns. To avoid injury, wear a full-face shield to prevent fluid contact with face and eyes. Wear chemically resistant gloves and clothing to prevent this acid from damaging skin and regular clothing.

Follow the instructions in the manufacturer's manual for battery boosted starting of engines (also known as jumping). Battery boosted starting is a two-person job. One person needs to be in the driver's seat so that the equipment is under control when the engine starts. Be sure that the voltage of the replacement battery is compatible with the power source. If a battery needs to be replaced, the old battery must be disposed of in accordance with environmental regulations.

WARNING!

Always disconnect the battery before welding on any machine.

4.6.7 Tire Maintenance

The stability and handling of the equipment can be dramatically affected by underinflated tires or those with inadequate ballast fill. As a minimum, the tires and wheels on the equipment need to be examined before each use. See the manufacturer's specifications for ballast and inflation requirements.

All tires must be serviced by a qualified tire service center or by an authorized person who has been properly trained in the procedures and equipment used during servicing. The types of rims and tires used on heavy equipment usually require that special care be taken during maintenance to prevent serious injury or death. Here are some important things to remember:

- Inflate tires from a distance using a self-attaching chuck. Always stand behind the tread when adjusting tire pressure.
- Do not inflate a tire with flammable gases or from a system using an alcohol injector.
- Never overinflate a tire—it can explode.
- Punctures that have allowed the ballast to leak out must be repaired and the ballast refilled before the equipment may be placed in operation.
- Never reinflate a tire that has been run flat or seriously underinflated without removing the tire from the wheel. Closely inspect the tire and rim for damage before remounting.
- Clean the area around all wheel lug nuts or bolts, and periodically check the torque per the manufacturer's instructions until the torque value stabilizes. Then check at regularly scheduled intervals.
- Never weld on a wheel rim with the tire in place. The tire could explode.
- Check for and remove any debris and rocks between dual tires.
- Check tire sidewalls for cuts or exposed metal belts.

Case History

An employee was using a MIG welder to repair a 100-gallon tank containing diesel fuel. The tank ruptured, and the fuel ignited and spilled onto the employee. The employee was engulfed in flames, suffering third- and fourth-degree burns from which he died.

The Bottom Line: Do not start a job until you have thought it through. Check material safety data sheets for information.

Source: Occupational Safety and Health Administration (OSHA)

Safety !

Warn trainees to always disconnect the battery before welding on any machine.

Classroom

Discuss problems that can occur when tires are underinflated or have inadequate ballast.

Stress the importance of checking the manufacturer's instructions and specifications for tire maintenance.

Point out the qualifications required to service tires.

Review the guidelines for servicing tires.

Teaching Tip/Notes

Classroom

Discuss qualifications required to work on air conditioning equipment.

Describe EPA restrictions on refrigerants.

! Safety

Stress the dangers of smoking or open flames near air conditioning/ refrigerant systems. Discuss the hazards of refrigerant gas fumes.

Classroom

Explain that bolts, fittings, and connections must be tightened to the proper torque after performing maintenance.

Point out that equipment should be tested for proper operation before returning it to service.

Homework

Have trainees review Sections 5.0.0–5.7.3.

Materials

Ensure that you have everything required for teaching this session.

Classroom

Describe typical hazards involved with trenching or excavating.

Audiovisual

Show Slide 36 (Figure 31).

4.6.8 Air Conditioning System Maintenance

Only qualified technicians may work on air conditioning equipment. When servicing air conditioning systems, refer to the refrigerant gas container for proper use. Keep in mind that the use of air conditioning refrigerant in the United States is governed by the Environmental Protection Agency (EPA), which strictly prohibits venting any type of refrigerant into the atmosphere. Therefore, EPA-approved recovery and recycling systems must be used to avoid accidental venting of refrigerant. To help maintain the air conditioning systems, keep the windows closed when it is in operation and check filters daily.

> **WARNING!**
>
> Do not smoke or have open flames in the area when servicing the air conditioning system—refrigerant gases may be present. Inhaling refrigerant gas through a cigarette or other smoking material or inhaling fumes released when refrigerant gas is burned can cause bodily harm or death. Also note that contact with the refrigerant stream can cause severe frost burns.

4.6.9 Post-Maintenance Repairs

Complete all servicing and repairs before releasing any equipment. Tighten all bolts, fittings, and connections to torques specified by the manufacturer. Reinstall all guards, covers, and shields after servicing. Replace or repair any damaged ones. Refill and recharge pressure systems only with manufacturer-approved or recommended fluids. Start the engine and check for leaks. Operate all controls to make sure that they are functioning properly. Road-test the equipment if necessary. After testing, shut down the engine and check the work you performed. Look for any missing hardware like cotter pins, washers, and locknuts. Recheck all fluid levels before releasing the equipment for operation.

5.0.0 Trenching and Excavation Safety

Heavy equipment is used for excavation. Safety is crucial during any excavation job. Excavations are done for a number of reasons, including laying sewer lines and during highway construction (*Figure 31*). During an excavation, earth is removed from the ground, creating a trench. A trench is a narrow excavation in which the depth is greater than the width and the width does not exceed 15 feet. The soil that is removed from the ground is called spoil. When earth is removed from the ground, extreme pressures may be generated on the trench walls. If the walls are not properly secured by shoring, sloping, or shielding, they will collapse.

> **GOING GREEN**
>
> **Refrigerant and the Law**
>
> It is unlawful to knowingly release CFCs, as well as other types of fluorocarbon refrigerants, into the atmosphere. If caught doing so, you can be subject to a stiff fine and possibly a prison term. Because of the damaging effects of these refrigerants on the environment, a new class of environmentally safe refrigerant blends, known as green refrigerants, has emerged.
>
> The US Environmental Protection Agency (EPA) requires that all persons who install, service, repair, or dispose of equipment containing a refrigerant possess a certification card. The certification card is obtained by passing a test for one or more categories of work as identified by the EPA. The categories are:
>
> - *Type I* – Small appliances containing less than five pounds of refrigerant, such as refrigerators and small air conditioners.
> - *Type II* – Appliances that use high-pressure refrigerants, such as R-22, R-500, and R-502.
> - *Type III* – Appliances such as centrifugal chillers that use low-pressure refrigerants.
> - *Type IV* – The universal certification for any of the above categories.

Teaching Tip/Notes

34 NCCER – *Heavy Equipment Operations Level One* 22102-12

Figure 31 Excavation sites.

The collapse of unsupported trench walls can instantly crush and bury workers. This type of collapse happens because not enough material is available to support the walls of an excavation or too much stress is placed on the ground surrounding the trench, such as the weight of an excavator.

5.1.0 Excavation Planning

Any excavation project requires a great deal of planning by specially trained workers before it can safely begin. One of these tasks is to arrange for buried utilities in the excavation area to be located and marked. Each state has its own agency to coordinate the location of buried utilities. They go by names such as One Call, Dig Line, and Dig Safely (*Figure 32*). There is now a nationwide 811 dig safely service that contractors and others can contact. The 811 service puts callers in touch with their state agency so they can arrange to have the job site surveyed and marked.

Your company must individually contact any utility companies that do not subscribe to One Call. Utility workers will come to your work site and mark their underground cables and pipes using the APWA underground color codes shown in *Table 4*.

The lead time for the One-Call agency to mark the job site varies by area and season. Do not start a digging job until the area has been cleared. Digging in uncleared areas can cause serious injuries, interruption of services, damage to the environment or equipment, or job delays.

5.2.0 Trenching Hazards

Working around excavations is one of the most hazardous jobs. When heavy equipment is used around trenches, safety precautions must be exercised at all times to prevent injury to yourself and others.

Classroom

Note that buried utilities must be located and marked before excavations begin.

Audiovisual

Show Slide 37 (Figure 32). Describe utility notification systems such as the 8-1-1, One Call, and Dig Line.

Show Slide 38 (Table 4). Discuss the use of color codes for buried utilities.

Classroom

Describe the different types of trenching hazards and their causes.

Teaching Tip/Notes

22102-12 Heavy Equipment Safety Module Two 35

Classroom

Discuss trench failure.

FRONT BACK

22102-12_F32.EPS

Figure 32 Dig safely card.

Table 4 APWA Underground Utility Color Codes

Color	Meaning
Red	Electric power lines, cables, conduit and lighting cables
Yellow	Gas, oil, steam, petroleum, or gaseous material
Orange	Communication, alarm or signal lines, cables or conduits
Blue	Potable water
Green	Sewers and drain lines
Purple	Reclaimed water, irrigation and slurry lines
White	Proposed excavation
Pink	Temporary survey markings

Some of the hazards that may be encountered during an excavation include the following:

- Cave-ins due to trench failure
- Flooding from broken water or sewer mains
- Electrical shock from striking electrical cable in the trench or striking overhead lines
- Toxic liquid or gas leaks from nearby facilities, pipes, or idling vehicles
- Falling dirt or rocks from an excavator bucket
- Oxygen-depleted environment

5.3.0 Trench Failure

A number of stresses and weaknesses can occur during excavations that affect the stability of trenches. The most common hazard during excavation is trench failure or cave-in. This hap-

Teaching Tip/Notes

pens when the walls of the trench are not strong enough to survive outside forces, causing the trench walls to collapse. Cave-ins can happen without warning and can bury workers in the trench.

5.4.0 Soil Hazards

Soil type is a major factor to consider in trenching operations. Only a company-assigned qualified person has the experience, training, and education to determine if the soil in and around a trench is safe and stable. However, it is still your responsibility to know the basics about soil and its associated hazards.

Soil is comprised of soil particles, air, and water in varying quantities. Surface soil often contains some amount of organic matter like decaying plant matter. The soil that is found on most sites is a mixture of many mineral grains coming from several kinds of rocks. Average soils are usually a mixture of two or three materials, such as sand and silt or silt and clay. The type of mixture determines the soil characteristics.

Each of the various soil types, depending on the condition of the soil at the time of the excavation, behaves differently. Sandy soil tends to collapse straight down. Wet clays and loams tend to slab off the side of the trench. Firm, dry clays and loams tend to crack. Wet sand and gravel tend to slide. These conditions are shown in *Figure 33.* Be aware of the type of soil you are working in and know how it behaves.

5.4.1 Type A Soil

Type A soil refers to solid soil with a good strength. Examples of cohesive soils are clay, silty clay, sandy clay, and clay loam. Excavations in Type A soil can have a maximum allowable slope of 53 degrees, as shown in *Figure 34.*

No soil can be considered Type A if any of the following conditions exist:

- The soil is fissured. Fissured means a soil material has a tendency to break along definite planes of fracture with little resistance, or a material that has cracks, such as tension cracks, in an exposed surface.
- The soil can be affected by vibration from heavy traffic, pile driving, or other similar effects.
- The soil has been previously disturbed. An exception to this rule occurs when the excavation is 12 feet deep or less and will remain open 24 hours or less.

Case History

Employees were laying sewer pipe in a 15-foot-deep trench. The sides of the trench, 4 feet wide at the bottom and 15 feet wide at the top, were not shored or protected to prevent a cave in. The trench was not protected from vibration caused by heavy vehicle traffic on the road nearby. To leave the trench, employees had to exit by climbing over the spoil. As they attempted to leave the trench, there was a small cave-in covering one employee to his ankles. When the other employee went to his co-worker's aid, another cave-in covered the first worker to his waist. The first employee died of a rupture of the right ventricle of his heart at the scene of the cave-in. The other employee suffered a hip injury.

The Bottom Line: Always use shoring, sloping, or shielding, or properly slope the sides of the excavation. In this case, there was no means of egress. Don't attempt a rescue; wait for professionals.

Source: The Occupational Safety and Health Administration (OSHA)

Classroom

Explain how the composition of soil affects the way soil will act in a trench.

Audiovisual

Show Slide 39 (Figure 33). Discuss soil behaviors.

Show Slide 40 (Figure 34). Describe the maximum allowable slope for various soils. Point out conditions that prevent a soil from being classed as Type A.

Teaching Tip/Notes

Classroom

Describe Type B soil and its compression strength.

Point out that Type C soil refers to gravel, loamy soil, and any soil from which water seeps freely.

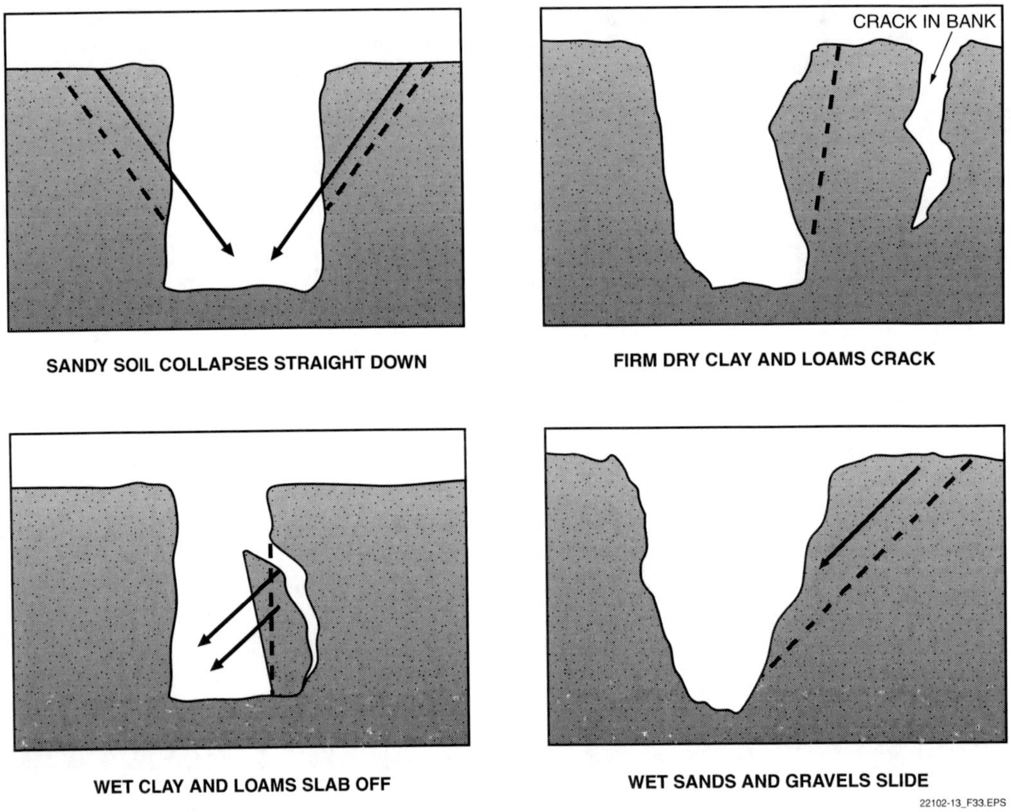

Figure 33 Soil behaviors.

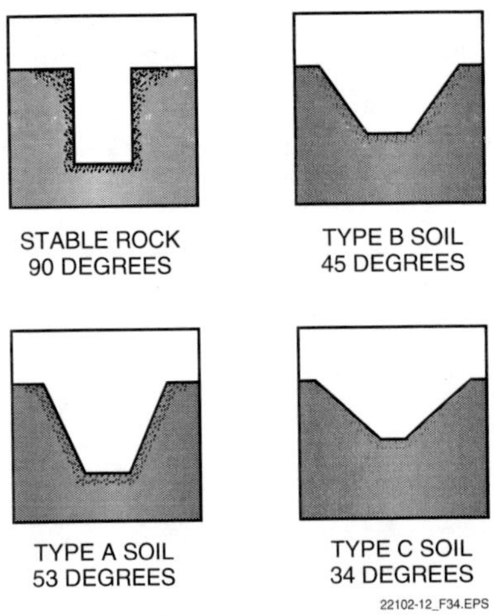

Figure 34 Maximum allowable slope for soil types.

5.4.2 Type B Soil

Type B soil refers to cohesive soils with compression strength of greater than 0.5 ton per square foot but less than 1.5 tons per square foot. It also refers to granular soils, including angular gravel (similar to crushed rock), silt, sandy loam, unstable rock, and any unstable or fissured Type A soils. Type B soils also include previously disturbed soils, except those that would fall into the Type C classification. Excavations made in Type B soils have a maximum allowable slope of 45 degrees, as shown in *Figure 34*.

5.4.3 Type C Soil

Type C soil is the most unstable soil type. Type C soil refers to cohesive soil with compression strength of 0.5 ton per square foot or less. Gravel, loamy soil, sand, any submerged soil or soil from which water is freely seeping, and unstable submerged rock are considered Type C soils. Exca-

Teaching Tip/Notes

Obtain samples of each soil type identified by OSHA and have them available for the trainees to examine.

vations made in Type C soils have a maximum allowable slope of 34 degrees, as shown in *Figure 34*.

When working in or around a trench, stay alert for changes in the trench. Watch for developing cracks, moisture, or small movements in the trench material. Alert the supervisor and co-workers to any changes. Small changes in trench walls may be an early sign of an impending cave-in.

Cave-ins are not the only cause of burial. Tons of dirt can be dumped onto workers when the excavated soil slides back into the trench. Such slides occur when the pile is place too close to the edge of the trench or when the ground beneath the pile gives way. OSHA requires a minimum of two feet between the trench wall and the spoils pile. This distance can vary depending on the type of soil and the height of the spoils pile. This area must also be kept free of heavy equipment, tools, and material.

5.5.0 Guidelines for Working Near a Trench

When working with heavy equipment around any excavation or trench, the operator is responsible for his or her personal safety, as well as that of others in the work zone. The following guidelines must be observed by heavy equipment operators as well as other workers in the area to ensure everyone's safety:

- Never move heavy equipment into an excavation area without the approval of the OSHA-approved competent person on site.
- A competent person must inspect the excavation area daily, at the start of each shift, or when conditions change. Conditions that can affect an excavation include changes in the environment, such as rain, frost, severe vibration from heavy equipment, rain, or change in soil conditions.
- Wear protective clothing and equipment, such as hard hats, safety glasses, work boots, and gloves. Use respirators and gas monitors if necessary.
- Do not walk under loads being handled by power shovels, cranes, or hoists.
- Stay clear of any vehicle that is being loaded.
- Be alert. Watch and listen for possible dangers.
- Do not work above or below a co-worker on a sloped or benched excavation wall.
- Barricade access to excavations to protect pedestrians and vehicles.
- Keep heavy equipment out of the area unless it is actively being used for work. Special attention should be given to cranes and crane matting near an excavation.
- Keep tools, equipment, and the excavated dirt (spoil) at least 2 feet from the edge of the excavation.
- Make sure shoring, trench boxes, benching, or sloping are used for excavations and trenches over 5 feet deep.
- Make sure means of exit, such as ladders or stairways, are placed within 25 feet of any employee in a trench 4 feet or deeper.
- Stop work and immediately exit the trench if there is any potential for a cave-in. Make sure any problems are corrected before starting work again.

5.6.0 Making the Trench Safe

There are several ways to make a trenching site a safer place to work. Trenches are often reinforced with shoring or shielding systems. Other times trench walls are cut at an angle away from the trench floor to relieve pressure and avoid cave-ins. This is called sloping. Shoring, shielding, and sloping are different methods used to protect workers and equipment. It is important that you recognize the differences between them.

CAUTION

Protective systems are designed for even loads of earth. Heaving and squeezing can place uneven loads on the shielding system and may stress particular parts of the protective system.

Case History

While installing a utility pole anchor in a city sidewalk, a St. Cloud, Minnesota construction company struck and ruptured a plastic gas pipeline. While utility workers and emergency response personnel were evaluating the situation and taking the initial steps to stop the gas leak, an explosion occurred. Because of the explosion, four people died; one person was seriously injured; and 10 people, including two firefighters and a police officer, received minor injuries. In addition, six buildings were destroyed.

The Bottom Line: Always contact the local One-Call office or local utility companies before starting any digging project.

Source: Occupational Safety and Health Administration (OSHA)

Classroom

Emphasize the importance of watching a trench for changes, such as fissures, moisture, or small movements.

Discuss the causes of cave-ins.

Review the safety guidelines for working near trenches.

Explain the purpose of trenching safety systems.

Teaching Tip/Notes

This exercise will familiarize trainees with various aspects of trenching and trenching safety. Trainees will need appropriate personal protective equipment, pencils, and paper. You will need to arrange for an opportunity to observe an excavation site. If possible, have the site safety manager give a tour of the area. Include discussion topics such as shoring, trench access, and planning. Allow 60 to 90 minutes for this exercise.

1. Ask the site safety manager to give a tour of the site and point out various safety features.
2. Have the trainees take notes and write down questions during the tour.
3. Ask the site safety manager to spend some time after the tour to answer questions. Have the trainees ask their questions.

Classroom

Describe ways to make trenches safer.

Explain what a trench shield is.

Describe safety guidelines for placing shielding systems.

Audiovisual

Show Slides 41 and 42 (Figures 35 and 36).

Classroom

Point out that there must be at least one method of entering and exiting any excavation over 4 feet deep.

Stress that ladders must be placed within 25 feet of each worker.

Describe other safety requirements for ladders.

Shoring in a trench is placed against the trench walls to support them and prevent their movement and collapse (*Figure 35*). Shoring does more than provide a safe environment for workers in a trench. Because it restrains the movement of trench walls, shoring also stops the shifting of surrounding soil, which may contain buried utilities or on which sidewalks, streets, building foundations, or other structures are built.

Trench shields, also called trench boxes (*Figure 36*), are placed in unshored trenches to protect personnel from trench wall collapse. They provide no support to trench walls or surrounding soil. But for specific depths and soil conditions, trench shields withstand the side weight of a collapsing trench wall to protect workers in the event of a cave-in.

NOTE

Shielding systems are designed to protect workers in an excavation. They are not designed to prevent cave-ins.

When helping to place shielding systems, follow these safety guidelines:

- Be sure that the vertical walls of a trench box extend at least 18 inches above the lowest point where the excavation walls begin to slope. If the trench box is over a pipe, the box must extend 24 inches above the pipe.
- Make sure shielding systems are placed against the sides of the trench.
- Never permit workers to enter the trench box during installation or removal.
- Backfill the excavation as soon as the trench box has been removed.
- If a trench box is to be used in a pit or a hole, all four sides of the trench box must be protected.

WARNING!

Be aware of pinch points between trench boxes.

5.6.1 Ladders

There must be at least one method of entering and exiting all excavations over 4 feet deep. Ladders are generally used for this purpose. Ladders must be placed within 25 feet of each worker.

When ladders are used, there are a number of requirements that must be met:

- Ladder side rails must extend a minimum of 3 feet above the landing.

22102-12_F35.EPS

Figure 35 Trench shoring.

- Ladders must have nonconductive side rails if work will be performed near equipment or systems using electricity.
- Two or more ladders must be used where 25 or more workers are working in an excavation in which ladders serve as the primary means of entry and exit or where ladders are used for two-way traffic in and out of the trench.
- All ladders must be inspected before each use for signs of damage or defects.
- Damaged ladders should be labeled Do Not Use and removed from service until repaired.

Case History

An employee was working in a trench 4 feet wide and 7 feet deep. About 30 feet away a backhoe was straddling the trench when the backhoe operator noticed a large chunk of dirt falling from the side wall behind the worker in the trench, he called out a warning. Before the worker could climb out, 6 to 8 feet of the trench wall had collapsed on him and covered his body up to his neck. He suffocated before the backhoe operator could dig him out. There were no exit ladders. No sloping, shoring or other protective system had been used in the trench.

The Bottom Line: OSHA's construction standards include several requirements, which, if they had been followed here, might have prevented this fatality. Provide protection from cave-ins by an adequate protective system and provide a means of egress within 25 feet of employees in a trench 4 feet or more deep, such as a ladder or stairway.

Source: The Occupational Safety and Health Administration (OSHA)

Teaching Tip/Notes

Figure 36 Trench box.

- Use ladders only on stable or level surfaces.
- Secure ladders when they are used in any location where they can be displaced by excavation activities or traffic.
- While on a ladder, do not carry any object or load that could cause you to lose your balance.
- Exercise caution whenever using a trench ladder.

5.6.2 Sloping Systems

A sloping system is a method of protecting workers from cave-ins. Sloping is accomplished by inclining the sides of an excavation. The angle of incline varies according to such factors as the soil type, environmental conditions of exposure, and nearby loads. There are three general classifications of soil types, and one type of rock. For each classification of soil type, OSHA defines maximum angles for the slope of the walls. MSHA uses the same guidelines for surface excavations but has special requirements for shafts and tunnels. The designation and selection of the proper sloping system is far more complex than described in this section. The depth of the trench, the amount of time the trench is to remain open, and other factors affect the maximum allowable slope. The important thing to understand is that a particular slope has been selected. When excavating slopes, follow the plans carefully.

5.7.0 Personal Safety

To stay safe, be aware of all hazards on the job site. When asked to list hazards of their profession, most construction workers would put heavy equipment, chemicals, and many others well ahead of hazards from soils, demolition dust, and sewage. While diseases from these sources are rare, they do occur. Although most of the time the diseases are short-lived and mild, they can sometimes be severe.

5.7.1 Soil and Demolition Dust

Excavation work generates a great deal of dust. The dust may contain animal droppings and other organisms that can cause diseases. Most of the illnesses that result from breathing this dust are mild, but a few can be serious. Fungi that occur naturally in the environment can cause serious illnesses when inhaled. Some diseases caused by

Case History

An operator working on a trenching job tried to position a backhoe to backfill an electrical trench on the top outer edge. As the operator steered into softer, excavated soil, the equipment tilted onto its side and slid downward, stopping only when it engaged the chain link fence 10 feet below the trench. The operator was wearing a seatbelt, and no injuries resulted from this incident.

The Bottom Line: If the employee had not been wearing his seatbelt and the fence had not been in place, this incident could have easily resulted in a very serious injury.

Source: The Occupational Safety and Health Administration (OSHA)

Classroom

Explain that a sloping system is a way of protecting workers from cave-ins.

Note the OSHA and MSHA definitions for surface excavations.

Emphasize the importance of being aware of hazards on the job site.

Describe some of the hazards found in ordinary soil and dust.

Teaching Tip/Notes

Classroom

Point out that dust from sources such as concrete can also be hazardous.

Safety

Stress the dangers of asbestos fibers.

Audiovisual

Show Slide 43 (Figure 37).

Classroom

Describe some of the diseases that can occur with exposure to sewage.

Point out some of the contaminants that may be in the soil where new construction or demolition is taking place.

Describe warning signs for contaminants.

bird droppings attack the lungs; these can be severe in people with pre-existing conditions such as asthma. Before demolition work is allowed to begin, a company safety inspector should have evaluated the site for materials that could be harmful. In some demolition situations, respirators and protective garments such as coveralls may be required.

> **WARNING!**
> Breathing asbestos fibers or getting them on your skin can cause serious illness. Proper PPE must be worn.

Asbestos was once commonly used in construction materials, such as insulation, ceiling tiles, siding, and flooring. Since that time, asbestos dust has been found to cause a variety of health problems ranging from simple skin irritation to cancer. Because of these hazards, asbestos use in this country is now extremely limited. Virtually all demolition work on asbestos material creates dust, thus exposing workers to this hazard. Asbestos removal may only be done by properly licensed companies. It requires specialized protective clothing, breathing apparatus, and containment equipment. Structures must be inspected by a qualified person for the presence of asbestos before demolition begins.

Another serious dust that may be encountered is concrete dust. Concrete dust can come from either the demolition of an existing structure or road (*Figure 37*), or from working in concrete-stabilized soil. The dust created when breaking up dry concrete has the same hazard as that of new cement. Concrete dust can cause a fatal, incurable lung disease called silicosis. Silicosis can occur after breathing high concentrations of concrete dust for a short period or low concentrations for a long period (10 or more years). Do not eat or drink in the area.

You can help to limit your exposure to soil contaminants and concrete dust by wearing a well-fitted respirator or proper PPE. Shower and change into clean clothes at your job site when possible. Avoid eating or drinking in the dusty area. Periodically spray the demolition area with water.

5.7.2 Sewage

Construction workers on an excavation site may uncover live sewers or septic tanks. Not only is this unpleasant, but contact with raw sewage can place workers at risk for disease. Stomach and intestinal diseases, including hepatitis, are common results. Wear latex or similar gloves when working around sewage. Wash your hands frequently and avoid touching your face. Do not eat or drink in the area. If you do become ill after working on a job where you came in contact with sewage, tell your employer and doctor. Employers must offer the hepatitis vaccine to employees.

22102-12_F37.EPS

Figure 37 Concrete demolition.

5.7.3 Soil Contamination

When working at a new construction site or performing demolition on a building, there is always the possibility of uncovering a contaminated site. Contaminants can be in the soil, buried underground, or stored in some container such as a barrel. Identifying contaminants is a complicated and time-consuming procedure that requires advanced education and training. Heavy equipment operators are not expected to identify contaminants, but are required to report any potential contaminations to a supervisor. Some warning signs of potential contamination are as follows:

- Puddles of fluid on the ground, especially fluid with an unusual color or odor
- Any fluid seeping out of the ground
- Unusual or foul odors
- Unmarked barrels or tanks—often buried or otherwise camouflaged

Teaching Tip/Notes

SUMMARY

Working in and around heavy equipment is dangerous. In addition to the physical hazards associated with heavy equipment, the equipment adds noise and confusion to the job site. Heavy equipment operators need to operate machinery safely and follow company and site safety procedures. One important safety responsibility of the equipment operator is to keep track of nearby workers. At the same time, other workers need to be sure they stay in plain sight of the heavy equipment operator while completing their assigned tasks. They must constantly be aware of the presence of the equipment.

The operator must be aware of the limitations of the equipment being operated. The operator's manual from the manufacturer provides instructions for safely operating the equipment, including the capabilities and limitations of the equipment. It is up to the operator to be familiar with the content of this manual and to operate the equipment within its intended parameters.

Equipment operators work outdoors, and are often exposed to the extremes of hot and cold weather, as well as other site hazards. It is therefore important to dress appropriately and wear appropriate PPE. Heavy equipment is used constantly on the edges of trenches and other excavations, so the operator must be familiar with the ways different soils behave. In addition, the operator must be conscious of the safety of anyone working in a trench or excavation.

Safety on the job is everyone's responsibility. Government organizations such as OSHA set requirements for employers to keep employees safe, but in the end, it is up to you to use the available information and equipment to keep yourself safe. Your safety is your responsibility.

Classroom

Summarize the major concepts presented in this module.

Teaching Tip/Notes

Classroom

Have the trainees complete the Review Questions, and go over the answers prior to administering the Module Examination.

Have the trainees complete the Trade Terms Quiz, and go over the answers prior to administering the Module Examination.

? Examination

Administer the Module Examination. Record the results on Training Report Form 200, and submit the results to the Training Program Sponsor.

✓ Performance Testing

Administer the Performance Test and fill out the Performance Profile Sheets for each trainee. If desired, trainee proficiency noted during laboratory sessions may be used to complete the Performance Test. Record the results on Training Report Form 200 and submit the results to the Training Program Sponsor.

Review Questions

1. Caution signs with a yellow background and black letters are used to inform workers of _____.
   a. dangerous conditions
   b. potential hazards or unsafe conditions
   c. hazardous materials in the area
   d. weather conditions that could become hazardous

2. Safety tags can be used in place of safety warning signs.
   a. True
   b. False

3. Barricades and barriers are used _____.
   a. only for traffic diversion on highway construction sites
   b. when there is not enough time to put up a sign
   c. to alert workers to potential hazards
   d. as a last resort to keep vehicles out of the work area

4. In areas where the ground is too soft to permit a machine to operate safely on a grade, it is a good idea to put up a _____.
   a. barrier
   b. flag
   c. caution sign
   d. danger sign

5. The spacing of temporary traffic control signs will depend on the road's _____.
   a. amount of traffic
   b. surface condition
   c. normal posted speed
   d. proximity to pedestrian traffic

6. Channeling devices are used in a temporary traffic control area _____.
   a. when flaggers are not available
   b. to guide traffic into a desired path
   c. to protect workers from speeding traffic
   d. to keep heavy equipment out of the road

7. The area next to the border that is designed to protect workers from closely passing vehicles is called the _____.
   a. buffer zone
   b. transition area
   c. pedestrian zone
   d. advance warning area

8. To use the Stop/Slow paddle, the flagger faces road users and _____.
   a. waves the flag rapidly across the lane of oncoming traffic
   b. steps into the lane of oncoming traffic and raises the flag
   c. repeatedly raises and lowers the flag to shoulder level
   d. extends the arm holding the paddle horizontally away from the body

9. For each hazardous substance in a shipment, OSHA and MSHA require that suppliers of hazardous chemicals provide _____.
   a. a respirator
   b. splash-proof goggles
   c. chemical-resistant gloves
   d. a material safety data sheet (MSDS)

10. The maximum amount of gasoline that OSHA allows to be stored outside an approved storage enclosure is _____.
    a. 5 gallons
    b. 10 gallons
    c. 25 gallons
    d. 50 gallons

11. What are the advanced symptoms of carbon monoxide poisoning?
    a. Dizziness
    b. Headaches
    c. Weakness and confusion
    d. Staggering and falling

12. Which of the following noise sources has the highest dBA level?
    a. Thunder
    b. Stock car races
    c. A jet plane taking off
    d. Shouting near the ear

Teaching Tip/Notes

13. Pale, clammy skin; heavy sweating with nausea and possible vomiting; a fast, weak pulse; and possible fainting are symptoms of _____.
    a. exposure to cold
    b. heat exhaustion
    c. hypothermia
    d. heat stroke

14. Workers may perform maintenance on a vehicle or equipment if they _____.
    a. are very knowledgeable about engine work
    b. are qualified and authorized by the employer
    c. are the only person present who knows what to do
    d. have thoroughly studied the manufacturer's manual

15. The only person who can decide if the soil in and around a trench is safe and stable is _____.
    a. a supervisor
    b. the engineer
    c. a company-assigned qualified person
    d. the most experienced crew member

Teaching Tip/Notes

Trade Terms Quiz

Fill in the blank with the correct term that you learned from your study of this module.

1. A person specially trained to direct traffic through and around a work zone is called a ________.
2. Cones, barrels, or markers that are used to move traffic from its normal lane to a desired lane are known as ________.
3. The point where all of an object's weight is evenly distributed is the ________.
4. The devices and plan used to safely divert traffic when the normal use of the road is disrupted by construction are called ________.
5. An area in which two moving equipment parts come together is known as a ________.

Trade Terms

Center of gravity
Channeling devices
Flagger
Pinch Point
Temporary traffic control (TTC)

Teaching Tip/Notes

Cornerstone of Craftsmanship

Mark S. Jones

Operations Support Manager
Saiia Construction Company LLC

Mark Jones is one of the lucky few who were able to parlay a summer job into a lifelong successful career. His is the classic story of a man who started at the bottom in a company and worked his way up to a senior management position.

How did you get started in the construction industry?
When I was at the University of Alabama studying corporate finance, I was fortunate to be able to spend summers working for Saiia Construction as a heavy equipment operator. When I graduated, I was offered a full-time job as operations coordinator. That was 16 years ago. Since then, I worked my way up to my current position as Operations Support Manager. I found a home at Saiia. I enjoyed the work and there was plenty of room for advancement for an employee who was willing to work hard and learn the business. One of the things I really appreciate about my employer is the company's commitment to training its employees.

What inspired you to enter the construction industry?
When I was working summers at Saiia Construction, my mentor was Donald Stansberry, who is Vice President of Operations at Saiia. I hadn't considered a career in construction, but his influence convinced me to give it a try.

What kinds of work have you done during your career?
During my career, I have worked on power plants, landfills, subdivisions, shopping centers, quarries, and automobile manufacturing plants. After graduation, I started out with Saiia Construction as Operations Coordinator. In that position, I was responsible for placing heavy equipment operators on jobs and for recruiting and hiring operators. As Operations Support Manager for Saiia, I am responsible for field operations. This includes responsibility for superintendents, pipe crews, and surveying crews. I was also one of the first NCCER heavy equipment instructors in the country.

What do you enjoy most about your job?
I enjoy working in the outdoors and the challenge of pulling a complex project together. There are so many things happening at once on a construction site and so many different skills making them happen. It takes a great deal of teamwork and it's exciting to me to be one of the people who make it all come together. I also get a great deal of satisfaction when I drive by a finished project that I worked on, especially one that I was responsible for. My work as a subject matter expert in the development of the Heavy Equipment training program has been very rewarding because it gives me the opportunity to bring my knowledge of the industry to people who are just starting out.

How do you feel about training in general and NCCER training specifically?
Training, especially safety training, is an important part of a successful career in construction. Anyone who wants to succeed in the construction industry needs to make a lifelong commitment to learning because new equipment, methods, and safety regulations are constantly evolving. Anyone who doesn't make an effort to keep up will fall behind. NCCER training is the industry standard. Their training courses are reviewed and developed by industry professionals and their credentials are accepted by companies all over the United States.

What kinds of training have you had?
In addition to my college education, I have completed MSHA instructor training, the OSHA 10-hour course, the NCCER Supervisor course, and the NCCER Project Management course. An in-depth knowledge of OSHA safety requirements is critical for anyone working in construction. In addition, anyone operating heavy equipment, especially in mining or quarrying work, must also know the MSHA safety regulations.

Teaching Tip/Notes

Would you recommend a career in construction to others?
Yes. There is a great demand for people who are skilled at operating heavy equipment. The pay is excellent and the work is challenging.

What advice would you give to someone just entering the heavy equipment field?
Have patience and always keep your mind on your work because distractions are the number one reason why people have accidents or are injured in the heavy equipment field. Listen to the people that have made their careers in the industry.

How would you define craftsmanship?
Craftsmanship is having a unique skill that enables you to perform a craft with a consistently high standard of quality.

Teaching Tip/Notes

Appendix

MSDS FOR DIESEL FUEL

Teaching Tip/Notes

Material Safety Data Sheet

MSDS ID NO.: 0116MAR019
Revision date: 12/07/2010

1. CHEMICAL PRODUCT AND COMPANY INFORMATION

Product name: Marathon No. 2 High Sulfur Fuel Oil Dyed 3000 ppm Sulfur Max
Synonym: Fuel Oil No. 2 Non-Road Use, Dyed; No. 2 Fuel Oil Dyed (0.3% Sulfur Max); No. 2 NR 3000 HS Fuel Oil; No. 2 Fuel Oil Dyed 0.3% Sulfur Max; HS No. 2 Fuel Oil, Non-Road Use, Dyed
Chemical Family: Petroleum Hydrocarbon
Formula: Mixture

Manufacturer:
Marathon Petroleum Company LP
539 South Main Street
Findlay OH 45840

Other information: 419-421-3070
Emergency telephone number: 877-627-5463

2. COMPOSITION/INFORMATION ON INGREDIENTS

No. 2 Fuel Oil is a complex mixture of paraffins, cycloparaffins, olefins and aromatic hydrocarbons having hydrocarbon chain lengths predominantly in the range of C11 through C20. May contain a trace amount of benzene (<0.01%). Can contain small amounts of dye and other additives (<0.15%) which are not considered hazardous at the concentrations used.

Note: May contain up to 5% Renewable Diesel, CASN 928771-01-1.

Product information:

Name	CAS Number	Weight %	ACGIH Exposure Limits:	OSHA - Vacated PELs - Time Weighted Ave	Other:
Marathon No. 2 Fuel Oil Dyed (0.3% Sulfur Max)	68476-30-2	100	Skin - potential significant contribution to overall exposure by the cutaneous route 100 mg/m³ TWA		

Component Information:

Name	CAS Number	Weight %	ACGIH Exposure Limits:	OSHA - Vacated PELs - Time Weighted Ave	Other:
Saturated Hydrocarbons	Mixture	54-85			

22102-12_A01A.EPS

Teaching Tip/Notes

Name	CAS Number	Weight %	ACGIH Exposure Limits:	OSHA - Vacated PELs - Time Weighted Ave	Other:
Aromatic Hydrocarbons	Mixture	15-45			
Unsaturated Hydrocarbons	Mixture	1-6			
Naphthalene	91-20-3	0.01-0.5	Skin - potential significant contribution to overall exposure by the cutaneous route 10 ppm TWA 15 ppm STEL	= 10 ppm TWA = 50 mg/m³ TWA = 15 ppm STEL = 75 mg/m³ STEL	

Notes: The manufacturer has voluntarily elected to reflect exposure limits contained in OSHA's 1989 air contaminants standard in its MSDS's, even though certain of those exposure limits were vacated in 1992.

MSDS ID NO.: 0116MAR019 **Product name:** Marathon No. 2 High Sulfur Fuel Oil Dyed 3000 ppm Sulfur Max **Page 2 of 12**

22102-12_A01B.EPS

Teaching Tip/Notes

22102-12 Heavy Equipment Safety Module Two 51

3. HAZARDS IDENTIFICATION

EMERGENCY OVERVIEW

CAUTION!

VAPORS, FUMES, OR MISTS MAY CAUSE RESPIRATORY TRACT IRRITATION
MAY BE HARMFUL OR FATAL IF SWALLOWED
MAY CAUSE LUNG DAMAGE
OVEREXPOSURE MAY CAUSE CNS DEPRESSION

MAY CAUSE CANCER BASED ON ANIMAL DATA
SEE TOXICOLOGICAL INFORMATION SECTION FOR MORE INFORMATION

COMBUSTIBLE LIQUID AND VAPOR
VAPOR MAY CAUSE FLASH FIRE
MATERIAL MAY ACCUMULATE STATIC CHARGE

STABLE

Inhalation:

Breathing high concentrations may be harmful.
May cause central nervous system depression or effects. Symptoms may include headache, excitation, euphoria, dizziness, incoordination, drowsiness, light-headedness, blurred vision, fatigue, tremors, convulsions, loss of consciousness, coma, respiratory arrest and death, depending on the concentration and duration of exposure.
Overexposure to this material may cause systemic damage including target organ effects listed under "Toxicological Information."

Ingestion:

Swallowing this material may be harmful.
May cause irritation of the mouth, throat and gastrointestinal tract. Symptoms may include salivation, pain, nausea, vomiting and diarrhea.
Aspiration into lungs may cause chemical pneumonia and lung damage. Exposure may also cause central nervous system symptoms similar to those listed under "Inhalation" (see Inhalation section).

Skin contact:

Contact may cause reddening, itching and inflammation. Effects may become more serious with repeated or prolonged contact. Skin contact may cause harmful effects in other parts of the body.

Eye contact:

Contact may cause pain and severe reddening and inflammation of the conjunctiva. Effects may become more serious with repeated or prolonged contact.

Carcinogenic Evaluation:

Product information:

Name	IARC Carcinogens:	NTP Carcinogens:	ACGIH - Carcinogens:	OSHA - Select Carcinogens:
Marathon No. 2 Fuel Oil Dyed (0.3% Sulfur Max) 68476-30-2	NE			

MSDS ID NO.: 0116MAR019 **Product name:** Marathon No. 2 High Sulfur Fuel Oil Dyed 3000 ppm Sulfur Max **Page 3 of 12**

22102-12_A01C.EPS

Teaching Tip/Notes

Notes:

The International Agency for Research on Cancer (IARC) has determined that there is inadequate evidence for the carcinogenicity of diesel fuel/fuel oil in humans. IARC determined that there was limited evidence for the carcinogenicity of marine diesel fuel in animals. Distillate (light) diesel fuels were not classifiable as to their carcinogenicity to humans (Group 3A).

IARC has determined that there is sufficient evidence for the carcinogenicity in experimental animals of diesel engine exhaust and extracts of diesel engine exhaust particles. IARC determined that there is only limited evidence for the carcinogenicity in humans of diesel engine exhaust. However, IARC's overall evaluation has resulted in the IARC designation of diesel engine exhaust as probably carcinogenic to humans (Group 2A) because of the presence of certain engine exhaust components.

The International Agency for Research on Cancer (IARC) has also determined that there is sufficient evidence for the carcinogenicity in experimental animals of light and heavy vacuum distillates, of light and heavy catalytically cracked distillates and of cracked residues (including heavy thermocracked distillates/residues) derived from the refining of crude oil.

Component Information:

Name	IARC Carcinogens:	NTP Carcinogens:	ACGIH - Carcinogens:	OSHA - Select Carcinogens:
Naphthalene 91-20-3	Monograph 82 [2002]	Reasonably Anticipated To Be A Human Carcinogen male rat-clear evidence; female rat-clear evidence; male mice-no evidence; female mice-some evidence	A4 - Not Classifiable as a Human Carcinogen	Present

Notes:

The International Agency for Research on Cancer (IARC) and the Environmental Protection Agency (EPA) have determined that naphthalene is a possible human carcinogen.

22102-12_A01D.EPS

Teaching Tip/Notes

22102-12 Heavy Equipment Safety Module Two 53

4. FIRST AID MEASURES

Eye Contact:

Flush immediately with large amounts of water for at least 15 minutes. Eyelids should be held away from the eyeball to ensure thorough rinsing. GET IMMEDIATE MEDICAL ATTENTION.

Skin Contact:

Immediately wash exposed skin with plenty of soap and water while removing contaminated clothing and shoes. Get medical attention if irritation persists. Place contaminated clothing in closed container until cleaned or discarded. If clothing is to be laundered, inform the person performing the operation of contaminant's hazardous properties.

Ingestion:

Do not induce vomiting. If spontaneous vomiting is about to occur, place victim's head below knees. If victim is drowsy or unconscious, place on the left side with head down. Never give anything by mouth to an unconscious person. Keep affected person warm and at rest. GET IMMEDIATE MEDICAL ATTENTION.

Inhalation:

Remove to fresh air. If not breathing, institute rescue breathing. If breathing is difficult, ensure airway is clear and give oxygen. If heart has stopped, immediately begin cardiopulmonary resuscitation (CPR). Keep affected person warm and at rest. GET IMMEDIATE MEDICAL ATTENTION.

NOTES TO PHYSICIAN:

INGESTION: If ingested this material represents a significant aspiration and chemical pneumonitis hazard. Induction of emesis is not recommended.

Medical Conditions Aggravated By Exposure:

skin,

5. FIRE FIGHTING MEASURES

Suitable extinguishing media: For small fires, Class B fire extinguishing media such as CO2, dry chemical, foam (AFFF/ATC) or water spray can be used. For large fires, water spray, fog or foam (AFFF/ATC) can be used. Fire fighting should be attempted only by those who are adequately trained and equipped with proper protective equipment.

Specific hazards: This product is not a combustible liquid per the OSHA Hazard Communication Standard, but will ignite and burn at temperatures exceeding the flash point. For additional fire related information, see NFPA 30 or the North American Emergency Response Guide 128.

Special protective equipment for firefighters: Avoid using straight water streams. Water spray and foam (AFFF/ATC) must be applied carefully to avoid frothing and from as far a distance as possible. Avoid excessive water spray application. Keep surrounding area cool with water spray from a distance and prevent further ignition of combustible material. Keep run-off water out of sewers and water sources.

Flash point: 130-190 F
Autoignition temperature: 637 F
Flammable limits in air - lower (%): 0.7

MSDS ID NO.: 0116MAR019 **Product name:** Marathon No. 2 High Sulfur Fuel Oil Dyed 3000 ppm Sulfur Max **Page 5 of 12**

22102-12_A01E.EPS

Teaching Tip/Notes

5. FIRE FIGHTING MEASURES

Flammable limits in air - upper (%): 5.0

NFPA rating:
Health: 1
Flammability: 2
Instability: 1
Other: -

6. ACCIDENTAL RELEASE MEASURES

Personal precautions: Keep public away. Isolate and evacuate area. Shut off source if safe to do so. Eliminate all ignition sources. Advise authorities and National Response Center (800-424-8802) if the product has entered a water course or sewer. Notify local health and pollution control agencies, if appropriate. Contain liquid with sand or soil. Recover and return free product to proper containers. Use suitable absorbent materials such as vermiculite, sand, or clay to clean up residual liquids.

7. HANDLING AND STORAGE

Handling:
Comply with all applicable EPA, OSHA, NFPA and consistent state and local requirements. Use appropriate grounding and bonding practices. Store in properly closed containers that are appropriately labeled and in a cool well-ventilated area. Do not expose to heat, open flames, strong oxidizers or other sources of ignition. Do not cut, drill, grind or weld on empty containers since they may contain explosive residues.

Avoid repeated and prolonged skin contact. Never siphon this product by mouth. Exercise good personal hygiene including removal of soiled clothing and prompt washing with soap and water.

8. EXPOSURE CONTROLS / PERSONAL PROTECTION

PERSONAL PROTECTIVE EQUIPMENT

Engineering measures: Local or general exhaust required when using at elevated temperatures that generate vapors or mists.

Respiratory protection: Use approved organic vapor chemical cartridge or supplied air respirators when material produces vapors that exceed permissible limits or excessive vapors are generated. Observe respirator assigned protection factors (APFs) criteria cited in federal OSHA 1910.134. Self-contained breathing apparatus should be used for fire fighting.

Skin and body protection: Neoprene, nitrile, polyvinyl alcohol (PVA), polyvinyl chloride and polyurethane gloves to prevent skin contact.

Eye protection: No special eye protection is normally required. Where splashing is possible, wear safety glasses with side shields.

Hygiene measures: No special protective clothing is normally required. Select protective clothing depending on industrial operations. Use mechanical ventilation equipment that is explosion-proof.

9. PHYSICAL AND CHEMICAL PROPERTIES:

Appearance: Red Liquid

22102-12_A01F.EPS

Teaching Tip/Notes

22102-12 Heavy Equipment Safety Module Two 55

9. PHYSICAL AND CHEMICAL PROPERTIES:

Physical state (Solid/Liquid/Gas):	Liquid
Substance type (Pure/Mixture):	Mixture
Color:	Red
Odor:	Slight Hydrocarbon
Molecular weight:	180
pH:	Neutral
Boiling point/range (5-95%):	400-640 F
Melting point/range:	Not determined.
Decomposition temperature:	Not applicable.
Specific gravity:	C.A. 0.8
Density:	6.76 lbs/gal
Bulk density:	No data available.
Vapor density:	4-5
Vapor pressure:	1-10 mm Hg @ 100 F
Evaporation rate:	No data available.
Solubility:	Negligible
Solubility in other solvents:	No data available.
Partition coefficient (n-octanol/water):	No data available.
VOC content(%):	10%
Viscosity:	1.9-3.4 @ 40 C

10. STABILITY AND REACTIVITY

Stability:	The material is stable at 70 F, 760 mm pressure.
Polymerization:	Will not occur.
Hazardous decomposition products:	Combustion produces carbon monoxide, aldehydes, aromatic and other hydrocarbons.
Materials to avoid:	Strong oxidizers such as nitrates, perchlorates, chlorine, fluorine.
Conditions to avoid:	Excessive heat, sources of ignition and open flames.

11. TOXICOLOGICAL INFORMATION

Acute toxicity:

Product information:

Name	CAS Number	Inhalation:	Dermal:	Oral:
Marathon No. 2 Fuel Oil Dyed (0.3% Sulfur Max)	68476-30-2	>2 mg/l for 4 hr [Dog]	>5 ml/kg [Rabbit]	9-16 ml/kg [Rat]

Toxicology Information:

MSDS ID NO.: 0116MAR019 **Product name:** Marathon No. 2 High Sulfur Fuel Oil Dyed 3000 ppm Sulfur Max **Page 7 of 12**

22102-12_A01G.EPS

Teaching Tip/Notes

MIDDLE DISTILLATES, PETROLEUM: Long-term repeated (lifetime) skin exposure to similar materials has been reported to result in an increase in skin tumors in laboratory rodents. The relevance of these findings to humans is not clear at this time.

MIDDLE DISTILLATES WITH CRACKED STOCKS: Light cracked distillates have been shown to be carcinogenic in animal tests and have tested positive with in vitro genotoxicity tests. Repeated dermal exposures to high concentrations in test animals resulted in reduced litter size and litter weight, and increased fetal resorptions at maternally toxic doses. Dermal exposure to high concentrations resulted in severe skin irritation with weight loss and some mortality. Inhalation exposure to high concentrations resulted in respiratory tract irritation, lung changes/infiltration/accumulation, and reduction in lung function.

ISOPARAFFINS: Studies in laboratory animals have shown that long-term exposure to similar materials (isoparaffins) can cause kidney damage and kidney cancer in male laboratory rats. However, in-depth research indicates that these findings are unique to the male rat, and that these effects are not relevant to humans.

NAPHTHALENE: Severe jaundice, neurotoxicity (kernicterus) and fatalities have been reported in young children and infants as a result of hemolytic anemia from overexposure to naphthalene. Persons with Glucose 6-phosphate dehydrogenase (G6PD) deficiency are more prone to the hemolytic effects of naphthalene. Adverse effects on the kidney have been reported in persons overexposed to naphthalene but these effects are believed to be a consequence of hemolytic anemia, and not a direct effect. Hemolytic anemia has been observed in laboratory animals exposed to naphthalene. Laboratory rodents exposed to naphthalene vapor for 2 years (lifetime studies) developed non-neoplastic and neoplastic tumors and inflammatory lesions of the nasal and respiratory tract. Cataracts and other adverse effects on the eye have been observed in laboratory animals exposed to high levels of naphthalene. Findings from a large number of bacterial and mammalian cell mutation assays have been negative. A few studies have shown chromosomal effects (elevated levels of Sister Chromatid Exchange or chromosomal aberrations) in vitro. Naphthalene has been classified as Possibly Carcinogenic to Humans (2B) by IARC, based on findings from studies in laboratory animals.

DIESEL EXHAUST: Chronic inhalation studies of whole diesel engine exhaust in mice and rats produced a significant increase in lung tumors. Combustion of kerosine and/or diesel fuels produces gases and particulates which include carbon monoxide, carbon dioxide, oxides of nitrogen and/or sulfur and hydrocarbons. Significant exposure to carbon monoxide vapors decreases the oxygen carrying capacity of the blood and may cause tissue hypoxia via formation of carboxyhemoglobin.

TARGET ORGANS: central nervous system, skin, respiratory system, lungs, kidney, liver, thymus, reproductive organs,

MSDS ID NO.: 0116MAR019 **Product name:** Marathon No. 2 High Sulfur Fuel Oil Dyed 3000 ppm Sulfur Max **Page 8 of 12**

22102-12_A01H.EPS

Teaching Tip/Notes

12. ECOTOXICOLOGICAL INFORMATION

Mobility:
May partition into air, soil and water.

Ecotoxicity:
Toxic to aquatic organisms.

Bioaccummulation:
Not expected to bioaccumulate in aquatic organisms.

Persistance/Biodegradation:
Readily biodegradable in the environment.

13. DISPOSAL CONSIDERATIONS

Cleanup Considerations:
This product as produced is not specifically listed as an EPA RCRA hazardous waste according to federal regulations (40 CFR 261). However, when discarded or disposed of, it may meet the criteria of an "characteristic" hazardous waste. This material could become a hazardous waste if mixed or contaminated with a hazardous waste or other substance(s). It is the responsibility of the user to determine if disposal material is hazardous according to federal, state and local regulations.

14. TRANSPORT INFORMATION

49 CFR 172.101:

DOT:

Transport Information: This material when transported via US commerce would be regulated by DOT Regulations.

Proper shipping name:	Fuel Oil, No. 2
UN/Identification No:	NA 1993
Hazard Class:	3
Packing group:	III
DOT reportable quantity (lbs):	Not applicable.

Proper shipping name:	Fuel Oil, No. 2
UN/Identification No:	NA 1993
Hazard Class:	3
Packing group:	III

15. REGULATORY INFORMATION

US Federal Regulatory Information:

MSDS ID NO.: 0116MAR019 **Product name:** Marathon No. 2 High Sulfur Fuel Oil Dyed 3000 ppm Sulfur Max **Page 9 of 12**

22102-12_A01I.EPS

Teaching Tip/Notes

58 NCCER – *Heavy Equipment Operations Level One* 22102-12

US TSCA Chemical Inventory Section 8(b): This product and/or its components are listed on the TSCA Chemical Inventory.

OSHA Hazard Communication Standard: This product has been evaluated and determined to be hazardous as defined in OSHA's Hazard Communication Standard.

EPA Superfund Amendment & Reauthorization Act (SARA):

SARA Section 302: This product contains the following component(s) that have been listed on EPA's Extremely Hazardous Substance (EHS) List:

Name	CERCLA/SARA - Section 302 Extremely Hazardous Substances and TPQs
Saturated Hydrocarbons	NA
Aromatic Hydrocarbons	NA
Unsaturated Hydrocarbons	NA
Naphthalene	NA

SARA Section 304: This product contains the following component(s) identified either as an EHS or a CERCLA Hazardous substance which in case of a spill or release may be subject to SARA reporting requirements:

Name	CERCLA/SARA - Hazardous Substances and their Reportable Quantities
Saturated Hydrocarbons	NA
Aromatic Hydrocarbons	NA
Unsaturated Hydrocarbons	NA
Naphthalene	= 100 lb final RQ = 45.4 kg final RQ

SARA Section 311/312 The following EPA hazard categories apply to this product:

Acute Health Hazard
Fire Hazard
Chronic Health Hazard

SARA Section 313: This product contains the following component(s) that may be subject to reporting on the Toxic Release Inventory (TRI) From R:

Name	CERCLA/SARA 313 Emission reporting:
Saturated Hydrocarbons	None
Aromatic Hydrocarbons	None
Unsaturated Hydrocarbons	None
Naphthalene	= 0.1 % de minimis concentration

State and Community Right-To-Know Regulations:
The following component(s) of this material are identified on the regulatory lists below:

Saturated Hydrocarbons

Louisiana Right-To-Know:	Not Listed
California Proposition 65:	Not Listed
New Jersey Right-To-Know:	Not Listed.
Pennsylvania Right-To-Know:	Not Listed.
Massachusetts Right-To Know:	Not Listed.
Florida substance List:	Not Listed.
Rhode Island Right-To-Know:	Not Listed
Michigan critical materials register list:	Not Listed.

MSDS ID NO.: 0116MAR019 **Product name:** Marathon No. 2 High Sulfur Fuel Oil Dyed 3000 ppm Sulfur Max **Page 10 of 12**

22102-12_A01J.EPS

Teaching Tip/Notes

Saturated Hydrocarbons	
Massachusetts Extraordinarily Hazardous Substances:	Not Listed
California - Regulated Carcinogens:	Not Listed
Pennsylvania RTK - Special Hazardous Substances:	Not Listed
New Jersey - Special Hazardous Substances:	Not Listed
New Jersey - Environmental Hazardous Substances List:	Not Listed
Illinois - Toxic Air Contaminants	Not Listed
New York - Reporting of Releases Part 597 - List of Hazardous Substances:	Not Listed
Aromatic Hydrocarbons	
Louisiana Right-To-Know:	Not Listed
California Proposition 65:	Not Listed
New Jersey Right-To-Know:	Not Listed.
Pennsylvania Right-To-Know:	Not Listed.
Massachusetts Right-To Know:	Not Listed.
Florida substance List:	Not Listed.
Rhode Island Right-To-Know:	Not Listed
Michigan critical materials register list:	Not Listed.
Massachusetts Extraordinarily Hazardous Substances:	Not Listed
California - Regulated Carcinogens:	Not Listed
Pennsylvania RTK - Special Hazardous Substances:	Not Listed
New Jersey - Special Hazardous Substances:	Not Listed
New Jersey - Environmental Hazardous Substances List:	Not Listed
Illinois - Toxic Air Contaminants	Not Listed
New York - Reporting of Releases Part 597 - List of Hazardous Substances:	Not Listed
Unsaturated Hydrocarbons	
Louisiana Right-To-Know:	Not Listed
California Proposition 65:	Not Listed
New Jersey Right-To-Know:	Not Listed.
Pennsylvania Right-To-Know:	Not Listed.
Massachusetts Right-To Know:	Not Listed.
Florida substance List:	Not Listed.
Rhode Island Right-To-Know:	Not Listed
Michigan critical materials register list:	Not Listed.
Massachusetts Extraordinarily Hazardous Substances:	Not Listed
California - Regulated Carcinogens:	Not Listed
Pennsylvania RTK - Special Hazardous Substances:	Not Listed
New Jersey - Special Hazardous Substances:	Not Listed
New Jersey - Environmental Hazardous Substances List:	Not Listed
Illinois - Toxic Air Contaminants	Not Listed
New York - Reporting of Releases Part 597 - List of Hazardous Substances:	Not Listed
Naphthalene	
Louisiana Right-To-Know:	Not Listed

MSDS ID NO.: 0116MAR019 **Product name:** Marathon No. 2 High Sulfur Fuel Oil Dyed 3000 ppm Sulfur Max **Page 11 of 12**

22102-12_A01K.EPS

Teaching Tip/Notes

60 NCCER – *Heavy Equipment Operations Level One* 22102-12

Saturated Hydrocarbons

California Proposition 65:	carcinogen, initial date 4/19/02
New Jersey Right-To-Know:	sn 1322
Pennsylvania Right-To-Know:	Environmental hazard
Massachusetts Right-To Know:	Present
Florida substance List:	Not Listed.
Rhode Island Right-To-Know:	Toxic; Flammable
Michigan critical materials register list:	Not Listed.
Massachusetts Extraordinarily Hazardous Substances:	Not Listed
California - Regulated Carcinogens:	Not Listed
Pennsylvania RTK - Special Hazardous Substances:	Not Listed
New Jersey - Special Hazardous Substances:	carcinogen
New Jersey - Environmental Hazardous Substances List:	SN 1322 TPQ 500 lb
Illinois - Toxic Air Contaminants	Present
New York - Reporting of Releases Part 597 - List of Hazardous Substances:	= 1 lb RQ land/water = 100 lb RQ air

Canadian Regulatory Information:

Canada DSL/NDSL Inventory: This product and/or its components are listed either on the Domestic Substances List (DSL) or are exempt.

Name	Canada - WHMIS: Classifications of Substances:	Canada - WHMIS: Ingredient Disclosure:
Naphthalene	B4, D2A	1 %

NOTE: Not Applicable.

16. OTHER INFORMATION

Additional Information: No data available.

Prepared by: Mark S. Swanson, Manager, Toxicology and Product Safety

The information and recommendations contained herein are based upon tests believed to be reliable. However, Marathon Petroleum Company LP (MPC) does not guarantee their accuracy or completeness nor shall any of this information constitute a warranty, whether expressed or implied, as to the safety of the goods, the merchantability of the goods, or the fitness of the goods for a particular purpose. Adjustment to conform to actual conditions of usage maybe required. MPC assumes no responsibility for results obtained or for incidental or consequential damages, including lost profits arising from the use of these data. No warranty against infringement of any patent, copyright or trademark is made or implied.

End of Safety Data Sheet

MSDS ID NO.: 0116MAR019 **Product name:** Marathon No. 2 High Sulfur Fuel Oil Dyed 3000 ppm Sulfur Max **Page 12 of 12**

22102-12_A01L.EPS

Teaching Tip/Notes

Trade Terms Introduced in This Module

Channeling device: Cones, barrels, or markers that are used to move traffic from its normal lane to a desired lane.

Flagger: A person who is specially trained to direct traffic through and around a work zone.

Pinch points: The area in which two moving equipment parts come together.

Center of gravity: The point where all of an object's weight is evenly distributed.

Temporary traffic control (TTC): Device and plan used to safely divert traffic when the normal use of the road is disrupted due to constriction.

Teaching Tip/Notes

Additional Resources

This module presents thorough resources for task training. The following resource material is suggested for further study.

Construction Safety. 1996. Jimmie Hinze. Englewood Cliffs, NJ: Prentice Hall.

Construction Safety Council, www.buildsafe.org.

Field Safety Participant Guide, 2003. NCCER. Upper Saddle River, NJ: Prentice Hall.

Handbook of OSHA Construction Safety and Health, Second Edition. Charles D. Reece and James V. Eidson: CRC Press.

HazCom for Construction. DVD. DuPont Sustainable Solutions – Training Solutions. Virginia Beach, VA.

Manual on Uniform Traffic Control Devices for Streets and Highways. Latest edition. Washington, DC: U.S. Department of Transportation, Federal Highway Administration

MSHA excavation guidelines, www.usbr.gov.

NAHB-OSHA Jobsite Safety Handbook, 1999. Washington, DC: Home Builder Press. Available online at www.osha.gov.

OSHA Standards 29 CFR, Part 1926 Safety and Health Regulations for the Construction Industry. Washington, DC: Occupational Safety and Health Administration (OSHA), US Department of Labor, US Government Printing Office.

Safety Orientation Pocket Guide, 2004. NCCER. Upper Saddle River, NJ: Prentice Hall.

Safety Technology Participant Guide, 2003. NCCER. Upper Saddle River, NJ: Prentice Hall.

The National Institute for Occupational Safety and Health (NOISH) Centers for Disease Control and Prevention. Atlanta, GA. Available online at www.cdc.gov.

United States Department of Labor, Occupational Safety and Health Administration Home Page, www.osha.gov.

Figure Credits

Reprinted courtesy of Caterpillar Inc., Module opener, Figures 27, 29–31, and 37

Dan Nickel, Figures 1, 3, and 20

Topaz Publications, Inc., Figures 12, 28 (boot), and 32

MUTCD, 2009 Edition, published by FHWA at http://mutcd.fhwa.dot.gov, Figures 15, 16, and 21

Courtesy of Justrite Mfg. Co. LLC, Figures 22 and 23

North Safety Products USA, Figures 24 (A–C) and 26

Becki Swineheart, Figure 24 (D)

Bacou-Dalloz, Figures 25 and 28 (safety glasses)

Bon Tool Company, Figure 28 (gloves)

Bullard, Figure 28 (hard hat)

Kundel Industries, Figures 35 and 36

Marathon Petroleum Company, LLC, Appendix

Teaching Tip/Notes

ANSWERS TO REVIEW QUESTIONS

Answer		Section Reference
1.	b	2.1.1
2.	b	2.1.2
3.	c	2.1.4
4.	a	2.1.4
5.	c	2.2.1
6.	b	2.2.1
7.	a	2.2.1
8.	d	2.2.2
9.	d	3.1.0
10.	c	3.1.7
11.	c	3.1.8
12.	c	3.2.0
13.	b	4.4.4
14.	b	4.6.0
15.	c	5.4.0

ANSWERS TO TRADE TERMS QUIZ

1. flagger
2. channeling devices
3. center of gravity
4. temporary traffic control (TTC)
5. pinch point

NCCER CURRICULA — USER UPDATE

NCCER makes every effort to keep its textbooks up-to-date and free of technical errors. We appreciate your help in this process. If you find an error, a typographical mistake, or an inaccuracy in NCCER's curricula, please fill out this form (or a photocopy), or complete the online form at **www.nccer.org/olf**. Be sure to include the exact module ID number, page number, a detailed description, and your recommended correction. Your input will be brought to the attention of the Authoring Team. Thank you for your assistance.

Instructors – If you have an idea for improving this textbook, or have found that additional materials were necessary to teach this module effectively, please let us know so that we may present your suggestions to the Authoring Team.

NCCER Product Development and Revision

13614 Progress Blvd., Alachua, FL 32615

Email: curriculum@nccer.org

Online: www.nccer.org/olf

❑ Trainee Guide ❑ AIG ❑ Exam ❑ PowerPoints Other ____________________

Craft / Level: Copyright Date:

Module ID Number / Title:

Section Number(s):

Description:

Recommended Correction:

Your Name:

Address:

Email: Phone:

22103-12

Identification of Heavy Equipment

Module Three

Materials

reminds you to gather materials needed for classes, labs, and testing.

Homework

suggests that you assign reading for the next session.

Classroom

prompts you to discuss a topic, stress a point, explain a concept, or provide examples.

Audiovisual

indicates an appropriate time to show a slide or transparency.

! Safety

used to emphasize safety issues.

Demonstration

directs you to show trainees how to perform tasks.

Laboratory

used when trainees are to practice performing tasks.

✓ Performance Testing

recommends administering a performance test, or a portion thereof.

PT / Laboratory

indicates that lab practice corresponds with a performance task. If desired, you can note the proficiency of the trainees during the lab to satisfy performance testing requirements.

Teaching Tip

indicates tips for practice exercises, Quick Quizzes, field trips, or other activities designed to encourage class participation.

? Examination

recommends administering the written module exam.

NCCER Standardized Craft Training Program

NCCER provides a standardized national program of accredited craft training. Key features of the program include instructor certification, competency-based training, and performance testing. The program provides trainees, instructors, and companies with a standard form of recognition through the National Registry. The program is described in full in the *Guidelines for Accreditation*, published by NCCER. For more information on standardized craft training, contact NCCER by writing to 13614 Progress Blvd., Alachua FL 32615; calling 1-888-NCCER20; or emailing **info@nccer.org**. More information is available at **www.nccer.org**.

How to Use this Annotated Instructor's Guide

Each page presents two sections of information. The larger section displays each page exactly as it appears in the Trainee Module. The narrow column ties suggested trainee and instructor actions to each page and provides icons (detailed on the left) to call your attention to material, safety, audiovisual, or testing requirements. The bottom of each page includes space for your notes, or when available, teaching tips that suggest activities or class exercises that will help your students learn the material.

Preparation

Before teaching this module, you should review the Objectives, Performance Tasks, Materials and Equipment List, and Module Outline. Be sure to allow ample time to prepare your own training or lesson plan and gather all required materials and equipment.

Instructors purchasing a single module receive the printed exam and performance tasks with the AIG module. Perfect-bound versions of this title no longer include the printed exam booklet and performance task sheets. Log on to the Instructor Resource Center at **www.NCCERIRC.com** to download these materials. Your access code is on the card bound into the front of this AIG. This code will allow you to access exams and performance tasks for all levels of this craft. Be sure to keep all testing materials in a secure place at all times.

Identification of Heavy Equipment
Annotated Instructor's Guide

Module 22103-12

Module Overview

This module introduces the most commonly used heavy equipment machines, including dump trucks, backhoes, excavators, and dozers. It also provides an overview of heavy equipment used in the mining industry. Coverage of drive systems and hydraulic systems is included.

Objectives

Upon completion of this module, the trainee will be able to do the following:

1. Identify the various types of heavy equipment and explain their primary uses.
2. Identify and explain the systems that make up the drive system used on heavy equipment.
3. Explain the basics of a hydraulic system and identify hydraulic components.

Performance Tasks

Under the supervision of the instructor, the trainee should be able to do the following:

1. Identify the various types of heavy equipment and their uses.
2. Identify the basic parts of each type of equipment and explain the differences in models of type of equipment.

Materials and Equipment

Markers/chalk
Pencils and paper
Whiteboard/chalkboard
Heavy Equipment Operations Level One PowerPoint® Presentation Slides (ISBN 978-0-13-292167-1)
Multimedia projector and screen
Computer
Appropriate personal protective equipment
Various pieces of heavy equipment
Sales brochures from heavy equipment distributors and manufacturers
Trade Terms Quiz*
Module Examinations**
Performance Profile Sheets**

* Located at the back of the Trainee Guide module

** Single-module AIG purchases include the printed exam and performance task sheet. If you have purchased the perfect-bound version of this title, download these materials from the IRC using your access code.

Safety Considerations

Ensure that the trainees are equipped with appropriate personal protective equipment and know how to use it properly. Review safety guidelines associated with working on or around heavy equipment. Emphasize the importance of proper housekeeping.

Additional Resources

This module presents thorough resources for task training. The following resource material is suggested for further study.

The Earthmover Encyclopedia, 2007. St. Paul, MN: MBI Publishing.

Clean Diesel Technology for Off-Road Engines and Equipment: Tier 4 and More, July 2011. Frederick, MD: Diesel Technology Forum.

Teaching Time for This Module

An outline for use in developing your lesson plan is presented below. Note that each Roman numeral in the outline equates to one session of instruction. Each session has a suggested time period of 2½ hours. This includes 10 minutes at the beginning of each session for administrative tasks and one 10-minute break during the session. Approximately 5 hours are suggested to cover *Identification of Heavy Equipment*. You will need to adjust the time required for hands-on activity and testing based on your class size and resources. Because laboratories often correspond to Performance Tasks, the proficiency of the trainees may be noted during these exercises for Performance Testing purposes.

Topic	**Planned Time**
Session I. Introduction; Heavy Equipment	
A. Introduction	____________
B. Construction Tractors	____________
C. Dump Trucks	____________
D. Roller/Compactors	____________
E. Scrapers	____________
F. Backhoe Loaders	____________
G. Excavators	____________
H. Dozers	____________
I. Loaders	____________
J. Forklifts	____________
1. Fixed-Mast Forklifts	____________
2. Telescoping-Boom Forklifts	____________
3. Articulating Forklifts	____________
K. Motor Graders	____________
L. Trenchers	____________
1. Pedestrian Trenchers	____________
2. Compact Trenchers	____________
3. Large Trenchers	____________
4. Utility Concerns	____________
M. Mining Equipment	____________
1. Aboveground Mining Equipment	____________
2. Belowground Mining Equipment	____________
N. PT/Laboratory	____________

Have trainees identify the various types of heavy equipment and their uses. This laboratory corresponds to Performance Task 1.

Session II. Vehicle Drive System; Hydraulic System; Review and Testing

A. Vehicle Drive System

1. Engines
2. Fuel System
3. Exhaust System
4. Cooling System
5. Electrical System
6. Lubrication System
7. Air Filtration System

B. Hydraulic System

1. Hydraulic Fundamentals
2. Hydraulic Brakes and Clutches
3. Hydraulic Power Couplers
4. Hydraulic Pumps

C. PT/Laboratory

Have trainees identify the basic parts of each type of equipment and explain the differences in models of type of equipment. This laboratory corresponds to Performance Task 2.

D. Review

E. Module Examination

1. Trainees must score 70% or higher to receive recognition from NCCER.
2. Record the testing results on Training Report Form 200 and submit the results to the Training Program Sponsor.

F. Performance Testing

1. Trainees must perform each task to the satisfaction of the instructor to receive recognition from NCCER. If applicable, proficiency noted during laboratory exercises can be used to satisfy the Performance Testing requirements.
2. Record the testing results on Training Report Form 200 and submit the results to the Training Program Sponsor.

22103-12

Identification of Heavy Equipment

Module Three

Homework

Assign the reading of this module.

Teaching Tip/Notes

Trainees with successful module completions may be eligible for credentialing through NCCER's National Registry. To learn more, go to **www.nccer.org** or contact us at **1.888.622.3720**. Our website has information on the latest product releases and training, as well as online versions of our *Cornerstone* newsletter and Pearson's product catalog.

Your feedback is welcome. You may email your comments to **curriculum@nccer.org,** send general comments and inquiries to **info@nccer.org**, or fill in the User Update form at the back of this module.

V.1 4/12

Teaching Tip/Notes

22103-12
Identification of Heavy Equipment

Objectives

When you have completed this module, you will be able to do the following:

1. Identify the various types of heavy equipment and explain their primary uses.
2. Identify and explain the systems that make up the drive system used on heavy equipment.
3. Explain the basics of a hydraulic system and identify hydraulic components.

Performance Tasks

Under the supervision of your instructor, you should be able to do the following:

1. Identify the various types of heavy equipment and their uses.
2. Identify the basic parts of each type of equipment and explain the differences in models of type of equipment.

Trade Terms

Articulated
Direct current (DC)
Dogged
Haul road
Hydraulic actuator
Pascal's law
Pneumatic
Ripper
Scarifier
Torque converter

Industry Recognized Credentials

If you're training through an NCCER-accredited sponsor you may be eligible for credentials from NCCER's Registry. The ID number for this module is 22103-12. Note that this module may have been used in other NCCER curricula and may apply to other level completions. Contact NCCER's Registry at 888.622.3720 or go to nccer.org for more information.

Materials

Ensure that you have everything required to teach the course. Check the Materials and Equipment list at the front of this module.

Classroom

Explain that terms shown in bold (blue) are defined in the Glossary at the back of this module.

Audiovisual

Show Slide 1, Objectives, and Slide 2, Performance Tasks. Review the goals of the module, and explain what will be expected of the trainees.

Teaching Tip/Notes

When you call on someone to answer a question, the rest of the class relaxes or even tunes out because they expect that the question and answer will take place only between you and the trainee you called on. Instead, use this technique to involve more trainees in answering questions and to keep them on their toes:

1. Ask trainees to define a term or explain a concept.
2. After one trainee has answered, ask a trainee seated nearby if the answer is right. Then ask whether a trainee in the back of the room agrees.
3. Ask trainees to explain why they think an answer is right or wrong.
4. Use the session to clear up incorrect ideas and encourage trainees to learn from their mistakes.

Contents

Topics to be presented in this module include:

Teaching Tip/Notes

Figures and Tables

Teaching Tip/Notes

1.0.0 Introduction

In the heavy equipment operations business, there are many different types of machines. Each machine was initially designed for a particular type of work, but many can be modified to perform other functions. Heavy equipment operators must know the capabilities and limitations of the machine being operated. Operators also must be familiar with the various attachments that can be used with any given machine, and know how to use them. With such levels of knowledge, the equipment operator will have many opportunities for advancement.

The purpose of this module is to familiarize you with heavy equipment. Today's construction equipment is designed for versatility. Many of the machines are designed to perform two or more tasks. Take the backhoe loader (*Figure 1*), for example. It is used to dig trenches and to move or load soil, gravel, stone, and other materials. Available accessories, such as a tool carrier attachment, **ripper,** broom, and hydraulic hammer, can turn it into the workhorse of small construction projects.

All of the machines covered in this module are available in a variety of sizes and types to fit all kinds of construction environments. Within each type classification, there are often models designed for specific applications. For example, there are certain types of excavators made for areas with a tight turning radius. Other models are designed specifically to reach heights or to stretch out a great distance to dig in areas where the ground in front of the excavator is wet or unstable.

The following types of machines are covered in this module:

- Construction tractors
- Dump trucks
- Compactors
- Scrapers
- Backhoe loaders
- Excavators
- Dozers
- Loaders
- Forklifts
- Motor graders
- Trenchers

Classroom

Explain that there are many types of heavy equipment and that several of them are designed to perform more than one function.

Audiovisual

Show Slide 3 (Figure 1). Point out that many of the machines, such as the backhoe, also use a variety of attachments.

22103-12 F01.EPS

Figure 1 Backhoe loader.

Teaching Tip/Notes

22103-12 Identification of Heavy Equipment Module Three 1

Classroom

Discuss the uses of tractors in construction.

Audiovisual

Show Slides 4 and 5 (Figure 2). Point out that tractors are generally equipped with a drawbar and power takeoff.

Classroom

Explain that dump trucks are commonly used to haul soil, rock, or debris.

Audiovisual

Show Slides 6 and 7 (Figures 3 and 4). Point out that some dump trucks are designed to operate on highways and others are for on-site use only.

Show Slide 8 (Figure 5). Explain that dump trucks are usually loaded by an excavator or loader.

(A) UTILITY TRACTOR

(B) DUAL-WHEEL ARTICULATING TRACTOR

22103-12_F02A.EPS

Figure 2 Tractors used in construction. (1 of 2)

2.0.0 Construction Tractors

The type of tractors normally used in agricultural work are used as utility vehicles on some job sites. The smaller tractors typically fall into the 60- to 100-horsepower range, but some of the larger ones have engines that range between 300 and 600 horsepower. They are usually equipped with a hitch, which allows them to pull trailers and other attachments that do not require any kind of drive power supplied by the tractor. Such tractors are also equipped with a power takeoff (PTO) and hydraulic systems that allow them to power a variety of construction-related implements that may be used for grading, sweeping, or even post-hole digging. *Figure 2* shows a variety of tractors used on construction sites.

Large, **articulated** tractors, like the one shown in *Figure 2*, are used for heavy duty tasks such as pulling scrapers, oversized rollers, and water tanks. These tractors are usually 375 horsepower and higher. Some companies may prefer to use rubber track vehicles for the heavy work. Although they tend to cost more than rubber tire machines, rubber track vehicles deliver greater pull in difficult terrain.

3.0.0 Dump Trucks

Construction dump trucks are used to haul soil or debris from one location to another. The two basic types of dump trucks are on-road and off-road. The on-road dump trucks (*Figure 3*) are designed for use on highways, and must be registered for that purpose. Anyone operating these trucks on the highway must have a commercial driver's license (CDL). The truck shown in *Figure 3A* is a tandem-axle truck, which means it has two rear axles under the load, along with the front axle under the cab. A single-axle dump truck has only one rear axle and carries a smaller load.

The truck shown in *Figure 3B* has two additional axles located in front of its two rear axles. These two auxiliary axles are carried in the raised position to save fuel and tire costs when the truck is empty or lightly loaded. The two extra axles and wheels are lowered when the truck is traveling with a load, in order to distribute the load and provide extra braking. A cab-operated air bag or spring-retraction system is used to raise and lower the axle assembly. Single-wheel tag axle versions without braking capability are also available.

Off-road dump trucks (*Figure 4*) are intended for hauling material on the site. They are usually larger than the on-road trucks. Off-road trucks are designed for traveling over the rough terrain normally found on a construction site. The articulated truck (*Figure 4A*) is the most popular of the off-road dump trucks. It consists of a trailer bed hitched to a large tractor. The tractor does not operate independently of the cargo bed, however. Rather, the truck is articulated so that it can make sharp turns. Articulated trucks have capacities of up to about 40 tons.

Rigid-frame trucks (*Figure 4B*) are used for high-volume operations. The capacities of these trucks start about where the articulated trucks leave off and go up to more than 300 tons. In recent years, crawler dump trucks on rubber tracks (*Figure 4C*) have become popular for use on wet construction sites.

A dump truck is usually loaded using an excavator or loader (*Figure 5*). Once loaded, the truck hauls the material to another location and dumps

Teaching Tip/Notes

Obtain heavy equipment sales brochures from distributors and manufacturers. As different machines are covered in class, have the trainees refer to the brochures to learn more about the features of the equipment and the available options.

(C) WHEELED ARTICULATING TRACTOR WITH SCRAPERS

(D) RUBBER TRACK TRACTOR

22103-12_F02B.EPS

Figure 2 Tractors used in construction. (2 of 2)

Teaching Tip/Notes

Classroom

Explain that roller/compactors are used to stabilize the ground before a structure or road is built.

Audiovisual

Show Slide 9 (Site Safety). Stress the importance of remaining alert when working around heavy equipment.

(A) DUMP TRUCK

(B) AUXILIARY AXLE DUMP TRUCK

22103-12_F03.EPS

Figure 3 On-road dump trucks.

it using its hydraulic bed-dumping mechanism. While most dump trucks use a hydraulic lifting mechanism to raise the front of the truck bed, some trucks are designed with a bottom dump feature. Still other dump trucks are equipped with a side-dump capability.

4.0.0 Roller/Compactors

The foundations of buildings and roads are placed on soil, which ultimately carries the weight of the entire structure and its contents. It is important, therefore, that the soil base be as smooth and solid as possible. In order to accomplish this, the

On Site

Site Safety

During the site development phase of a project, it is common to find many machines operating at the same time, as well as people who may be working near the machines. Operators must remain alert to their surroundings at all times in order to avoid accidents that could kill or injure personnel or damage expensive equipment.

22103-12_SA01.EPS

Teaching Tip/Notes

(A) ARTICULATED TRUCK

(B) RIGID-FRAME DUMP TRUCK

(C) CRAWLER

22103-12_F04.EPS

Figure 4 Off-road dump trucks.

soil must be compacted. Compaction is the process in which the soil is rolled, tamped, vibrated, and/or pressed to make it more dense. The denser the soil, the more likely it is to support the weight of the structure without sinking. If the soil gives way under the weight, the structure is likely to crack.

Roller/compactors are used to smooth and compact earth, gravel, and asphalt. They compress the material to a desired density so it can support specified loads without excessive settling. There are four basic types of roller/compactors, as shown in *Figure 6*:

- Sheep's foot soil compactor
- Two-drum, steel wheel compactor
- Vibratory steel wheel soil compactor
- **Pneumatic** tire compactor

5.0.0 Scrapers

A scraper is used to remove dirt and other material from a site by scraping it from the surface into the bowl as it moves along. Scrapers are also used to spread dirt and other material, which can be placed in the bowl by a loader. There are four basic types of scrapers, as shown in *Figure 7*:

- Standard or self-propelled scraper
- Elevating scraper
- Tandem-powered scraper
- Pull-type scraper

A standard scraper (*Figure 7A*) is a self-propelled unit consisting of a bowl pulled by a single tractor. The cutting edge at the bottom of the bowl can be levered to dig to two feet or more. The bowls of some units can hold more than 40 cu-

Audiovisual

Show Slide 10 (Figure 6). Discuss the different types of roller/compactors and how they are used.

Classroom

Explain that scrapers are used to both remove and deposit soil and are often used in place of dump trucks to haul soil.

Audiovisual

Show Slide 11 (Figure 7). Discuss the different types of scrapers and how they are used.

Teaching Tip/Notes

22103-12 Identification of Heavy Equipment Module Three 5

Classroom

Discuss the uses of backhoe loaders and the types of attachments available for them.

Audiovisual

Show Slide 12 (Figure 8). Point out the major parts of a backhoe.

22103-12_F05.EPS

Figure 5 An excavator loading an off-road rigid-frame dump truck.

bic yards of earth. The width of the cut can range up to 13 feet, also depending on the model in use. The bowl contains a mechanism for ejecting the material. The forward wall of the bowl is a movable section called the apron. Standard scrapers are often pushed by a dozer in difficult terrain and on slopes.

The bowl of the elevating scraper (*Figure 7B*), also called a paddle wheel scraper, contains a conveyor system that moves the material toward the rear of the bowl as it is picked up from the ground. The capacity of these units is not as great as that of the standard scraper because of the space taken by the elevator system. This type of scraper is used for fine soils and sand. Elevating scrapers are used primarily for fine grading work.

Tandem-powered (push-pull) scrapers (*Figure 7C*) have two engines, one in the tractor and a second at the rear of the scraper unit. The advantage offered by these units is that they can work on steeper grades and rougher terrain. Tandem-powered scrapers are also available in a push-pull model that is equipped with a cushioned plate on the front and a hook at the rear of the bowl. When operating in a cut, the second unit pushes the first unit. When the first unit is full, it then uses its engines to pull the second scraper through the cut. Push-pull scrapers are designed for heavier loads, typically 50-percent greater than a standard tandem scraper.

There is an auger-type attachment available for self-propelled and tandem scrapers. These systems have an independently powered auger at the center of the bowl. The auger is designed to evenly distribute the material that flows over the blade. Auger units have the same capacity as standard units, but are able to load and eject at a much greater rate.

The pulled, or towed, scraper (*Figure 7D*) is pulled along by a heavy duty tractor. These types of scrapers have large capacities and can be towed in tandem to pick up twice the material in a single run.

6.0.0 Backhoe Loaders

A backhoe loader (*Figure 8A*), often just called a backhoe, consists of three separate parts: a tractor, a loader attachment, and a backhoe attachment. The loader attachment is mounted on the front of the tractor while the backhoe attachment is mounted at the rear of the tractor. Other attachments are available to replace the loader bucket, such as the tool/material carrier shown in *Figure 8B*.

The backhoe attachment consists of three components known as the boom, the stick, and the bucket (*Figure 8C*). The backhoe is used to dig and clean ditches, excavate hard-to-reach areas,

Teaching Tip/Notes

(A) SHEEP'S FOOT SOIL COMPACTOR

(B) TWO-DRUM, STEEL WHEEL ROLLER (COMPACTOR)

(C) VIBRATORY SOIL COMPACTOR

(D) PNEUMATIC TIRE ROLLER (COMPACTOR)

22103-12_F06.EPS

Figure 6 Roller compactors.

and hoist light loads. One common use is digging trenches for pipes and other utilities. Many backhoes, including the one shown, have a boom that can slide left and right, which adds versatility to the machine. The backhoe buckets on larger backhoe loaders are able to lift over 400 pounds, or 14 cubic feet of material, and can be extended more than 20 feet.

The loader attachment is used primarily to move soil, gravel, stone, and similar material. It is also used to fill in ditches after the pipes or other items have been installed. Like its larger cousin the loader, the loader attachment on a backhoe loader can lift material from the ground and dump it into the bed of a dump truck. The largest loader buckets can lift more than a yard and a half of soil.

Backhoes are equipped with stabilizer legs (outriggers) that carry the weight when the backhoe is digging. Stabilizers distribute the lateral loads created by the bucket as it digs. They keep the backhoe from tipping or rolling. They widen the stance of the machine and increase its capacity.

The operator faces the loader end when moving the machine and when operating the loader. The backhoe has separate controls and the operator simply turns the seat around to operate that attachment.

Classroom

Stress the importance of using the stabilizer legs when digging with a backhoe.

Teaching Tip/Notes

22103-12 Identification of Heavy Equipment Module Three 7

Classroom

Discuss the various ways that excavators are used.

Audiovisual

Show Slide 13 (Figure 9). Point out the different types of excavators.

Show Slide 14 (Giant Shovel). Compare this machine to a common excavator.

Figure 7 Scrapers.

On Site

Giant Shovel

This monster of an excavator is used in mine and quarry operations. It can lift 50 cubic yards of material in one scoop.

22103-12_SA02.EPS

7.0.0 Excavators

The excavator is the workhorse on most construction sites. The excavator is used for the following types of work:

- Digging trenches and foundations
- Loading soil, stone, gravel, and other materials into dump trucks and scrapers
- Demolishing structures
- Clearing land
- Lifting and placing tools and materials

The majority of excavators are track vehicles, such as the medium construction excavator shown in *Figure 9A*. Some manufacturers also make a few wheeled excavators. The one shown in *Figure 9B* is a wheeled version of the excavator shown in *Figure 9A*. The compact excavator (*Figure 9C*) has become popular, especially for work on sites where there is very little working clearance. Telescoping excavators are preferred for some applications. There are both track and wheel versions of the telescoping excavator.

Teaching Tip/Notes

Figure 8 Backhoe loaders.

There are many types of excavator attachments and many variations within the types. The bucket (*Figure 10*) is the most common attachment, and each excavator comes equipped with a bucket. One excavator manufacturer lists five different types of buckets for tasks such as trenching, excavation, rock ripping, utility work, and ditch cleaning. Each type of bucket comes in a range of sizes to match the excavator capacity. Most buckets have teeth on the lower edge. The number and type of teeth is a function of the bucket width and purpose. Typically excavators have from three to seven teeth. Buckets made for ditch cleaning do not have teeth.

Excavators can be used for many tasks, including demolition. In some cases, the bucket attachment can be used to knock over or break up a structure. In other cases, specialized attachments such as powerful hydraulic shears and hammers are needed (*Figure 11*). Excavators can also be used to knock over and remove trees in clearing a site for construction. To collect and hold demolition debris, roots and stumps, or other such materials, excavator buckets can be equipped with an extra part often called a thumb.

Like the backhoe, an excavating assembly consists of a boom and stick (*Figure 12*). There are many combinations of these two components that are used to satisfy various needs. The boom and stick are matched to the application. Some have a short reach of less than 10 feet; others can reach 50 feet or more. Some are designed for an extended reach, while others are designed for work in areas where a there is a short swing radius. Different amounts of counterweight at the back of the excavator are used to balance the machine. The longer the reach and the greater the load capacity, the more counterweight is needed.

Audiovisual

Show Slide 15 (Figure 10). Describe the different types of buckets used on excavators.

Show Slide 16 (Figure 11). Discuss the ways that excavators can be used for demolition work.

Show Slide 17 (Figure 12). Describe the boom and stick of an excavating assembly.

Teaching Tip/Notes

22103-12 Identification of Heavy Equipment Module Three 9

Classroom

Discuss the uses of dozers and point out the differences between wheeled and track dozers.

Audiovisual

Show Slides 18 and 19 (Figure 13 and 14).

Classroom

Discuss common dozer attachments.

(A) HYDRAULIC CRAWLER EXCAVATOR

(B) HYDRAULIC WHEEL EXCAVATOR

(C) COMPACT EXCAVATOR

22103-12_F09.EPS

Figure 9 Excavators.

8.0.0 Dozers

A dozer is a heavy-duty tractor with a pusher blade mounted on the front. Depending on the manufacturer, these machines may be called tractors, dozers, or bulldozers.

The blade of a dozer can be hydraulically raised, lowered, or positioned at an angle. Most dozers are tracked vehicles, although there are some wheel versions, including heavy-duty four-wheel-drive dozers (*Figure 13A*).

Dozers are used for a variety of purposes including the following:

- Clearing land, including pulling out tree stumps
- Moving soil, rock, and other material
- Spreading material dumped on the site
- Rough and finish grading
- Maintaining **haul roads**

Dozers are configured in different ways to serve different purposes. For example, there are two types of crawler dozers. The low-track machine (*Figure 13B*) is used primarily as a grading machine. The high-track type (*Figure 13C*) is a high-powered machine used primarily as a pusher. There are also compact versions of these machines that can be used for lighter grading work.

Many dozers are equipped with a powered winch (*Figure 14A*) so they can pull stumps, drag logs, pull other machines out of mud, and even pull themselves out of trouble. Another common attachment is the ripper, which is used to break up hard soil, rock, and other materials, such as concrete and asphalt (*Figure 14B*).

Teaching Tip/Notes

(A) EXCAVATOR BUCKET WITH TEETH

(B) DITCH-CLEANING EXCAVATOR BUCKET WITHOUT TEETH

(C) EXCAVATOR WITH REPLACEABLE BUCKET TEETH

22103-12_F10.EPS

Figure 10 Excavator buckets.

On Site

Wheeled Dozers

This Caterpillar 854G wheel dozer has a blade that can push up to 58 cubic yards of material. It weighs over 200,000 pounds and has an 800-horsepower diesel engine.

22103-12_SA03.EPS

Audiovisual

Show Slide 20 (Wheeled Dozers).

Teaching Tip/Notes

Classroom

Describe the ways loaders are used on a job site. Point out that wheel loaders are more common than track loaders.

Audiovisual

Show Slide 21 (Figure 15).

Classroom

Explain what a skid steer is and how it got its name.

Audiovisual

Show Slide 22 (Figure 16). Discuss the fixed-mast forklift.

Show Slide 23 (Figure 17). Discuss the advantages of a telescoping-boom forklift.

(A) ROCK HAMMER

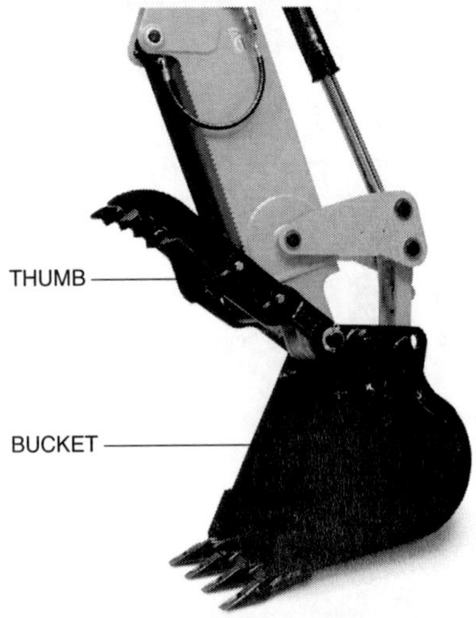

(B) BUCKET WITH THUMB

(C) SHEARS

22103-12_F11.EPS

Figure 11 Demolition attachments.

9.0.0 Loaders

Loaders are used for a variety of purposes, including the following:

- Loading trucks
- Stockpiling soil, gravel, and other material
- Rough grading
- Digging
- Hauling material and equipment

Like many of the other types of equipment that have been covered, loaders come in crawler and wheel configurations. Contrary to what was shown earlier for excavators and dozers, the wheel version of the loader (*Figure 15A*) is more common and has a wider range of models and sizes than the track version (*Figure 15B*).

The skid steers (*Figures 15C* and *15D*) are a compact version of the loaders. They are equipped with either wheels or tracks. The tracked vehicles have better traction. Regardless of how a skid steer is equipped (wheels or tracks), they are steered by pushing or pulling on two levers in the cab. If the levers are moved in opposite directions, the machine will counter-rotate. This ability to turn in small spaces is where the term *skid steer* came from. Like the backhoe loader, the skid steer is available with a range of attachments, such as pallet forks, augers, brooms, and grapple buckets.

10.0.0 Forklifts

Forklifts are used to move equipment and materials. They are also used extensively to unload materials from trucks and trailers. The forklifts used in construction are generally diesel-powered, rough-terrain forklifts. These forklifts have a higher ground clearance and larger tires than warehouse forklifts. The two basic types of forklifts are fixed-mast and telescoping boom.

10.1.0 Fixed-Mast Forklifts

The upright member along which the forks travel is called the mast. On a fixed-mast forklift (*Figure 16*), the forks can be raised as high as the limits of the mast allow. The mast may be able to be tilted forward 15 to 20 degrees and backward by a few degrees. These forklifts must be able to get very close to the pickup or landing point because of the limited movement range of the mast.

10.2.0 Telescoping-Boom Forklifts

A telescoping-boom forklift (*Figure 17*) is able to lift material much higher than fixed-mast forklifts and has much more flexibility in positioning

Teaching Tip/Notes

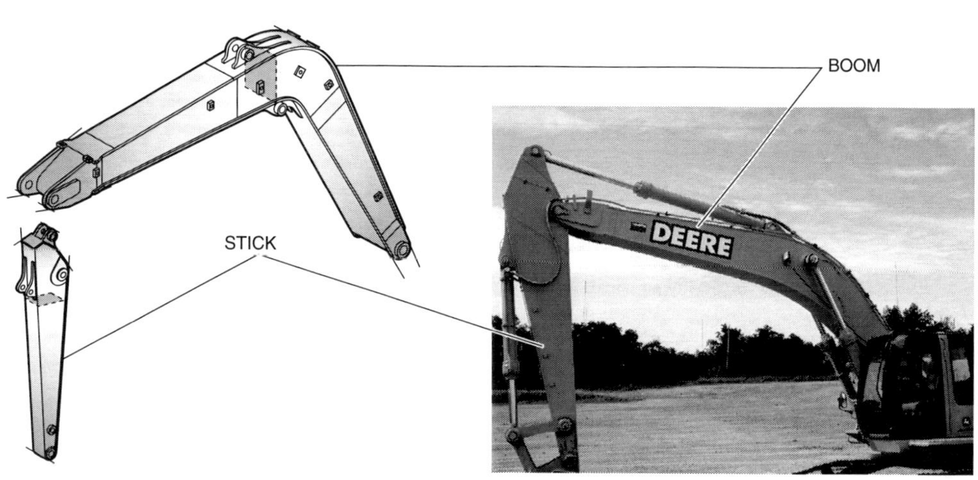

Figure 12 Stick and boom.

Figure 13 Dozers.

Teaching Tip/Notes

Audiovisual

Show Slide 24 (Figure 18). Point out that the articulating forklift has a movable fork carriage.

Show Slide 25 ("Bull" Dozer). Discuss the evolution of dozers.

Classroom

Discuss the uses of motor graders.

(A) WINCH

(B) RIPPER

22103-12_F14.EPS

Figure 14 Dozer attachments.

a load. Some telescoping-boom forklifts have a level-reach fork carriage, which is often called a squirt boom. A squirt boom allows the fork carriage to be moved in the horizontal plane while the boom remains stationary.

10.3.0 Articulating Forklifts

Articulating forklifts (*Figure 18*) are built with a fork carriage that is hinged so that it pivots left or right of center line. The operator uses hydraulic controls and cylinders to move the fork carriage. Most articulating forklifts are designed for operations within narrow, confined spaces such as the narrow aisles inside a warehouse. They are also used inside truck trailers.

11.0.0 Motor Graders

A motor grader is a machine with a centrally located blade, also called a moldboard, that can be angled to push material to the side. On most motor graders, the moldboard can be positioned several feet outside the rear wheels and angled to work on slopes. The grader is used to establish a finish grade, but can also be used for cut and fill work.

At one time, motor graders were made with a rigid frame. Modern motor graders have a hydraulically controlled articulating frame that pivots at a point just forward of the engine. The articulated frame provides greater maneuverability and a tighter turn radius. For example, articulation gives the machine the ability to work a slope with

On Site

"Bull" Dozer

The original bulldozer was not a machine at all, and it had nothing to do with bulls. Rather, it was a blade pushed by mules. The first step in mechanizing this tool was accomplished when a tractor was converted to a continuous chain tread, or track crawler, format early in the twentieth century. The next step was to provide independent control for each track. The bulldozer came about when the blade was attached to a tractor a few years later. The final step occurred when a power control for the blade was added. Although the principle is still the same, dozers have come a long way since the early days.

22103-12_SA04.EPS

Source: about.com

Teaching Tip/Notes

(A) WHEEL LOADER

(B) CRAWLER LOADER

(C) WHEELED SKID STEER WITH BUCKET

(D) TRACKED SKID STEER WITH AUGER

22103-12_F15.EPS

Figure 15 Loaders.

On Site

Attachments Used with a Skid Steer

A skid steer can be used to perform the functions of many other machines when equipped with attachments available from the manufacturer. This skid steer is equipped with a pallet forks attachment, which allows it to be used as a forklift. Many different attachments are made by the manufacturer for use with this machine, including backhoe, auger, and trenching attachments.

22103-12_SA05.EPS

Audiovisual

Show Slide 26 (Attachments Used with a Skid Steer). Discuss the versatility of the skid steer.

Teaching Tip/Notes

22103-12 Identification of Heavy Equipment Module Three 15

Audiovisual

Show Slide 27 (Figure 19). Point out that modern motor graders have an articulating frame that gives them greater maneuverability.

Show Slide 28 (Figure 20). Describe how GPS systems are used to control motor graders.

Show Slide 29 (Figure 21). Discuss the types of attachments used with motor graders.

22103-12_F16.EPS

Figure 16 Fixed-mast forklifts.

22103-12_F17.EPS

Figure 17 Telescoping-boom forklift.

22103-12_F18.EPS

Figure 18 Articulating forklift.

the front wheels and moldboard on the slope and the rear wheels on the roadway for greater stability and power. The moldboard can be raised and lowered, turned in the horizontal plane, and have one end raised to grade slopes (*Figure 19*).

Regardless of how the grader is being used, getting its blade or moldboard set to the proper angle is most critical. Graders are now equipped with a global positioning system (GPS) that allows the position of the moldboard to be set by a computer system. GPS antennas (one or two) are mounted near the outer tips of the moldboard (*Figure 20*). A GPS base station (surveying system) is set up near where the grader is working. The base station works with the GPS devices on the grader to control the positioning of the moldboard. Motor grader operators must learn how to work with these GPS devices. The GPS base station shown in *Figure 20* is an Ashtech ProFlex 500 GNSS receiver (with antenna) mated with a radio receiver.

The attachments commonly used on motor graders are a **scarifier** and a ripper (*Figure 21*). Scarifiers are attached to either the front of the grader or just in front of the moldboard (blade). Scarifiers rip up the ground so that the moldboard has an easier job of breaking up and moving the dirt.

A ripper mounted on the rear of a grader breaks up the ground as the grader passes over a given section of ground. The grader's moldboard may or may not be down and cutting the dirt at the same time the ripper is being used.

16 NCCER – *Heavy Equipment Operations Level One* 22103-12

Teaching Tip/Notes

Figure 19 Grading a slope.

12.0.0 Trenchers

Trenchers are machines used to dig trenches for the installation of underground services. There are different styles of trenchers available depending upon the scale of the work to be performed (*Figure 22*). There are essentially three styles of trenchers. The lighter-weight one is called a pedestrian trencher because the operator walks behind it. A heavier, but yet small one is called a compact trencher and the operator rides along on the back end of it. The third style is more of a heavy construction model used to dig large pipelines on large construction jobs. With the large trenchers, the operator sits in an operator cab position on top of the machine. These large construction trenchers ride on tracks, similar to the tracks of a dozer, so that it has much more traction. Some companies make trencher attachments that mount on the front of a skid steer machine. Such skid steer attachments operate like the compact trenchers in that the operator sits inside the skid steer and operates the trenching attachment from additional controls installed in the skid steer cab.

Trenchers use either a gasoline or diesel engine to drive a hydraulic system that provides power to the drive wheels and digging chain mounted on the digging boom. The digging chain and boom are very similar to the blade and chain of a chainsaw. When activated, the chain moves forward along the top of the boom, around the end of the boom, and returns to the mounting end of the boom. When the boom is lowered onto the ground, the chain begins to dig into the ground. As the digging progresses, the dirt (and rock) being removed is kicked out the sides. Also, as the digging progresses, the operator makes the digging boom go farther and farther into the ground until the desired depth of the new trench is reached. As the desired depth is reached, the trenching machine is moved on to dig another section of the trench. On some trenchers, the trenching is done as the trencher machine is backing up. If that is the case, the dirt is kicked back toward the machine. Most large trenchers move forward and kick the dirt out and away from the machine.

12.1.0 Pedestrian Trenchers

Operating the pedestrian trenchers requires the operator to manually control the machine with handles. The operator must have good upper body strength to control the machine. The operator must also be very careful to maintain stable footing while operating the machine. Most pedestrian trenchers have a kill handle that must be held down to keep the machine running. If the

On Site

Laser-Guided Machines

Laser-controlled graders are being used more and more because they provide very precise grading and significantly reduce grading errors and rework. The laser transmitter is set up to establish the desired grade. It then transmits a laser beam to sensors attached to a mast mounted on the grader blade. If the blade is off grade, the sensor detects the difference. In a fully automatic system, the sensor sends a signal to the hydraulic controls to adjust the blade. In a semi-automatic system, the sensor signal is sent to a cab display unit that tells the operator how to adjust the blade position.

In addition to graders, this technology is being used on dozers to control cut and fill, as well as on excavating equipment to control the depth and slope of trenches.

Classroom

Discuss the different types of trenching machines.

Describe how trenchers are used.

Compare trenchers to excavators and backhoes for digging trenches.

Audiovisual

Show Slide 30 (Figure 22).

Classroom

Compare the pedestrian trencher to the compact trencher.

Audiovisual

Show Slide 31 (Laser-Guided Machines). Explain how lasers are used to control motor graders.

Teaching Tip/Notes

Figure 20 Motor graders with GPS.

operator loses control of the machine and releases the kill handle, the machine stops.

12.2.0 Compact Trenchers

As noted earlier, the operator rides on the rear of a compact trencher. Most compact trenchers allow for trenching forward or backward. Like any other trencher, the compact trencher kicks the dirt out the side as it progresses along the way. Operators of these compact trenchers must be very careful about the grades over which they travel with these trenchers. With the operator standing on the rear end of the trencher, there is

Teaching Tip/Notes

A SCARIFIER ON THE FRONT OF A MOTOR GRADER

A RIPPER ON THE REAR OF A MOTOR GRADER

22103-12_F21.EPS

Figure 21 Common motor grader attachments.

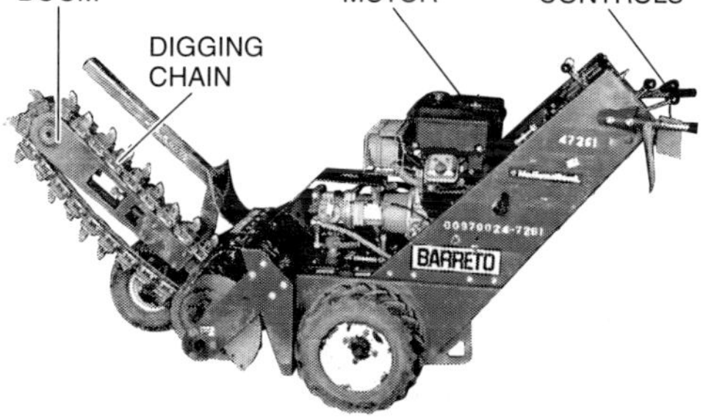

PEDESTRIAN TRENCHER

LARGE CONSTRUCTION TRENCHER

SKID STEER WITH TRENCHER ATTACHMENT

22103-12_F22.EPS

Figure 22 Examples of different trenchers.

Teaching Tip/Notes

22103-12 Identification of Heavy Equipment Module Three 19

Safety

Point out the hazards of operating a compact trencher on a slope.

Classroom

Discuss the uses of large trenchers.

Describe how buried utilities should be marked before trenching activity.

Point out that MSHA safety practices must be followed when working in mines and quarries.

Audiovisual

Show Slide 32 (Figure 23). Discuss the heavy equipment used in surface mining.

Classroom

Explain how the equipment used in belowground mining differs from that used in surface mining.

Audiovisual

Show Slide 33 (Figure 24).

Safety

Discuss confined space and other safety issues unique to belowground mining.

PT/Laboratory

Under your supervision, have trainees identify the various types of heavy equipment and their uses. This laboratory corresponds to Performance Task 1.

Homework

Have trainees review Sections 14.0.0–15.4.0.

Materials

Ensure that you have everything required for teaching this session.

additional weight on that end of the machine. If the compact trencher is digging backward and is allowed to go over an incline, it may flip backward with the additional weight of the operator on it. If it is digging forward and moves across an incline, the weight of the digging bar and chain can make the machine flip forward, even with the operator's weight on the rear. Operators of compact trenchers need to be constantly aware of the trencher's balance.

12.3.0 Large Trenchers

Large trenchers are simply large construction vehicles. They are operated much like an excavator in that controlling their tracks controls the direction in which they move. Their digging boom and chain assembly is a lot like the boom, stick, and bucket of an excavator. It is raised for travel and lowered for trenching. The operator inside the cab must stay alert to his or her surroundings. The operator's concentration is on the trenching equipment, but still must stay alert to anything going on around the trenching process.

12.4.0 Utility Concerns

Because any kind of trenching requires that the trenching machinery dig into the ground, trenching operations must be well planned prior to any trenching work. Underground utility lines must be well marked by the utility companies owning them. Trenching operators need to walk the area to be trenched and verify that all underground utility lines are well marked and easily seen from the operator's position either behind, on, or in the trenching machine being used.

13.0.0 Mining Equipment

Heavy equipment operators may find themselves working in mining. Mining is a form of earthmoving work that is carried out both above ground and below ground. In addition to the usual safety guidelines required on any construction site, personnel doing mining work must also comply with the guidelines set forth by the Mine Safety and Health Administration (MSHA). MSHA is charged with guarding the safety of miners to prevent death, disease, and injury resulting from mining work.

13.1.0 Aboveground Mining Equipment

Aboveground mining involves digging down through the surface of the earth to reach particular minerals, such as coal, mineral ore, or stone. Construction materials, such as sand, fill dirt, and gravel, are taken from open-pit quarries located all over the country. The equipment used for aboveground mining is usually the large dozers, excavators, loaders, and rigid-framed trucks. *Figure 23* shows a few of the heavy equipment machines often found in aboveground mining operations.

13.2.0 Belowground Mining Equipment

Belowground mining involves digging a tunnel into Earth and then working underground to extract the desired materials. Any time a tunnel is dug into the ground, the ground above the work area must be either strong enough to support any weight above the open area, or be supported by fabricated bracing that prevents a collapse of the ground above the opening. To help reduce the risk of a collapse, the ceiling or overhead surfaces above the underground work areas are kept low, which means that the machines used in underground mining are built low to the ground. The machines that are primarily used in underground mining are the underground mining loaders and underground mining trucks, but normal excavators and loaders may be used for higher-ceiling areas. *Figure 24* shows such machines often used for underground mining.

Because underground mining is done in a confined space, additional attention must be paid to the air in the area. Workers and the machines they are using, must be able to breathe the air without problems. The exhausts that the machines produce must not pollute the air to the point it hurts the workers.

Underground mining means digging through the dirt, rock, and minerals that make up Earth. That dirt, rock, and minerals may at times include gas pockets or veins of water. Such potential hazards should have already been identified and eliminated by mining safety engineers before the heavy equipment operators start their work in the mines. Heavy equipment operators must review all MSHA guidelines, and all company safe-operating procedures, before doing any underground mining work.

14.0.0 Vehicle Drive System

The mechanical operation of heavy equipment relies on two separate systems—the vehicle drive system and the hydraulic system. This section focuses on the vehicle drive system and the hydraulic system is addressed later.

Teaching Tip/Notes

Show Slide 34 (Figure 25). Explain that diesel engines are more common than gasoline engines in heavy equipment because they deliver more power.

Figure 23 Aboveground mining equipment.

14.1.0 Engines

The engine of a machine provides the power that drives the machine. The two primary types of engines used in construction machines are gasoline-fired and diesel. But more and more companies are building hybrid-powered machines that are strong enough to be used in heavy construction. Although such hybrid-powered machines may be on the market, this module focuses on the normal combustion engines that are still used in most construction machines.

Gasoline engines are used in some smaller machines, but most machines operated on construction sites use the diesel engine. Diesel engines (*Figure 25*) operate at roughly twice the compression ratio of gasoline engines, and therefore deliver much more power to the drive mechanism.

Teaching Tip/Notes

22103-12 Identification of Heavy Equipment Module Three 21

Classroom

Describe the differences between diesel and gasoline engines.

Figure 24 Underground mining equipment.

Both the gasoline and diesel engines are internal combustion engines. That is, their drive comes from the explosion of a fuel-air mixture in a piston chamber. The main difference between diesel and gasoline engines is in the way they are fired. In the gasoline engine, air and gasoline are mixed, and then injected into the piston chamber where it is compressed. A spark plug then emits a spark that causes the compressed air-fuel mixture to explode. The explosion pushes the piston down in the chamber, and this motion is transmitted to the crank shaft and then to the drive train.

In a diesel engine, air and diesel fuel are not mixed before being injected, and there is no spark plug. Air and fuel are injected directly into the piston chamber and compressed. The extreme compression creates enough heat to cause the mixture to ignite.

> **CAUTION**
>
> Do not put gasoline in a diesel engine. It will damage the engine.

Teaching Tip/Notes

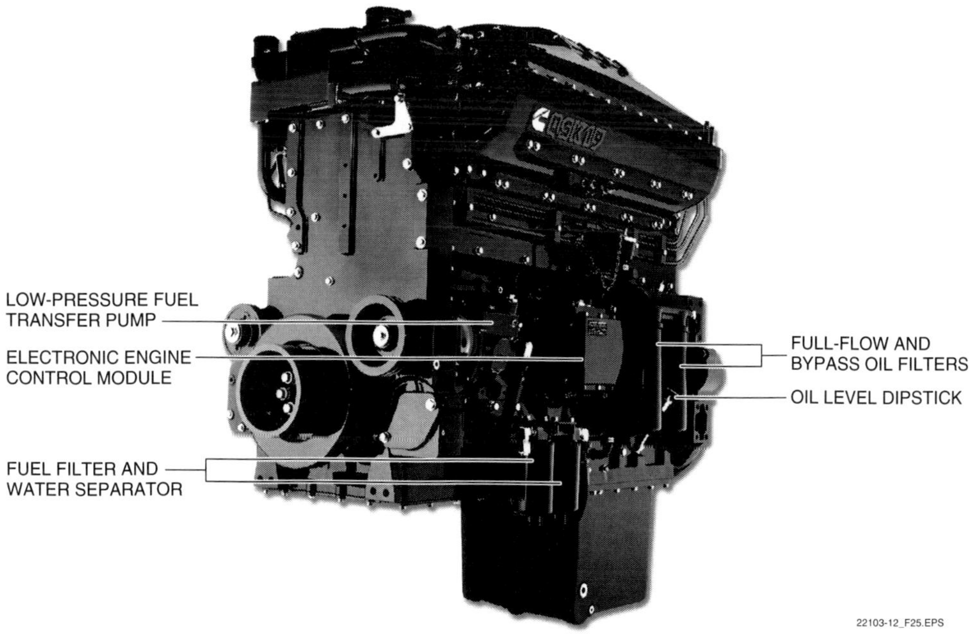

Figure 25 Diesel engine.

All new engines must be built to comply with the clean engine standards that have been established by environmental agencies all over the world. Those standards especially apply to all new diesel-powered vehicles built for both on-road and off-road uses.

Environmental protection agencies in different parts of the world have their own guidelines for all the different types of vehicles. Manufacturers of such vehicles must build machines that comply with the standards of each country. In the United States and Canada, the environmental standards that apply to new diesel engines used in off-road equipment are referred to as Tier-4 standards. While the Tier-4 standards have been in existence for several years, they are being more strictly applied to all new off-road (construction) equipment being built during and after 2011. The whole point of the Tier-4 standards is to lower the amount of nitrogen oxides (NO_x) and particulate matter (PM) being emitted from engines. Particulate matter is the small solid or liquid matter that remains in gas or liquid emissions being released into the atmosphere. Nitrogen oxides and particulate matter are pollutants that must be controlled.

To meet the Tier-4 standards, the new diesel engines must be built to burn cleaner ultra-low sulfur diesel fuel. That fuel can contain no more than 15 parts per million (ppm) of sulfur. To burn that cleaner fuel more efficiently, the new engines are equipped with computer-controlled components that better manage the fuel and air mixture coming into the engine, and the combustion processes taking place inside the engine. All those efforts are to optimize engine performance while minimizing the harmful emissions. On the exhaust side of the engines, changes are being made to the exhaust systems themselves. Even the design and location of the exhaust pipes is being changed to better control the exhausts being released into the atmosphere. Inside the exhaust systems themselves, improved technologies are being used to better clean the exhaust gases passing through the exhausts.

14.2.0 Fuel System

The fuel for a diesel engine is injected under high pressure into the combustion cylinder (*Figure 26*). Some engines use high-pressure fuel injection pumps (20,000 to 28,000 psi). Other engines use lower-pressure pumps and cam-operated injectors to pressurize the fuel to the pressure required during injection. To reduce pollution and increase efficiency, many newer engines control the injec-

Classroom

Explain the Tier-4 environmental standards and how they affect the design of heavy equipment engines and exhaust systems.

Audiovisual

Show Slide 35 (Figure 26). Explain how fuel is injected into a diesel engine.

Teaching Tip/Notes

22103-12 Identification of Heavy Equipment Module Three 23

Audiovisual

Show Slide 36 (Figure 27). Describe how a turbocharger increases power by increasing the amount of air delivered to the engine.

Show Slide 37 (Sustainability). Explain that hybrid engines are becoming more common.

tion time using electronic controls that monitor a variety of engine functions and conditions during operation.

Some engines are also equipped with turbochargers. Turbochargers are designed to increase the amount of air delivered to each of the engine's cylinders. This increases the compression ratio and produces more power. When the air is admitted into the cylinder, the cylinder piston compresses the air. Then the fuel is injected and compression continues until a detonation takes place when the piston is near the end of its compression stroke. The detonation creates the piston's power stroke, which in turn rotates a crankshaft, producing crankshaft torque to the engine's load (see *Figure 27*).

One of the ways to reduce the formation of NO_x in an engine is the use of exhaust gas recirculation (EGR) equipment. The EGR equipment takes a portion of the exhaust gases and runs them through a cooler before returning them to the air supply going back into the combustion chamber

On Site

Sustainability

The Caterpillar Company is producing a D7E dozer that is being powered by electric drive motors. The dozer still uses a combustion engine, but the engine now drives a generator that produces electricity, and that electricity in turn powers the electric drive motors. According to Caterpillar, the hybrid dozer can move 25 percent more material per gallon of fuel than a conventional D7R dozer. These hybrid-powered machines also emit less exhaust into the atmosphere, and thus are more environmentally friendly. Other companies are building similar systems for other types of construction equipment.

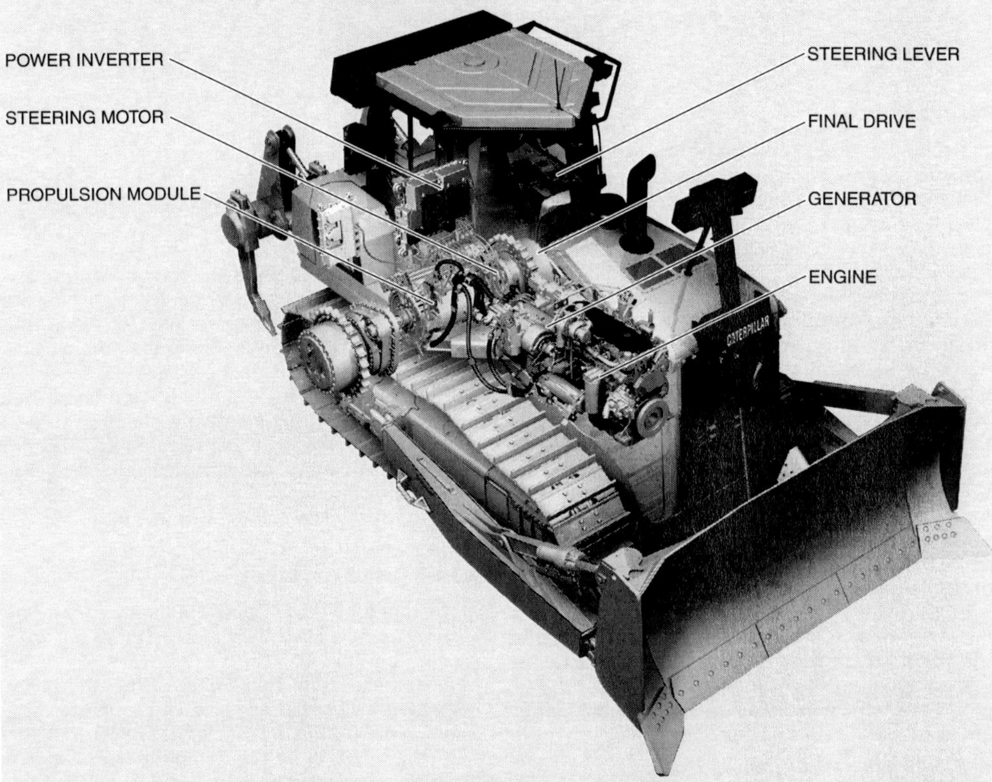

22103-12_SA07.EPS

24 NCCER – *Heavy Equipment Operations Level One* 22103-12

Teaching Tip/Notes

Figure 26 Diesel engine cutaway.

of the engine. Inducing the cooler exhaust gases helps lower the combustion temperature inside the combustion chamber, which in turn reduces the formation of NO_x. Engines equipped with EGR equipment have additional manifolds and extra plumbing around the engine. *Figure 28* shows a side view of an engine with EGR equipment installed.

WARNING! Always wear the proper eye, face, and hand protection when working on diesel engines. High-pressure fuel leaks can cause severe injury. Only properly trained and qualified mechanics are permitted to work on diesel engines.

14.3.0 Exhaust System

When combustion engines operate, they emit gases that exit the vehicle through its exhaust system. All modern vehicles have fuel systems that are directly related to the vehicle's exhaust system. Sensors in the exhaust system monitor the exhaust gases and tell the fuel system controller if the fuel is burning correctly or not. If not, the fuel system controller tries to adjust the combustion process to clean up the exhaust fumes as much as it can.

Three of the major pollutants coming out of a combustion engine are hydrocarbons (HC), carbon monoxide, and nitrogen oxides or NO_x. The exhaust systems on modern vehicles are designed to minimize the release of such pollutants. One of the devices installed in exhaust systems is a catalytic converter. Catalytic converters contain materials that chemically affect the gasses passing through the converter. The

On Site

The Diesel Engine

The diesel engine was patented by Rudolph Diesel in 1898. This engine revolutionized industry, which until that time had used less efficient steam engines. Today, locomotives, heavy construction equipment, trucks, and many other applications depend on the diesel engine.

Audiovisual

Show Slide 38 (Figure 28). Describe how an exhaust gas recirculation system reduces emissions.

Safety !

Stress the importance of wearing eye protection and other PPE when working on or around a diesel engine. Explain that only properly trained mechanics may work on diesel engines.

Classroom

Explain how modern exhaust systems monitor and minimize the release of pollutants.

Teaching Tip/Notes

Classroom

Describe how a selective catalytic reduction system works.

Audiovisual

Show Slide 39 (Figure 29). Discuss the equipment operator's duties with respect to the DEF storage tank, as well as the SCR and EGR systems.

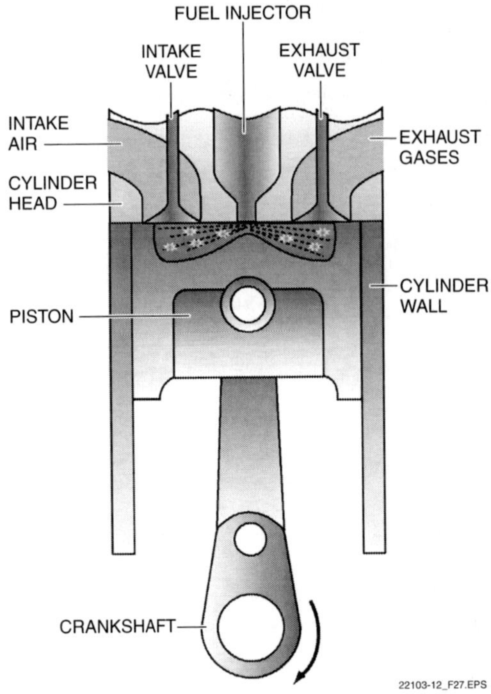

Figure 27 Cylinder and crankshaft.

conversion results in less pollutants exiting the exhaust system. One of the technologies being used to reduce the levels of exhaust NO_x is called a selective catalytic reduction (SCR) system. SCRs are installed in the machine's exhaust system. *Figure 29* shows a side view of an engine with SCR equipment installed.

Before the exhausts of the engine reaches the SCR device, a dosing system injects periodic sprays of a mist of a chemical reagent called a diesel exhaust fluid (DEF). The DEF solution is a nontoxic solution made up of approximately 68 percent purified water and 32 percent urea. Urea is an organic product created from animal wastes. When injected into the exhaust system and heated, the DEF creates ammonia that flows through the SCR catalyst and reacts with the engine gases to form harmless nitrogen and water vapor that make up the engine's exhaust.

SCR-equipped vehicles are equipped with a DEF storage tank. The vehicle operator refills the DEF tank as needed. Because of the ongoing efforts to reduce pollutants from all combustion engines, the heavy equipment used in construction will eventually be equipped with similar SCR systems using DEF. Operators are also responsible for visually inspecting the EGR and SCR systems installed on the engines of their machines.

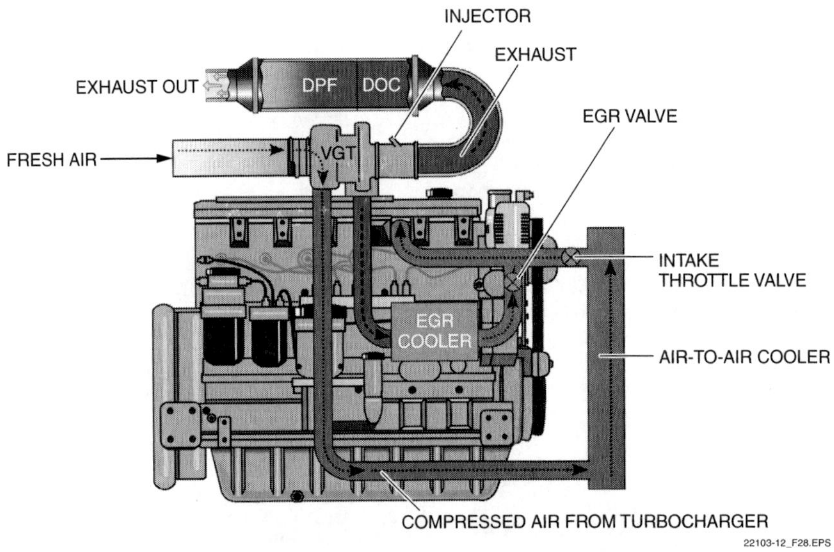

Figure 28 Location of EGR equipment on a diesel engine.

Teaching Tip/Notes

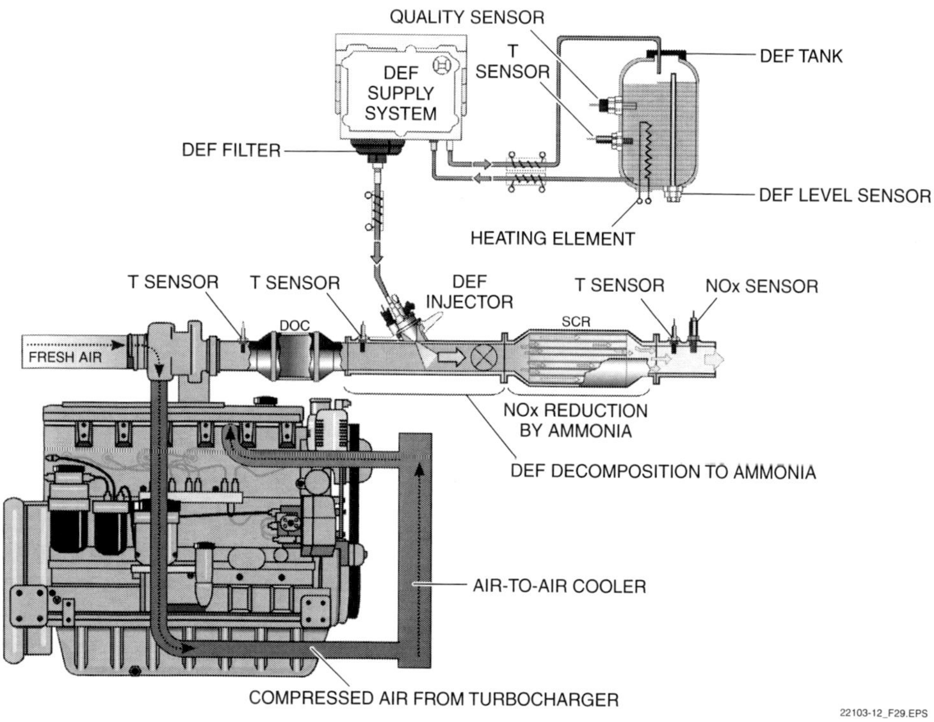

Figure 29 Location of SCR equipment on an engine's exhaust.

14.4.0 Cooling System

Depending on the design of the system, a diesel engine is either air-cooled or water-cooled. If the engine is water-cooled, it will have a system similar to the one shown in *Figure 30*, which removes the excess heat generated during combustion.

The components of this system are typical of most water-cooled engines. The coolant, radiator, fan, pressure cap, engine-cooling circuit, connecting hoses, and thermostat work together to keep the engine at the proper operating temperature. This is accomplished by the circulation of pressurized coolant (antifreeze) with a water pump.

The water pump is normally air-driven or turned by a belt connected to the crankshaft of the engine. As the coolant circulates, it is drawn from the bottom of the radiator by the pump. The pump forces the coolant into and through the water jacket of the engine. Coolant flowing through the water jacket controls the temperature of the engine. A thermostat regulates the flow of the coolant through the water jacket. After the coolant exits the water jacket and thermostat, it flows to the top of the radiator. The coolant flows down through thin tubes inside the radiator. The tubes are constructed with thin cooling fins mounted along their outer surfaces. Air passing over the tubes and the fins extracts heat from the coolant. A cooling fan is used to increase the air flowing through the radiator. The more air moving past the radiator tubes results in better cooling for the engine.

Each day, before starting operations, the machine's engine belts, hoses, and coolant level should be checked. A problem in this area could result in expensive repairs if the engine overheats. The coolant level should only be checked when

GOING GREEN

Antifreeze

When antifreeze is used as a coolant, identify the source of any leaks and repair the leaks. Also, try to control any spills. Antifreeze can harm animals and may pollute any soil it contacts. It must be kept away from any streams or water sources. Check local regulatory guidelines for instructions on how to handle coolant spills. Remove and properly dispose of any contaminated soil.

Classroom

Describe the engine cooling system and explain how it works.

Audiovisual

Show Slide 40 (Figure 30). Identify the cooling system components and explain their functions.

Classroom

Point out the importance of inspecting the cooling system every day, including belts and hoses, coolant level, and operating temperature.

Teaching Tip/Notes

22103-12 Identification of Heavy Equipment Module Three 27

Safety

Explain that workers must wear protective equipment when working on a hot cooling system.

Point out that cold water or coolant must never be added to a hot cooling system and that the radiator cap should never be removed while the engine is hot.

Audiovisual

Show Slide 41 (Figure 31). Describe the components and functioning of a diesel engine electrical system.

Classroom

Explain how to make sure an electrical system is functioning properly.

Safety

Point out the hazards associated working around batteries. Emphasize the need for proper PPE, including eye protection.

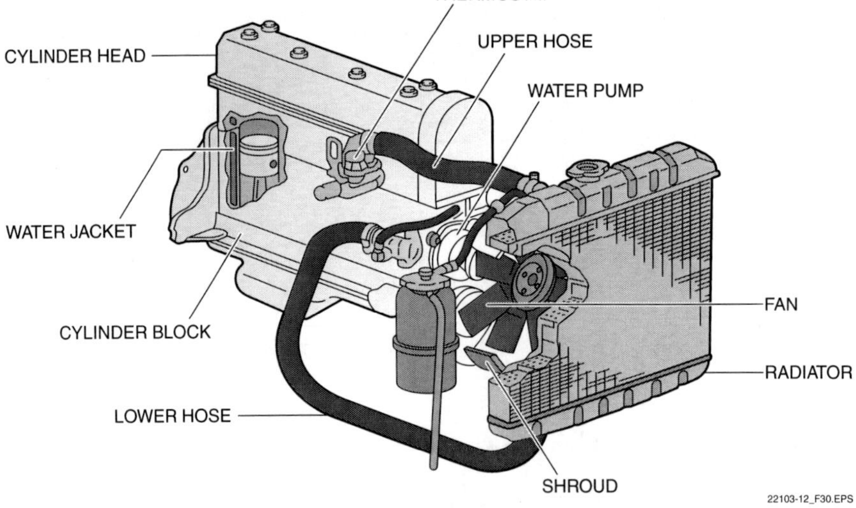

Figure 30 Typical cooling system.

the system is cool. Opening a closed cooling system when the coolant is hot can cause personal injuries. Always allow the system to cool before adding coolant. Verify that the level of coolant is at the level recommended by the manufacturer.

After an engine has been operating for 30 minutes or so, always check the temperature of the coolant. Check the temperature often as long as the engine is running. If an engine is showing signs of overheating, verify that the cooling fan is operating and the air paths into and out of the radiator are clear. If the fan is running and the air paths are unblocked, the next thing to do is to verify that the radiator has the proper level of coolant. Verify that the radiator and cooling system are not leaking coolant before attempting to check the level of coolant inside the radiator. Always allow an overheating engine to cool before checking or adding coolant.

WARNING!

When working on liquid coolant systems, make sure that the engine is stopped and wear proper eye, face, and hand protection. Hot coolant systems operate under pressure. To prevent an eruption of steam and coolant from the radiator, slowly depressurize the radiator before removing a radiator cap.

Do not attempt to add cold water to an overheated liquid coolant system. Allow the system to cool before depressurizing the system and adding coolant and/or water. Besides the danger of opening an overheated system, damage to the engine can be caused by adding cold coolant and/or water.

14.5.0 Electrical System

The typical charging and starting systems consist of a 12V or 24V starting system and an associated 12V **direct current (DC)** alternator, as shown in *Figure 31*. Direct current electrical systems are systems in which the electrical current flows in only one direction.

These systems provide either 12V or 24V battery power to rotate a high-torque starter, which starts the diesel engine. The systems shown in *Figure 31* can be a parallel or series-parallel systems. In the series-parallel system, the starting relay, when energized, configures the batteries in series to start the engine. After the engine starts, the relay is de-energized and the batteries are configured in parallel for charging and running purposes. In addition, 12V power is used for auxiliary systems, such as lights, computers, the load-indicating system, and engine instrumentation. The best way to determine that this system is operating properly is by observing ammeter or voltmeter readings to make sure that they are as specified by the manufacturer.

WARNING!

When working on electrical systems and lead-acid batteries, wear proper eye and face protection. When charging, batteries release explosive hydrogen gas and any spark or flame can cause accumulated gas or the batteries to explode. If battery acid gets on your skin, immediately flush the area with water and seek medical attention.

Teaching Tip/Notes

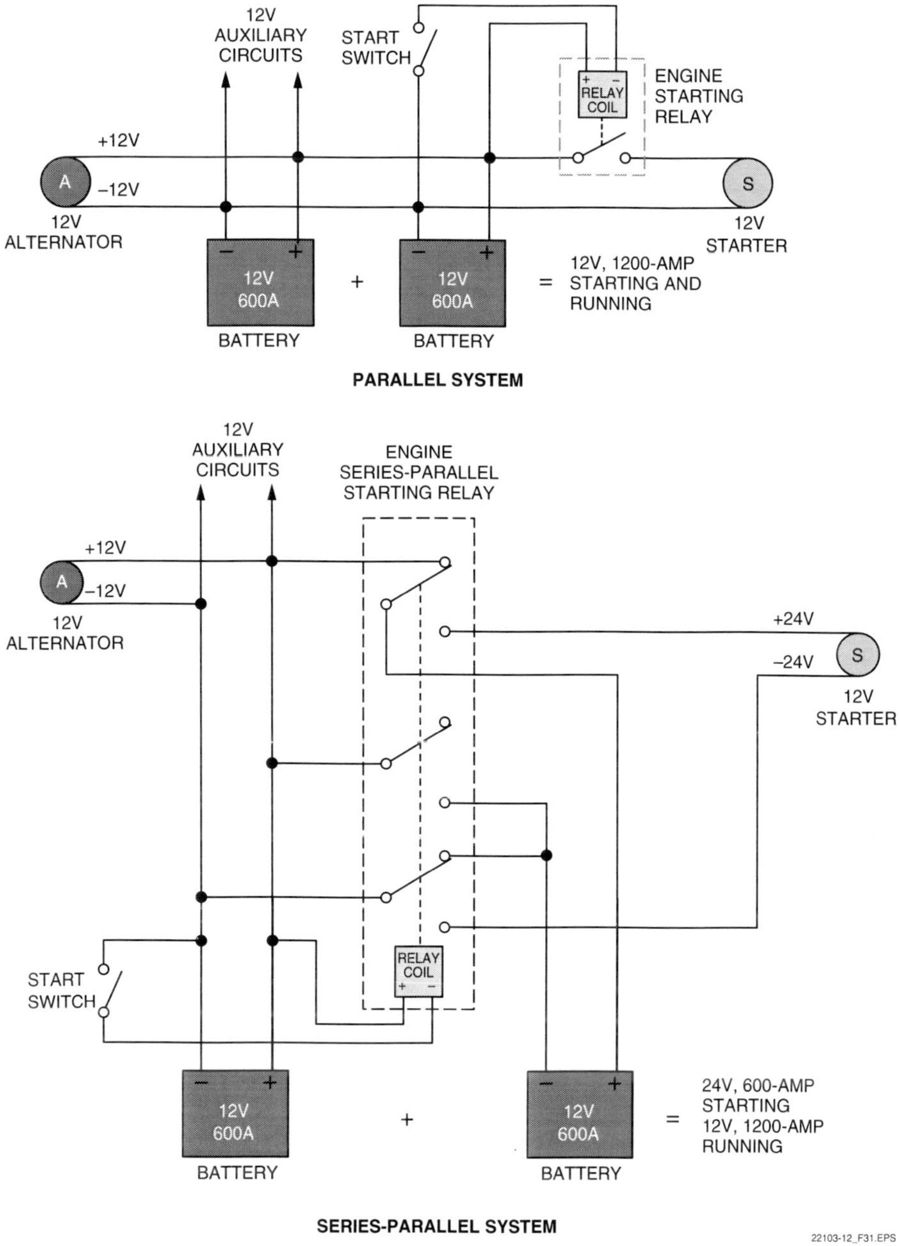

Figure 31 Typical charging and starting systems.

Teaching Tip/Notes

Classroom

Point out that electrical systems are not all the same. For example, some use a positive ground and others a negative ground.

Discuss the proper method for jump-starting a vehicle and stress the associated safety issues.

Audiovisual

Show Slide 42 (Figure 32). Describe the lubrication system for a diesel engine. Explain the difference between full-flow and bypass lubrication systems.

! Safety

Emphasize the importance of turning off the engine and wearing PPE when checking lubricants.

Audiovisual

Show Slide 43 (Figure 33). Describe the purpose and operation of an air filtration system.

Classroom

Describe the purpose of the hydraulic system.

Jump-starting one street vehicle from another is done often, but can be dangerous if improperly done. When dealing with the engines in larger heavy equipment machines, the danger increases because some of those machines require multiple batteries to start. There is also a danger because different machine manufacturers wire their machines differently. Some use a positive ground for their electrical systems while others use a negative ground. Always refer to the manufacturer's operations and maintenance manual to see how the machine is wired.

The type of batteries used in these heavy machines also differs from year to year. Some older batteries give off hydrogen sulfide gas. That gas will sometimes accumulate inside battery compartments on hot days. If a spark happens near that accumulated gas, the gas can explode causing possible injury to anyone near the explosion. Such sparks can be created by anything from a dropped tool to an improperly attached jumper cable.

Because most heavy equipment is located outside, the weather plays a major role in getting a large engine to start, especially cold weather. Any engine is going to be hard to start when cold because the lubricants inside the engine are still stiff from the cold. Companies often install block heaters on equipment engines operating in cold environments. The block heaters help keep the lubricants warmed and thinned, which makes starting an engine easier. If the starting batteries have been frozen and need to be jumped, they may explode as a charge is introduced into the cells. Check with your supervisor on how to deal with frozen batteries.

For jump-starting procedures, refer to the manufacturer's operations and maintenance manual for each machine. Also, check company policies about jump starting heavy equipment because jump starting may not be allowed in some companies due to safety concerns.

> **WARNING!**
>
> Use extreme care if jump starting is required. Improperly connected jumper cables can cause severe battery damage, starter damage, or an explosion. Make sure that the ground cable is connected to the chassis ground and not to the discharged battery ground.

14.6.0 Lubrication System

Another important system associated with the diesel engine is the lubrication system. This system provides lubrication and cooling to the engine bearings and other friction surfaces internal to the engine. This system can be one of two types: full flow and bypass (*Figure 32*). In some cases, both systems may be used on an engine. These systems consist of a pump, an oil sump, a pressure-regulating valve, and filter(s). In the bypass system, the pump pulls oil from the oil sump and sends it to the engine bearings under pressure. Some of the oil is also sent continuously through a filter to remove contaminants. The pressure-regulating valve bypasses a small amount of pressurized oil from the system back to the sump if the system pressure gets too high. In the full-flow system, all the oil is sent through a filter to the engine bearings. The filter is equipped with a bypass-relief valve in case the filter clogs.

> **WARNING!**
>
> When checking the oil in a hot engine, or when changing oil or filters, make sure the engine is stopped. Wear the proper eye, face, or hand protection as required because the oil, dipstick, and filters will be very hot.

14.7.0 Air Filtration System

The air intake system to the engine requires monitoring by the operator. This system allows the engine to draw in filtered air to mix with the fuel for combustion purposes. The air is filtered to prevent foreign material, which may cause damage to the engine, from entering the combustion chamber. *Figure 33* illustrates typical air filter systems. Some filters are equipped with dust cups or dust-unloading valves to rid the filter of some of the accumulated dust. Centrifugal action of the incoming air sweeps the dust into a dust cup or unloader valve.

15.0.0 HYDRAULIC SYSTEM

While the engine and drive train are used to control the movement of a machine, the hydraulic system is used to control the positioning of the blades, buckets, and other tools used on most heavy construction equipment. A hydraulic sys-

> **GOING GREEN**
>
> **Oil and Grease**
>
> Lubrication oils and greases are either petroleum-based or synthetic. Heavy equipment is often serviced in the field. To protect the environment, make every effort to keep any oils or greases from getting onto the surrounding soils, and make sure to keep them away from any streams or bodies of water.

Teaching Tip/Notes

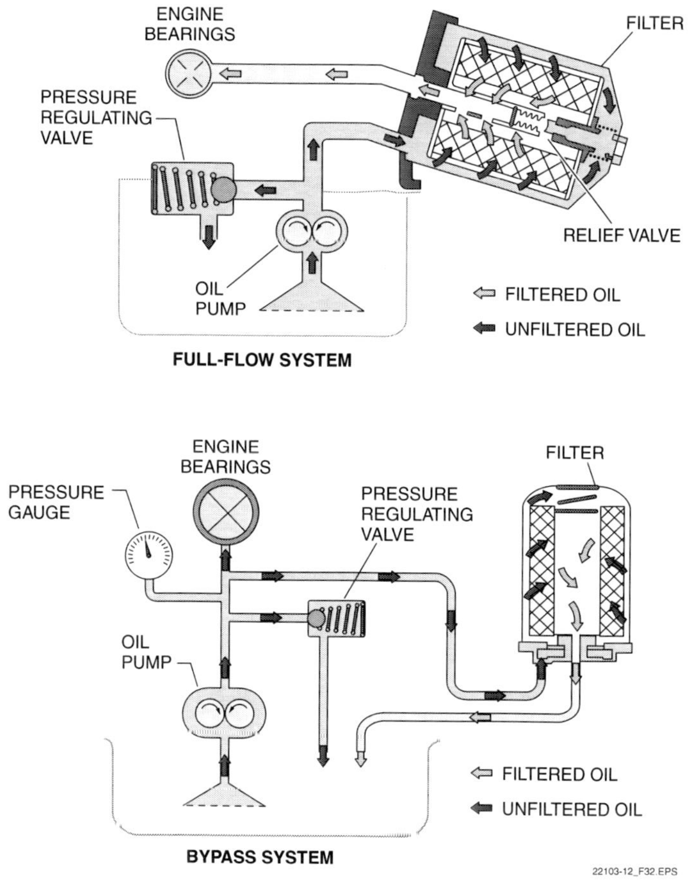

Figure 32 Lubrication systems.

tem uses hydraulic pressure on hydraulic fluids to do its work. Be prepared to follow all safety precautions when working on hydraulic systems. Because it is necessary to release hydraulic pressure during system maintenance or repair, it may be necessary to have hoist motors **dogged.**

WARNING!

When checking, filling, or working on a hydraulic system, wear proper eye, face, and hand protection. Hydraulic oil may be very hot if the machine has been operating. Also, ensure that any hydraulic cylinders are retracted as much as possible and that any hoist motors are dogged before releasing pressure on the system or any pressurized reservoir tank. Hydraulic system pressures can be very high if the system is under static load. Retracting any hydraulic cylinders will prevent overfilling of the reservoir tank.

15.1.0 Hydraulic Fundamentals

The transmitting of power in a hydraulic system requires an understanding of **Pascal's law** of hydraulics. This law states that any pressure applied to a fluid in a closed system is applied to all points

GOING GREEN

Hydraulic Fluid

Hydraulic systems notoriously leak. The source of such leaks must be identified and stopped. Hydraulic spills must be prevented. Spilled hydraulic fluids harm the environment. If unable to prevent a hydraulic spill, follow MSDS instructions and local laws for dealing with hydraulic fluid spills. Contaminated soils must be removed and disposed of in accordance with local guidelines.

Safety !

Point out the safety hazards associated with hydraulic fluid under pressure.

Classroom

Explain the principles of Pascal's law.

Teaching Tip/Notes

22103-12 Identification of Heavy Equipment Module Three 31

Audiovisual

Show Slide 44 (Figure 34).

Show Slide 45 (Figure 35). Describe the components of a hydraulic system and explain how the system works.

Classroom

Explain how hydraulics are used in brakes and clutches.

Explain how a torque converter replaces a conventional clutch.

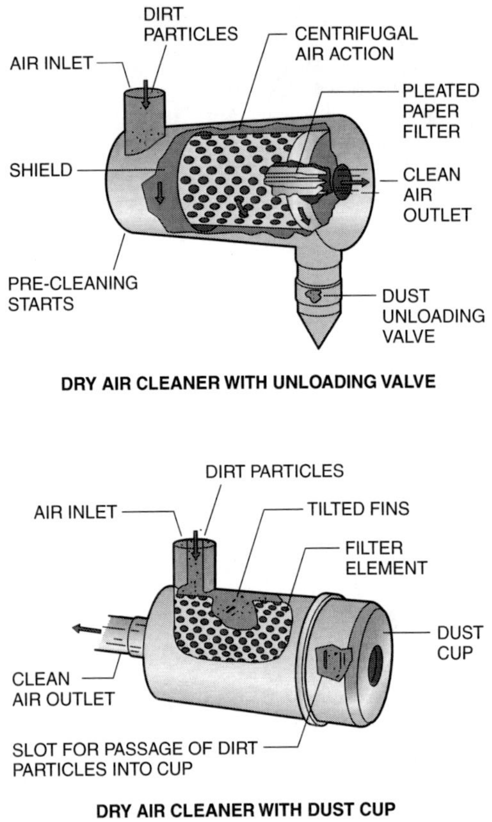

22103-12_F33.EPS

Figure 33 Typical air filters.

and surfaces in the system equally. Take the example shown in *Figure 34*. It shows a smaller piston having a force of 10 pounds applied to it, transmitting a force to the larger piston, which has five times the surface area of the smaller piston. This results in a force of 50 pounds generated by the larger piston. However, the larger piston was only moved one-fifth the distance of the smaller piston.

A typical hydraulic system consists of four basic components: a hydraulic pump, a reservoir with a strainer, a valve, and a **hydraulic actuator** (*Figure 35*). The pump moves the hydraulic fluid through the system. The reservoir is the source of the hydraulic fluid to be pumped into the system. It also allows for the return of the fluid. The valve controls the fluid as it does its work by porting the pressurized fluid to the proper place. The actuator is the device that performs the work. The actuator can be a cylinder that provides for linear work or a hydraulic motor for rotating work.

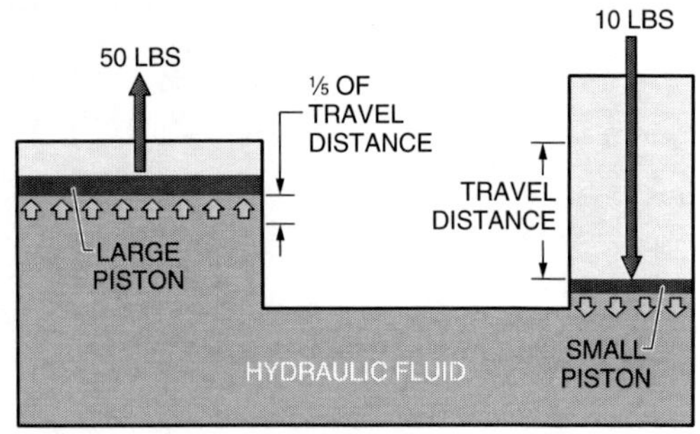

FORMULAS FOR PASCAL'S LAW

$$\text{PRESSURE (PSI)} = \frac{\text{FORCE (LBS)}}{\text{AREA (IN}^2\text{)}}$$

$$\text{FORCE (LBS)} = \text{PRESSURE (PSI)} \times \text{AREA (IN}^2\text{)}$$

22103-12_F34.EPS

Figure 34 Pascal's law.

Because hydraulics increase a system's mechanical advantage, they are frequently used to provide power assistance for many of the mechanical systems that a crane operator must position manually, such as brakes and clutches.

15.2.0 Hydraulic Brakes and Clutches

Power assist brakes are supplied on some machines. This system uses a small reservoir of brake fluid with a piston. As the operating pedal is pushed, it transmits the pressure to a brake cylinder, which in turn applies the brake pad to the drum or disc. This same type of system may be used for other braking systems, such as the swing brake and hoist brake on some types of cranes.

The clutch system uses a similar layout. This system typically involves a pedal operating a piston in a reservoir. The pressure applied when the pedal is pushed is transmitted to a cylinder that moves to overcome the spring pressure of the pressure plate. This allows the engine power to be disconnected from the power distribution system.

15.3.0 Hydraulic Power Couplers

A system that is being used more often is the **torque converter.** A torque converter, also known as a fluid coupler, is a small, self-contained hydraulic system. It is used to connect the main power source to the transmission or gearbox. The torque converter replaces a conventional clutch system. The principle of operation is simple. It is an oil-filled, donut-shaped device that has pump vanes connected to the engine surrounding tur-

Teaching Tip/Notes

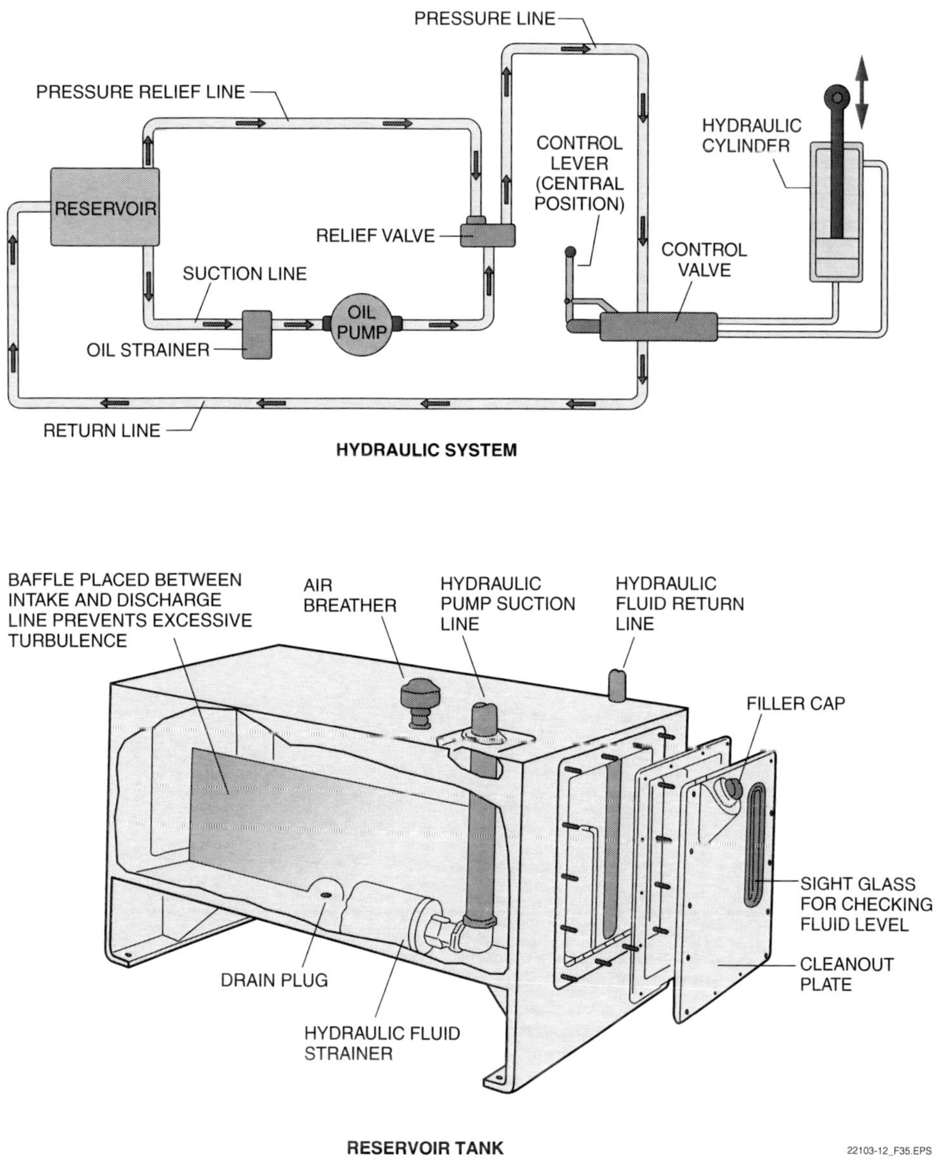

Figure 35 Typical hydraulic system and reservoir tank.

bine vanes. The vanes are connected to the transmission (*Figure 36*). As the pump is turned by the engine, it forces fluid into the turbine vanes, making them move. This movement causes the transmission gearing to transfer the torque created by the movement to the power distribution system.

15.4.0 Hydraulic Pumps

The pump is a major component of the hydraulic system. The pump is driven from the output of the transmission or gearbox to supply the fluid flow necessary for operation of the hydraulic

Audiovisual

Show Slide 46 (Figure 36).

Classroom

Explain the function of the pump in a hydraulic system.

Teaching Tip/Notes

22103-12 Identification of Heavy Equipment Module Three 33

Classroom

Point out the relationship between engine speed and hydraulic pressure.

PT/Laboratory

Under your supervision, have trainees identify the basic parts of each type of equipment and explain the differences in models of type of equipment. This laboratory corresponds to Performance Task 2.

Classroom

Summarize the major concepts presented in this module.

Figure 36 Torque converter.

power plant. Its speed is also dependent on engine speed. If the engine speed is kept too low, the hydraulic system will have insufficient flow to operate the system. This reduced flow can cause overheating of the hydraulic oil and prohibit the operation of some or all of the hydraulic actuators. For this reason, it is important that the equipment operator maintain sufficient engine speed to allow for proper operation of the hydraulic system.

On Site

Hydraulic Pressure and Systems

A pinpoint leak in a hydraulic system can be very dangerous. In some cases, the pressure from such a leak can be great enough to puncture the skin.

A good example of a hydraulic system is the braking system on your car. When you press the brake pedal, hydraulic fluid in the brake lines is pressurized. The pressure forces the abrasive brake lining or brake pad against the rotor, which causes the wheels to stop turning. The amount of pressure applied to the brakes is proportional to the amount of pressure you apply to the pedal. That is why you can use the brake to simply slow down or to come to a screeching stop. The same principle is used to position buckets and blades on heavy equipment.

SUMMARY

Heavy equipment is used in every aspect of site development, from clearing the land to grading a building site and roads. Heavy equipment is used to move material and equipment, demolish structures, excavate foundations and trenches, and level the site in preparation for new structures, roads, and parking lots.

Most heavy equipment can be used for more than one purpose simply by adding attachments. Manufacturers of heavy equipment have developed a huge range of equipment configurations, including both wheel and track versions, to meet the wide variety of site needs and conditions. The range extends from compact machines used on small sites to the mega-machines used in mining and quarry operations. Proper maintenance is required to keep all heavy equipment functioning safely and efficiently.

Teaching Tip/Notes

Review Questions

1. The only use for a tractor on a construction site is for pulling trailers.
   a. True
   b. False

2. Articulated trucks normally have capacities of up to _____.
   a. 20 tons
   b. 40 tons
   c. 100 tons
   d. 300 tons

3. The machine shown in *Figure 1* is a _____.
   a. wheel loader
   b. wheel dozer
   c. tractor
   d. roller

4. The machine shown in *Figure 2* is a(n) _____.
   a. scraper
   b. excavator
   c. skid steer
   d. motor grader

5. Which of these machines is equipped with stabilizers?
   a. Roller
   b. Dozer
   c. Motor grader
   d. Backhoe loader

6. The number of teeth on an excavator bucket is a function of the _____.
   a. bucket's width
   b. capacity of the bucket
   c. capacity of the excavator
   d. material being excavated

7. The machine shown in *Figure 3* is a(n) _____.
   a. dozer
   b. excavator
   c. wheel loader
   d. backhoe loader

8. The machine shown in *Figure 3* is used primarily as a(n) _____.
   a. pusher
   b. loader
   c. grading machine
   d. excavating machine

22103-12_RQ01.EPS

Figure 1

22103-12_RQ02.EPS

Figure 2

22103-12_RQ03.EPS

Figure 3

Classroom

Have the trainees complete the Review Questions, and go over the answers prior to administering the Module Examination.

Have the trainees complete the Trade Terms Quiz, and go over the answers prior to administering the Module Examination.

Examination ?

Administer the Module Examination. Record the results on Training Report Form 200, and submit the results to the Training Program Sponsor.

Performance Testing ✓

Administer the Performance Test and fill out the Performance Profile Sheets for each trainee. If desired, trainee proficiency noted during laboratory sessions may be used to complete the Performance Test. Record the results on Training Report Form 200 and submit the results to the Training Program Sponsor.

Teaching Tip/Notes

22103-12 Identification of Heavy Equipment Module Three 35

9. The machine shown in *Figure 4* is a(n) _____.
   a. motor grader
   b. excavator
   c. skid steer
   d. scraper

Figure 4

10. The machine shown in *Figure 5* is a(n) _____.
    a. wheeled loader
    b. fixed-mast forklift
    c. wheeled excavator
    d. telescoping boom forklift

Figure 5

11. In a diesel engine, what causes an explosion of the fuel and air mixture?
    a. The fuel
    b. An agitator
    c. Extreme compression
    d. An electrical spark

12. Some diesel engines use high-pressure fuel injection pumps that create pressures in the range of 20,000 psi to 28,000 psi.
    a. True
    b. False

13. On some diesel engines, the air to the engine fuel is delivered under high pressure directly into the combustion cylinder by a _____.
    a. crankshaft
    b. turbocharger
    c. mechanical actuator
    d. fuel injection pump

14. When working on a cooling system, the operator should _____.
    a. add coolant before depressurizing the system
    b. allow the system to cool before adding coolant
    c. add cold water before depressurizing the system
    d. remove the radiator cap to allow the system to depressurize

15. The device used on some air filters to rid the filters of accumulated dust is a _____.
    a. dust shaker shaft
    b. dust unloading valve
    c. small vacuum system
    d. small cyclone separator

Teaching Tip/Notes

Trade Terms Quiz

Fill in the blank with the correct term that you learned from your study of this module.

1. A(n) ________ is an attachment with teeth used on motor graders to loosen soil.
2. A compacted dirt road used to move material and equipment on and off the site is known as a(n) ________.
3. A(n) ________ is a device that uses fluid, usually oil, to transmit torque from one shaft to another.
4. An electrical current that flows in only one direction is called ________.
5. ________ is the basis of hydraulic theory.
6. A(n) ________ tire is inflated with air.
7. Two parts connected by a joint are ________, and move independently.
8. The fluid energy of the hydraulic system is converted into mechanical energy by a(n) ________.
9. A(n) ________ is an attachment used by dozers, motor graders, and other machines to loosen heavily compacted soil and soft rock.
10. ________ means to use a mechanical device to hold or fasten something tightly.

Trade Terms

Articulated
Direct current (DC)
Dogged
Haul road
Hydraulic actuator
Pascal's law
Pneumatic
Ripper
Scarifier
Torque converter

Teaching Tip/Notes

Cornerstone of Craftsmanship

Paul James

Department Head
Bridgerland Applied Technology College

Some people who earn their spurs in the construction trades migrate to the educational side. Paul James is a good example. After many years in the construction trades, he took his knowledge and experience to Bridgerland Applied Technology College in Utah and eventually became head of his department at the college.

How did you choose a career in the heavy equipment field?
I have always loved working in the construction field. Within that field, heavy equipment has always been one of the most interesting to me.

Who inspired you to enter the industry?
My father got me into the construction and maintenance field through his plumbing business. I started out my career as an apprentice plumber and had some great journeymen mentors, learning and working in the construction and maintenance field. After 10 years, I took a job with Thiokol Corporation in construction and maintenance. Through my on-the-job training and the related instruction in the apprentice training program, I developed my job into a rewarding and successful career.

What types of training have you been through?
After attending Utah State University for one year, I entered and completed a four-year apprenticeship program in plumbing while working for Donald H. James Plumbing and Heating and doing commercial, industrial, and residential plumbing. I also have completed OSHA and MSHA courses as well as other safety courses.

How important do you think education and training are in construction?
I have found that there is no substitute for good training. If you have the proper training it will pay big dividends in the future. Safety training is the most important. A person working in construction, especially a heavy equipment operator, must know and apply good safety practices every hour of every working day. You can never relax when it comes to safety.

How important are NCCER credentials to a career in construction?
NCCER has been the main curriculum for several of our programs at BATC, including Electrical, HVAC, and Heavy Equipment Operator training. NCCER relies on industry experts to make sure their training programs reflect the real world and meet industry standards. They are always willing to listen and are continually looking to improve the materials. Their programs have national recognition and widely accepted credentials. I am proud to serve as a subject matter expert for the development of the Heavy Equipment Operator program.

How has training/construction impacted your life?
Construction and training are my life. They have provided me with a great career so that I can provide for my family.

What kinds of work have you done in your career?
I have had 30-plus years in residential, commercial, and industrial construction, maintenance and installation of plumbing, industrial processes, underground utilities, and recently organized training in all of those fields.

Tell us about your present job.
I am currently the Department Head over the Heavy Equipment Operator, Professional Truck Driving, and Apprenticeship programs at Bridgerland Applied Technology College in Logan, Utah. This job requires me to plan courses, hire and manage instructors, and supervise the training facilities and equipment used to conduct the training courses. We currently have six apprenticeship programs functioning at BATC – Electrical, Plumbing, HVAC, Machinist, Industrial Maintenance, and Sign Fabrication.

Teaching Tip/Notes

What do you enjoy most about your job?
I really get a lot of satisfaction from helping others find employment. I enjoy seeing the students working in their chosen career field and knowing that the training we provided helped them get there.

What factors have contributed most to your success?
I have had great family, friends, and coworkers who have encouraged and supported me in my career. The training I have received, both on the job and in the classroom, have been major factors in my success.

Would you suggest construction as a career to others? Why?
Construction is a great career. One of my favorite things is the satisfaction of seeing the direct result of your efforts. I really like driving around our community and pointing out the buildings and projects that I have helped to construct.

What advice would you give to those new to the heavy equipment field?
My advice would be to take advantage of all the training and experience that you can. Make sure it is documented so you can show your employers or customers your credentials. A major benefit of the NCCER training is that completion credentials and transcripts are available to the students so that prospective employers can see what they have accomplished.

How do you define craftsmanship?
Craftsmanship is having the personal pride and knowledge to perform tasks and jobs to the best of your abilities. True craftsmen would never be afraid to have their names associated with a job because they know they gave it their best.

Teaching Tip/Notes

Trade Terms Introduced in This Module

Articulated: Two parts connected by a joint so as to move independently.

Direct current (DC): An electrical current that flows in only one direction.

Dogged: To hold or fasten tightly with a mechanical device.

Haul road: A compacted dirt road used to move material and equipment on and off the site.

Hydraulic actuator: A device that converts the fluid energy of the hydraulic system into mechanical energy using a hydraulic cylinder or motor.

Pascal's law: A scientific law stating that pressure at any one point in a closed static fluid system will be applied to all points in the system in every direction and will exert equal forces on equal areas; the basis of hydraulic theory, developed by Blaise Pascal (1623–1662), a French philosopher and mathematician.

Pneumatic: Inflated with compressed air.

Ripper: A towed attachment with teeth used on dozers, motor graders and other machines to loosen heavily compacted soil and soft rock.

Scarifier: An attachment with teeth used on motor graders to loosen soil.

Torque converter: A device that uses fluid, usually oil, to transmit torque from one shaft to another. Also known as a fluid coupler.

Teaching Tip/Notes

Additional Resources

This module presents thorough resources for task training. The following resource material is suggested for further study.

The Earthmover Encyclopedia, 2007. St. Paul, MN: MBI Publishing.

Clean Diesel Technology for Off-Road Engines and Equipment: Tier 4 and More, July 2011. Frederick, MD: Diesel Technology Forum.

Figure Credits

Reprinted courtesy of Caterpillar Inc., Module opener, Figures 2 (D), SA01, 4 (A, B), 5–9 (A, B), 10, 11 (B), 12, 13 (A, C), 14, 15 (A, B), 19, 20 (top and bottom left photos) 21, 22 (bottom photo) 23, 24, 26, 36, and SA03–SA07

Deere & Company, Figures 1, 9 (C), 13 (B), and 15 (C, D)

CNH America LLC, Figure 2 (A)

Images courtesy of Case IH, Figure 2 (B, C)

Images courtesy of Ray Walker Trucking, Figure 3 (A)

International Truck and Engine Corporation, Figure 3 (B)

Image courtesy of Dominion Equipment Parts, LLC, Figure 4 (C)

Komatsu America Corp., SA02

Stanley LaBounty, Figure 11 (A, C)

Sellick Equipment Limited, Figure 16

Manitou North America, Figure 17

Landoll Corporation, Figure 18

Image courtesy of Coastal Instrument & Supply Company, Figure 20 (bottom right photo)

Topaz Publications, Inc., Figure 22 (top photo)

The Charles Machine Works, Inc., Manufacturer of Ditch Witch products, Figure 22 (middle photo)

Cummins, Inc., Figure 25

US Department of Energy, Energy Efficiency & Renewable Energy, Figure 27

Image courtesy of John Deere and Diesel Technology Forum (DTF), Figures 28 and 29

US Navy, Figures 30, 32, 33, and 35

Teaching Tip/Notes

ANSWERS TO REVIEW QUESTIONS

Answer		Section Reference
1.	b	2.0.0
2.	b	3.0.0
3.	a	4.0.0; Figure 6
4.	a	5.0.0; Figure 7A
5.	d	6.0.0
6.	a	7.0.0
7.	a	8.0.0; Figure 13B
8.	c	8.0.0
9.	c	9.0.0; Figure 15C
10.	b	10.0.0; Figure 16
11.	c	14.1.0
12.	a	14.2.0
13.	b	14.2.0
14.	b	14.4.0
15.	b	14.7.0

ANSWERS TO TRADE TERMS QUIZ

1. scarifier
2. haul road
3. torque converter
4. direct current (DC)
5. Pascal's law
6. pneumatic
7. articulated
8. hydraulic actuator
9. ripper
10. dogged

NCCER CURRICULA — USER UPDATE

NCCER makes every effort to keep its textbooks up-to-date and free of technical errors. We appreciate your help in this process. If you find an error, a typographical mistake, or an inaccuracy in NCCER's curricula, please fill out this form (or a photocopy), or complete the online form at **www.nccer.org/olf**. Be sure to include the exact module ID number, page number, a detailed description, and your recommended correction. Your input will be brought to the attention of the Authoring Team. Thank you for your assistance.

Instructors – If you have an idea for improving this textbook, or have found that additional materials were necessary to teach this module effectively, please let us know so that we may present your suggestions to the Authoring Team.

NCCER Product Development and Revision

13614 Progress Blvd., Alachua, FL 32615

Email: curriculum@nccer.org

Online: www.nccer.org/olf

❑ Trainee Guide ❑ AIG ❑ Exam ❑ PowerPoints Other ____________________

Craft / Level: Copyright Date:

Module ID Number / Title:

Section Number(s):

Description:

____

____

____

____

Recommended Correction:

____

____

____

____

Your Name:

Address:

____

Email: Phone:

22104-12

Basic Operational Techniques

Module Four

Materials

reminds you to gather materials needed for classes, labs, and testing.

Homework

suggests that you assign reading for the next session.

Classroom

prompts you to discuss a topic, stress a point, explain a concept, or provide examples.

Audiovisual

indicates an appropriate time to show a slide or transparency.

! Safety

used to emphasize safety issues.

Demonstration

directs you to show trainees how to perform tasks.

Laboratory

used when trainees are to practice performing tasks.

✓ Performance Testing

recommends administering a performance test, or a portion thereof.

PT / Laboratory

indicates that lab practice corresponds with a performance task. If desired, you can note the proficiency of the trainees during the lab to satisfy performance testing requirements.

Teaching Tip

indicates tips for practice exercises, Quick Quizzes, field trips, or other activities designed to encourage class participation.

? Examination

recommends administering the written module exam.

NCCER Standardized Craft Training Program

NCCER provides a standardized national program of accredited craft training. Key features of the program include instructor certification, competency-based training, and performance testing. The program provides trainees, instructors, and companies with a standard form of recognition through the National Registry. The program is described in full in the *Guidelines for Accreditation*, published by NCCER. For more information on standardized craft training, contact NCCER by writing to 13614 Progress Blvd., Alachua FL 32615; calling 1-888-NCCER20; or emailing **info@nccer.org**. More information is available at **www.nccer.org**.

How to Use this Annotated Instructor's Guide

Each page presents two sections of information. The larger section displays each page exactly as it appears in the Trainee Module. The narrow column ties suggested trainee and instructor actions to each page and provides icons (detailed on the left) to call your attention to material, safety, audiovisual, or testing requirements. The bottom of each page includes space for your notes, or when available, teaching tips that suggest activities or class exercises that will help your students learn the material.

Preparation

Before teaching this module, you should review the Objectives, Performance Tasks, Materials and Equipment List, and Module Outline. Be sure to allow ample time to prepare your own training or lesson plan and gather all required materials and equipment.

Instructors purchasing a single module receive the printed exam and performance tasks with the AIG module. Perfect-bound versions of this title no longer include the printed exam booklet and performance task sheets. Log on to the Instructor Resource Center at **www.NCCERIRC.com** to download these materials. Your access code is on the card bound into the front of this AIG. This code will allow you to access exams and performance tasks for all levels of this craft. Be sure to keep all testing materials in a secure place at all times.

Basic Operational Techniques
Annotated Instructor's Guide

Module 22104-12

Module Overview

In this module, the trainee will learn about the pre-operational checks and operator maintenance tasks for heavy equipment. The trainee will learn basic startup procedures and will be introduced to basic operation of the various heavy equipment machines. The purpose of the module is to provide the trainee with an opportunity for some hands-on time.

Objectives

Upon completion of this module, the trainee will be able to do the following:

1. Describe basic prestart activities for heavy equipment machinery.
2. Describe basic safety measures associated with operating heavy equipment.
3. Explain how to properly start, operate, and shut down the following types of heavy equipment: utility tractors, dozers, loaders, backhoes, excavators, compaction equipment, motor graders, scrapers, on-road dump trucks, off-road dump trucks, forklifts, skid steers, and trenchers.

Performance Tasks

Under the supervision of the instructor, the trainee should be able to do the following:

1. Perform basic prestart inspection, startup, operational movement, and shutdown for the following types of heavy equipment:
   - Utility tractors
   - Dozers
   - Loaders
   - Backhoes
   - Excavators
   - Compaction equipment
   - Motor graders
   - Scrapers
   - On-road dump trucks
   - Off-road dump trucks
   - Forklifts
   - Skid steers
   - Trenchers

Materials and Equipment

Markers/chalk
Pencils and paper
Whiteboard/chalkboard
Heavy Equipment Operations Level One PowerPoint® Presentation Slides (ISBN 978-0-13-292167-1)
Multimedia projector and screen
Computer
Appropriate personal protective equipment
Operator's manuals
Heavy equipment machines including:
- Utility tractors
- Dozers
- Loaders
- Backhoes
- Excavators
- Compaction equipment
- Motor graders
- Scrapers
- On road dump trucks
- Off road dump trucks
- Forklifts
- Skid steers
- Trenchers

Trade Terms Quiz*
Module Examinations**
Performance Profile Sheets**

* Located at the back of the Trainee Guide module

** Single-module AIG purchases include the printed exam and performance task sheet. If you have purchased the perfect-bound version of this title, download these materials from the IRC using your access code.

Safety Considerations

Ensure that the trainees are equipped with appropriate personal protective equipment and know how to use it properly. Review safety guidelines associated with working on or around heavy equipment. Emphasize the importance of proper housekeeping.

Additional Resources

This module presents thorough resources for task training. The following resource material is suggested for further study.

The Earthmover Encyclopedia, 2007. St. Paul, MN: MBI Publishing.

Construction Safety. 1996. Jimmie Hinze. Englewood Cliffs, NJ: Prentice Hall.

Field Safety Participant Guide, 2003. NCCER. Upper Saddle River, NJ: Prentice Hall.

Handbook of OSHA Construction Safety and Health, Second Edition, 2006. Charles D. Reece and James V. Eidson. Boca Raton, FL: CRC Press.

HazCom for Construction. DVD. DuPont Sustainable Solutions–Training Solutions. Virginia Beach, VA.

Manual on Uniform Traffic Control Devices for Streets and Highways. 2009 edition. Washington, DC: US Department of Transportation, Federal Highway Administration.

OSHA Standards 29 CFR, Part 1926 Safety and Health Regulations for Construction. Washington, DC: Occupational Safety and Health Administration (OSHA), US Department of Labor, US Government Printing Office.

Teaching Time for This Module

An outline for use in developing your lesson plan is presented below. Note that each Roman numeral in the outline equates to one session of instruction. Each session has a suggested time period of 2½ hours. This includes 10 minutes at the beginning of each session for administrative tasks and one 10-minute break during the session. Approximately 27½ hours are suggested to cover *Basic Operational Techniques*. You will need to adjust the time required for hands-on activity and testing based on your class size and resources. Because laboratories often correspond to Performance Tasks, the proficiency of the trainees may be noted during these exercises for Performance Testing purposes.

Topic	Planned Time
Session I. Introduction; Safety; Prestart Activities	
A. Introduction	____________
B. Safety	____________
1. Work Area Familiarization	____________
2. Mounting the Vehicle	____________
C. Prestart Activities	____________
1. Structural Hardware Inspections and Checks	____________
2. Power Train Inspection and Checks	____________
3. Hydraulic System Inspection and Checks	____________
4. Electrical and Control Inspection and Checks	____________
5. Attachments	____________
6. Operator Cab	____________
7. Prestart Activities for Trucks	____________

Session II. Startup Activities

A. Startup Activities ____________

1. Indicator Checks ____________
2. Starting Aids ____________
3. Starting ____________
4. Warmup ____________

Sessions III–X. Equipment Operations; Shutdown Activities

A. Equipment Operations ____________

1. Construction and Steering ____________
2. Basic Operations ____________
3. Machine-Specific Operations ____________

B. Shutdown Activities ____________

C. PT/Laboratory ____________

Have trainees perform basic inspection, startup, operational movement, and shutdown of specified heavy equipment. This laboratory corresponds to Performance Task 1.

Session XI. Review and Testing

A. Review ____________

B. Module Examination ____________

1. Trainees must score 70% or higher to receive recognition from NCCER.
2. Record the testing results on Training Report Form 200 and submit the results to the Training Program Sponsor.

C. Performance Testing ____________

1. Trainees must perform each task to the satisfaction of the instructor to receive recognition from NCCER. If applicable, proficiency noted during laboratory exercises can be used to satisfy the Performance Testing requirements.
2. Record the testing results on Training Report Form 200 and submit the results to the Training Program Sponsor.

22104-12

Basic Operational Techniques

Module Four

Teaching Tip/Notes

Homework

Assign the reading of this module.

Trainees with successful module completions may be eligible for credentialing through NCCER's National Registry. To learn more, go to **www.nccer.org** or contact us at **1.888.622.3720**. Our website has information on the latest product releases and training, as well as online versions of our *Cornerstone* newsletter and Pearson's product catalog.

Your feedback is welcome. You may email your comments to **curriculum@nccer.org**, send general comments and inquiries to **info@nccer.org**, or fill in the User Update form at the back of this module.

V.1 4/12

Teaching Tip/Notes

22104-12
Basic Operational Techniques

Objectives

When you have completed this module, you will be able to do the following:

1. Describe basic prestart activities for heavy equipment machinery.
2. Describe basic safety measures associated with operating heavy equipment.
3. Explain how to properly start, operate, and shut down the following types of heavy equipment: utility tractors, dozers, loaders, backhoes, excavators, compaction equipment, motor graders, scrapers, on-road dump trucks, off-road dump trucks, forklifts, skid steers, and trenchers.

Performance Task

Under the supervision of your instructor, you should be able to do the following:

1. Perform basic prestart inspection, startup, operational movement, and shutdown for the following types of heavy equipment:
   - Utility tractors
   - Dozers
   - Loaders
   - Backhoes
   - Excavators
   - Compaction equipment
   - Motor graders
   - Scrapers
   - On-road dump trucks
   - Off-road dump trucks
   - Forklifts
   - Skid steers
   - Trenchers

Trade Terms

Blade
Bucket
Float
Glow plugs
Joystick
Moldboard
Pitch
Power takeoff (PTO)
Rollover protective structure (ROPS)
Scarifying
Stabilizers
Throttle

Industry Recognized Credentials

If you're training through an NCCER-accredited sponsor you may be eligible for credentials from NCCER's Registry. The ID number for this module is 22104-12. Note that this module may have been used in other NCCER curricula and may apply to other level completions. Contact NCCER's Registry at 888.622.3720 or go to nccer.org for more information.

Materials

Ensure that you have everything required to teach the course. Check the Materials and Equipment list at the front of this module.

Classroom

Explain that terms shown in bold (blue) are defined in the Glossary at the back of this module.

Audiovisual

Show Slide 1, Objectives, and Slide 2, Performance Tasks. Review the goals of the module, and explain what will be expected of the trainees.

Teaching Tip/Notes

When you call on someone to answer a question, the rest of the class relaxes or even tunes out because they expect that the question and answer will take place only between you and the trainee you called on. Instead, use this technique to involve more trainees in answering questions and to keep them on their toes:

1. Ask trainees to define a term or explain a concept.
2. After one trainee has answered, ask a trainee seated nearby if the answer is right. Then ask whether a trainee in the back of the room agrees.
3. Ask trainees to explain why they think an answer is right or wrong.
4. Use the session to clear up incorrect ideas and encourage trainees to learn from their mistakes.

Contents

Topics to be presented in this module include:

Teaching Tip/Notes

Figures and Tables

Teaching Tip/Notes

Figures and Tables (continued)

Teaching Tip/Notes

1.0.0 Introduction

Learning to operate a piece of heavy equipment is something like learning to drive and take care of a car. One of the first things an operator must learn is how to check the vehicle prior to starting it. Visual inspections of the vehicle are critical. The tires should appear to be inflated and undamaged. The lights and horn should be functional. All vehicle body parts need to be in place and usable. After the vehicle has been checked out, the operator should get into the operator's seat, fasten the seat belt, and become familiar with all the instruments and controls.

After verifying that the vehicle is safe to use, and after becoming familiar with the controls, the operator can start the vehicle, move it forward, change its direction, stop it, and move it backwards before parking it. Heavy equipment operators must learn to do such basic actions before they can move on to operating any attachments installed on the heavy equipment.

This module covers very basic information for the operation of heavy equipment under the guidance of an experienced instructor. An overview of basic daily preventive maintenance and basic operational guidelines is provided.

2.0.0 Safety

Most of the heavy equipment machines being used on a work site are large enough to kill someone if the person is run over by the machine. Operators must be constantly aware of their surroundings while operating one of these machines. Anyone on the ground and near a heavy equipment machine must keep away from the machine, whether it is moving or not. Always assume that the operator cannot see you. Some of the larger machines have a closed-circuit camera system that allows the operator to see blind spots that cannot normally be seen from the operator cab.

2.1.0 Work Area Familiarization

Operators must familiarize themselves with the day's operations. They must review any job-related paperwork to make sure that they know what is supposed to be done. Check for any warnings or notes in the job sheet. Look for any markers that show where to dig or fill. Walk around the area in which the work will be done. Check the ground for large holes, large rocks, soft or muddy areas, sudden dropoffs, and curbs. Driving over uneven terrain with a heavy load can cause the vehicle to tip. Be aware of the terrain, and plan how to move the equipment safely. Sharp objects can damage tires. If possible move them or mark them. Look above the site for overhead power lines, trees, and other overhangs that could contact a raised bucket. Be aware of any low clearances in the operational area.

2.2.0 Mounting the Vehicle

Getting into the vehicle can be dangerous. Slips and falls are common hazards, but they can be avoided. Clean the steps and handholds. Oil, grease, snow, ice, and mud can cause slips and falls. Make sure work shoes do not have any excessive amounts of mud or grease.

Mount the machine only at locations that are equipped with steps and/or handholds. Face the machine when mounting or dismounting. Maintain three points of contact with the machine at all times. Observe the following safety rules when mounting heavy equipment:

- Do not use controls as handholds.
- Do not mount the machine while carrying tools or supplies.
- Do not mount a moving machine.
- Do not jump off a machine or allow others to do so.

Once in the operator cab, remove all personal items or other objects from the operator's area. Secure loose items or remove them from the machine. Loose items can become flying objects and cause equipment damage or personal injury. Adjust the seat as needed to reach all the controls comfortably. Verify that the seat is locked into the correct position. Take time to become familiar with all the controls, warning devices, and gauges.

Check that all of the safety equipment is securely in place and in operating condition. Clean and adjust the mirrors. Verify that the fire extinguisher is fully charged. Make sure that any signs, shields, or guards are in place. Check the service and parking brakes. Always fasten the seat belt before operating the machine. OSHA and MSHA require that approved seatbelts and **rollover protective structures (ROPS)** be installed on virtually all heavy equipment. Check the ROPS bolts to make sure they are not loose or damaged. Old equipment must be retrofitted.

> **WARNING!**
> Do not use any heavy equipment that is not equipped with approved seatbelts and an ROPS.

Classroom

Emphasize that heavy equipment work is hazardous and must be performed in a safe and responsible manner.

Point out that compliance with safety requirements will be carefully scrutinized during the study of this module.

Reiterate the safety practices that apply to heavy equipment work.

Stress the importance of inspecting the work site for hazards before starting work on the site.

Describe the proper method for mounting a machine.

Stress the importance of three-point contact when mounting and dismounting.

Emphasize the importance of securing the cab and fastening the seat belt before starting the machine.

Teaching Tip/Notes

Classroom

Explain that an inspection of heavy equipment machinery must be performed at the start of each shift.

Audiovisual

Show Slide 3 (Figure 1). Discuss the machine areas that must be inspected.

3.0.0 Prestart Activities

Several of the prestart activities performed on heavy equipment machinery are the same on all machines. This section addresses the generic prestart activities that should be performed on any heavy equipment vehicle prior to starting it. Most companies have inspection checklists that specify which things need to be checked on a given machine. *Figure 1* shows an example of a shift inspection form similar to some used by construction companies. This type of inspection needs to be performed on all equipment.

The following are common areas that must be checked:

- Structural hardware (undercarriage, tires, tracks, and body parts)

Shift Inspection List

Date ________ Submitted by ________
Unit No./S.N. ________
Make/Model ________
Hour/Odometer Reading ________
Today's Use: Hours Machine...
- Operated ________
- Down (due to failure) ________
- Idle (machine not needed) ________

Use the daily inspection sheet as a guide to fill in the following information (if applicable). Explain any defects on the back side.

	OK F	OK R	Amount Added or *Needs Attention
Engine Coolant	___	___	________
Engine Oil	___	___	________
Transmission Oil	___	___	________
Pump Drive Oil	___	___	________
Pivot Shaft Oil	___	___	________
Hydraulic Oil	___	___	________
Fuel	___	___	________
Coolant Temp. Recorder Label	___	___	________
Engine Air Restriction Indicator	___	___	________
Ground Engaging Tools	___	___	________
Tires/Undercarriage	_/_	_/_	________
Fire Protection	___	___	________
Seat Belt	___	___	________
EMS/Gauges	___	___	________
Governor Control	___	___	________
Brakes	___	___	________
Transmission	___	___	________
Steering	___	___	________
Implements	___	___	________
Lights	___	___	________
Horn	___	___	________
Backup Alarm	___	___	________
Wipers/Washers	___	___	________

Inform supervisor of any malfunction before operating machine

Repairs Needed ________

Additional Comments ________

Signature ________

22104-12_F01.EPS

Figure 1 Example of a shift inspection form.

Teaching Tip/Notes

- Engine and power train
- Hydraulic system
- Electrical system for controls and instrumentation (gauges)
- Attachments

> **NOTE**
> For specific details on what needs to be checked on any given machine, refer to the vehicle-specific Operations and Maintenance (O&M) manual. The O&M manuals, also called the operator's manual, give specific procedures for any operator-performed maintenance activities. Most O&M manuals also give detailed pictures of all control panels, indicators, and control devices.

3.1.0 Structural Hardware Inspections and Checks

Heavy construction equipment is normally mounted on an undercarriage riding on either metal tracks or rubber tires. Structural hardware includes the undercarriage and all things mounted to it, as well as the machine's body.

3.1.1 Tires

Some machines ride on inflatable tires while others ride on solid rubber tires. Check tires for bulges, cuts, punctures, and signs of excessive wear. Check the wheels to insure that lug nuts are in place and properly torqued. Verify that the inflatable tires are properly inflated. Tire inflation pressures are often posted near each tire. *Figure 2* shows examples of both inflatable and solid rubber tires found on construction equipment.

Figure 2 Construction equipment tires.

3.1.2 Tracks

Machines with metal tracks are a bit more complicated to inspect due to their construction. Metal tracks are made up of track shoes fastened to a track chain (*Figure 3A*). The track chain is a drive chain that wraps around the drive sprocket (at the rear of the undercarriage) and the idler sprocket (closer to the front of the undercarriage). Between the sprockets, the chain is supported by rollers. The sprockets and rollers all ride on bearings that must be lubricated. Due to the environment in which most tracks operate, rocks, roots, and other debris may become trapped in between the track chain and the sprockets or rollers. Preoperational checks need to include making sure that all debris is removed from the tracks. Also, look for loose, damaged, or missing bolts or pins that hold the track shoes to the chain; check proper track tension per the manufacturer's instructions; and look for loose or damaged track shoes.

Some lighter-weight construction vehicles are equipped with rubber tracks. The rubber tracks are essentially one large rubber belt with molded cleats. Mechanically, the rubber tracks function the same as the metal tracks. A rubber track wraps around the drive and idler sprockets and is supported by rollers installed on the undercarriage frame. *Figure 3B* shows an undercarriage with rubber tracks. Inspections of a rubber track should include making sure that the track appears to be properly installed and tensioned per the manufacturer's instructions; is not excessively worn; and is not ripped or torn.

Classroom

Explain that the operator manual for the machine provides specific inspection criteria.

Describe the components that make up the structural hardware.

Discuss the inspection criteria for tires.

Audiovisual

Show Slide 4 (Figure 2).

Classroom

Point out the inspection and lubrication points for tracks.

Explain the importance of proper tensioning for both metal and rubber tracks.

Audiovisual

Show Slides 5 and 6 (Figure 3). Explain the differences between metal and rubber tracks.

Teaching Tip/Notes

Provide a copy of an operator manual for the trainees to examine. Point out the prestart inspection section.

22104-12 Basic Operational Techniques Module Four 3

Classroom

Point out that rotating parts have bearings that must be greased to prevent them from wearing out.

Discuss the importance of checking all machine fluids.

Describe how to fill fluid reservoirs and lubricate pivot points.

Audiovisual

Show Slide 7 (Figure 4).

Classroom

Stress the importance of keeping surfaces such as ladders and body panels clean.

A. METAL TRACKS

22104-12_F03A.EPS

Figure 3 Metal and rubber tracks. (1 of 2)

3.1.3 Rotating Parts

It is important to keep rotating parts properly lubricated in order to get the most use out of the parts. Remember that all rotating parts are riding on some form of bearing. Bearings must be either greased or bathed in some kind of lubricant (oil). Refer to the O&M manual for lubrication instructions on any undercarriage components. If bearings need to be greased, make sure that the grease fittings are cleaned before attempting to add grease to them. Also, make sure that the right kind of lubricant is being used. *Figure 4* shows a grease gun being used to grease a fitting on the bottom end of a hydraulic cylinder. All pivot points, such as the end of the hydraulic cylinder, have a grease fitting.

B. RUBBER TRACKS

22104-12_F03B.EPS

Figure 3 Metal and rubber tracks. (2 of 2)

3.1.4 Frame and Body Panels

The frame of a heavy equipment machine is supported above the undercarriage of the vehicle. Keeping the main protective structure around

22104-12_F04.EPS

Figure 4 Example of grease fittings.

Teaching Tip/Notes

and over the operator in good condition is critical to the safe operations of the machine. Anyone inspecting the machine's frame and body panels should be looking for cracks in the metal (especially the frames), loose connections (bolts or other types of fasteners), broken or cracked welds, and excessive corrosion.

3.1.5 Machine Surface Areas

Dirt, mud, and moisture collect on the surfaces of the machine's body panels, ladders, and walking surfaces. Make sure ladders and walking surfaces are clean.

3.1.6 Safety Equipment

Check all installed or mounted safety equipment on the vehicle. This includes mirrors, seat belts, horn(s), and the fire extinguisher. Verify that lights are properly mounted and not damaged. If the machine is so equipped, ensure that all brush guards are properly mounted and undamaged. Make sure windows are clean and undamaged.

3.2.0 Power Train Inspections and Checks

The power train of a vehicle starts with the engine. Heavy equipment machines are powered by either gasoline or diesel engines. The vehicle's engine is tied through a transmission or torque converter to the machine's drive axle. Regardless of engine type, the operator needs to visually check the machine's fuel supply system, coolant system, engine, transmission, drive axle(s), and hydraulic system for signs of leaks.

3.2.1 Fuel Supply

Engines must have the right type of fuel to operate. Gas engines must have only an approved grade of gasoline. Diesel engines also must have only an approved grade of diesel fuel. Never use gas in a diesel engine, or diesel fuel in a gas engine. Most fuel tanks are clearly marked for gas or diesel fuel.

> **WARNING!**
> Never fill the fuel tank or work on any part of the fuel system with the engine running, while smoking, or near an open flame. Doing so may cause a fire or an explosion. Make sure a fire extinguisher is nearby.

Before the fuel reaches the engine from the fuel tank, it usually passes through some kind of fuel filter. Those fuel filters must be changed at times. Fuel filters on heavy equipment machines are usually mounted near the engine's fuel injector (or carburetor), as shown in *Figure 5*. Check for visible water in the fuel-water separator and drain it into an approved container (*Figure 6*).

Figure 5 Fuel filter change.

Figure 6 Fuel-water separator.

> **GOING GREEN**
> **Lubricants, Coolants, and Fuels**
> Lubricants, coolants, and fuels harm the environment. Be extremely careful to not spill coolants and fuels when servicing a vehicle. As for lubricants, make sure that all leaks are contained, reported, and stopped. When oil must be changed in an engine or transmission, make sure that the used oil is contained and recycled or disposed of properly.

Classroom

Point out the need to check body parts for loose fasteners and damage, such as cracks or corrosion.

Discuss the components that make up the machine's power train.

Stress the importance of using the proper fuel.

Safety !

Point out that refueling must never be done with the engine running, around an open flame, or while anyone nearby is smoking.

Audiovisual

Show Slides 8 and 9 (Figures 5 and 6). Point out that fuel filters must be periodically cleaned or replaced.

Teaching Tip/Notes

22104-12 Basic Operational Techniques Module Four 5

Classroom

Discuss the coolant system components.

Audiovisual

Show Slides 10 and 11 (Figures 7 and 8).

! Safety

Point out that the radiator cap should never be removed while the engine is hot.

Classroom

Explain the importance of checking the engine oil and periodically replacing the oil and oil filter.

Audiovisual

Show Slide 12 (Figure 9).

3.2.2 Coolant System Check

Start the power train inspections at the engine's radiator. Make sure that the air paths into, through, and out of the radiator are free of debris. Some radiators have electric fans that increase the air flow over the fins of the radiator tubes. *Figure 7* shows a swingout electric fan on a large heavy equipment machine.

The hoses between the engine and the radiator must be securely fastened and in good condition. If the engine is cool, remove the radiator cap (*Figure 8*) and check the fluid level in the radiator. Some machines have coolant storage tanks for overflow coolant. If in doubt about any fluid level, review the O&M manual for the machine being inspected. Add coolant as needed.

> **CAUTION**
> Use the specified coolant. Do not mix coolant types (colors).

> **WARNING!**
> To prevent personal injury, never open the radiator cap on a hot engine. Always allow it to cool before relieving the pressure inside the radiator and opening the cap.

> **NOTE**
> Different machines have their radiators located in different places. The radiator cap and coolant storage tank are usually protected inside the body of the machine.

3.2.3 Engine Oil Check

Engines have dipsticks that are used to measure the amount of oil in the engine (*Figure 9*). Dipsticks have markers on them that show where the oil level should be, and a level where oil needs to be added. Pull the dipstick out of its storage area and check the oil level of the engine. If oil must be added, make sure that only the oil called for in the machine's O&M manual (or company procedures) is used. Do not overfill the oil. Make sure that the dipstick is securely replaced.

While checking the oil, also locate and check the engine's oil filter(s). Some companies have their maintenance workers mark the oil filter(s) with a date when the filter was last changed. Maintenance records should also have that date. Try to determine if the machine's oil filter needs to be changed. Make sure the machine is serviced in accordance with the manufacturer's maintenance schedule.

3.2.4 Engine Air Filter Check

Due to the environments in which most heavy construction equipment is used, getting clean air

Figure 7 Radiator cooling fan.

Teaching Tip/Notes

Figure 8 Radiator cap and coolant storage tank.

Figure 9 Oil dipstick and engine oil fill area.

to the engines is most critical. Most heavy equipment engines have large air filters housed inside of metal canisters mounted somewhere near the engine. In most cases, a lid can be removed from the filter housing and the filter can be slipped out for visual inspection. In some cases, a filter can be cleaned if it is not too oily and dirty. If in doubt, replace any dirty filter. *Figure 10* shows an air filter being checked and replaced.

3.2.5 Transmission and Drive Axle Checks

Transmissions and drive axles, regardless of type, need lubricants to protect the gears inside them. Any transmission, other than a standard shift transmission, also requires a specific transmission fluid in order to function as designed. Always check the O&M manual for the specific type of transmission fluid needed in the machine being inspected. Some transmissions have sight glasses that indicate the level of fluid in the transmission. Others may have a dip stick. Drive axles usually have inspection ports that can be opened to check the level of lubricant inside them.

WARNING!

Be careful when checking the fluid levels in transmissions and drive axles. If checking the fluid while it is hot, be careful to not get any of the hot fluid on exposed skin.

Some modern machines may have temperature and level sensors in or on both the transmission and drive axle(s). Monitoring the operating temperature of a transmission is critical to the continued operation of the machine.

Audiovisual

Show Slide 13 (Figure 10). Explain the importance of checking and periodically cleaning or replacing the engine air filter.

Classroom

Explain the inspection and lubrication requirements for the transmission and drive axle.

Safety !

Warn trainees to use caution when checking fluid levels while fluid is hot.

Teaching Tip/Notes

22104-12 Basic Operational Techniques Module Four 7

Classroom

Discuss the components of a hydraulic system.

Audiovisual

Show Slide 14 (Figure 11).

Classroom

Explain the function of the hydraulic reservoir and pump.

Audiovisual

Show Slides 15 and 16 (Figures 12 and 13).

Safety

Explain the hazards of hydraulic fluid under pressure. Describe the proper way to check for leaks.

Classroom

Discuss the methods for checking hydraulic hoses.

22104-12_F10.EPS

Figure 10 Air filter replacement.

3.3.0 Hydraulic System Inspections and Checks

Almost all heavy equipment vehicles use some form of hydraulic system to control their attachments. Remember that a hydraulic system is a closed-loop system consisting of the reservoir, a pump, filters and strainers, control valves, hoses, and actuators (cylinders or motors). Heavy construction equipment may also use hydraulic power to drive cooling fans and lubrication pump motors. *Figure 11* shows a hydraulic power management system that may be found on some heavy equipment.

3.3.1 Hydraulic System Reservoir and Pump

The heart of a hydraulic system is its reservoir (storage tank) and the pump that moves the hydraulic fluid. Most hydraulic reservoirs have a sight glass that allows the level of fluid to be checked without opening the reservoir. If hydraulic fluid needs to be added, it can be poured in through the fill port located on the reservoir. Most fill ports have some kind of strainer/filter inside them to help ensure that the fresh hydraulic fluid does not have trash in it. *Figure 12* shows the location of the reservoir's sight glass and fill port.

The temperature of hydraulic fluid rises as it is forced through the hoses and control valves to make the actuators (cylinders or motors) perform work. After hydraulic fluid has done its job as an actuator, the fluid returns to the reservoir. It may or may not pass through a filter before returning to the reservoir. Internally, reservoirs are designed to cool the returning hydraulic fluid. In some cases, a heat exchanger may be used before the reservoir to provide additional cooling for the returning hydraulic fluid.

The size of the heavy equipment often determines the size and complexity of its hydraulic system. Hydraulic pumps may be driven by the same belt(s) used to run the engine's water pump and coolant fan, or they may be directly driven off a cam or shaft inside the machine's engine. *Figure 13* shows a hydraulic system pump mounted onto the side of the machine's engine.

GOING GREEN

Oil Filter Disposal

To protect the environment, make sure that used and removed oil filters are disposed of in accordance with all environmental guidelines. Used oil filters do not get thrown into a landfill. Most communities have specific collection containers for used oil filters and empty oil containers.

WARNING!

Do not use bare hands to check for fluid leaks on a vehicle. Daily exposure to chemicals in this manner can cause health hazardous.

3.3.2 Hydraulic System Hoses

When inspecting a hydraulic system pump, pay particular attention to the metal connectors that connect the hydraulic hoses to the body of the pump. Ensure that the hose connectors are se-

Teaching Tip/Notes

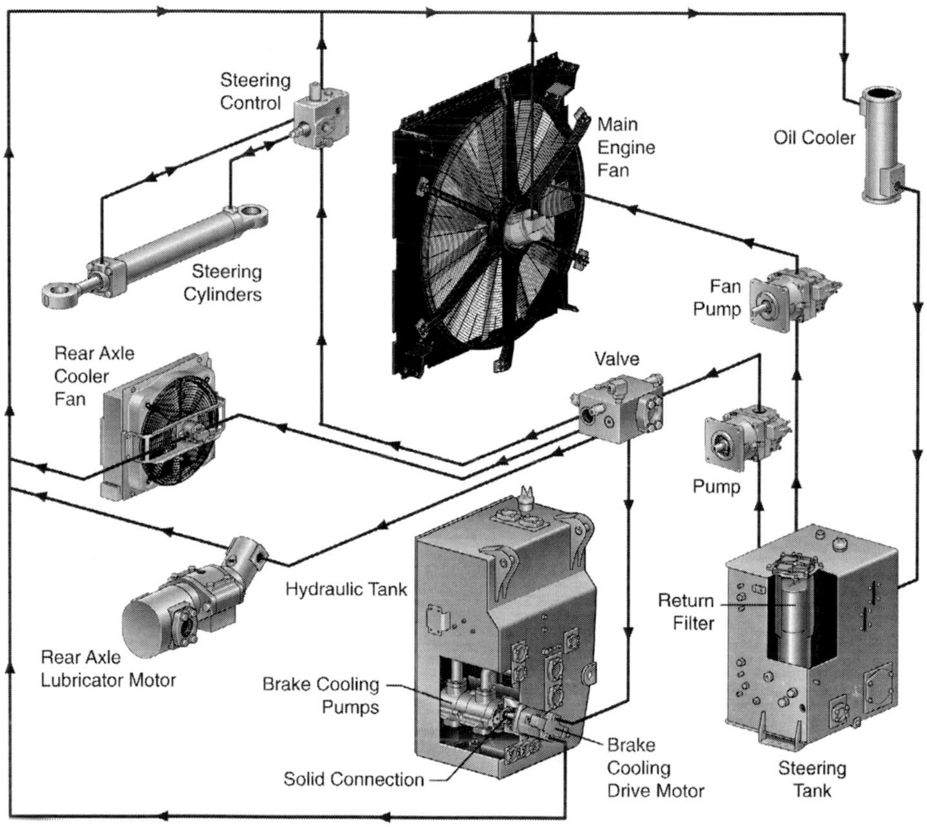

Figure 11 Hydraulic power management system components.

curely fastened to the pump's body. Always check the hoses for cracks and wear spots. Hydraulic hoses tend to move as the hydraulic fluid flows back and forth through them. If they happen to get too close to something else, even another hydraulic hose, the rubbing can wear a hole in the hose.

In most cases, hydraulic fluid that must be moved from one end of the machine to the other is run through rigid lines securely attached to the machine's frame. Flexible hoses are used to connect the rigid lines to devices such as the pump and the control valves for the actuators. The points where the flexible hoses and the rigid lines connect should always be checked for leaks. *Figure 14* shows both flexible and rigid hydraulic lines.

The hydraulic lines and hoses mounted inside the machine's protective housing are less likely be exposed to potential damage than those installed out on the boom and stick of a backhoe or excavator. When inspecting the hydraulic system hoses, check every hose or line that can be seen. The hoses and lines on the movable attachments are more likely to get damaged. *Figure 15* shows how the hydraulic hoses and lines on the boom and stick of an excavator are exposed.

GOING GREEN

Hydraulic Leaks

Hydraulic systems tend to leak. To protect the environment, make sure that any parked vehicles are not leaking onto the ground. Clean any leaking hydraulic fluid from the machine and dispose of the cleaning cloths properly. Report leaks so they can be repaired before the machine is used again.

Teaching Tip/Notes

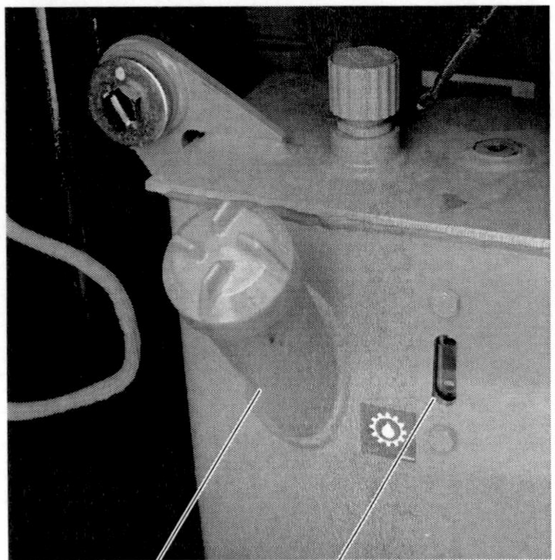

Figure 12 Typical hydraulic reservoir and its associated parts.

Figure 13 Hydraulic pump on side of an engine.

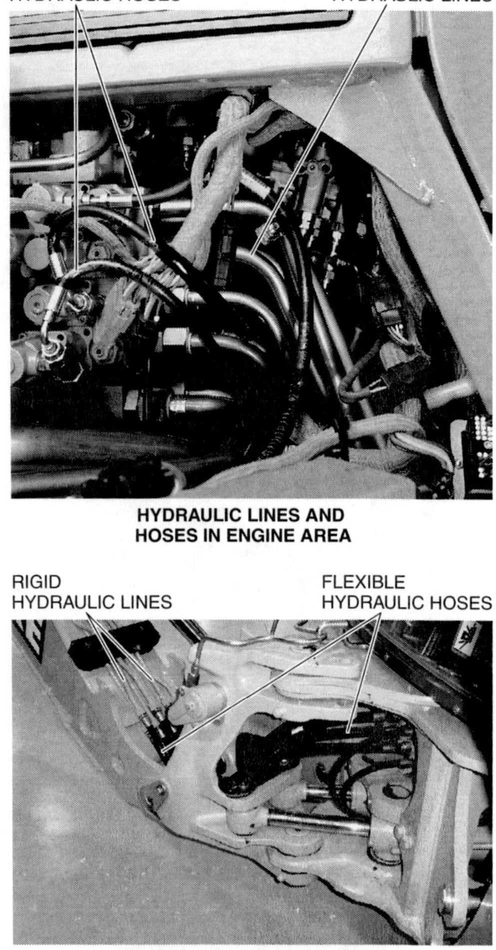

Figure 14 Hydraulic lines.

3.4.0 Electrical and Control Inspections and Checks

Modern heavy equipment vehicles have become very dependent on electrical sensors, controllers (computers), and indicators. The control panels in the operator's cab area are made up of switches, pushbuttons, analog gauges (gauges that have rotating pointers to show different values), digital readouts, and different colored indicator lights. Some control panels also have audio (sound) de-

Teaching Tip/Notes

Audiovisual

Show Slide 19 (Figure 16). Point out the key indicators on the control panels.

Figure 15 Exposed hydraulic hoses and lines.

vices that give off different sounds for different situations.

Without turning the machine's On/Off switch On, it is difficult to tell if any of the control panel devices are operational or malfunctioning. Visually check the control panels for broken glass or gauges that appear to be broken. If allowed, turn the On/Off switch to On and recheck all the gauges and indicators on the control panel being inspected. *Figure 16* shows examples of indicator panels.

NOTE

Every machine has a slightly different control panel. Refer to the machine's O&M manual for details about its control panels.

GOING GREEN

Hydraulic Lines

Hydraulic lines, especially those on construction equipment, are more likely to be damaged through daily use. Hydraulic lines, under intense pressure, occasionally rupture. If a line breaks, for whatever reason, the hydraulic pump continues to pump until shut off. Most environmentally conscious companies have fluid containment kits located either on the machines or nearby. These containment kits include various devices and absorbent materials used to clean up oil spills. Always report any oil spill.

Teaching Tip/Notes

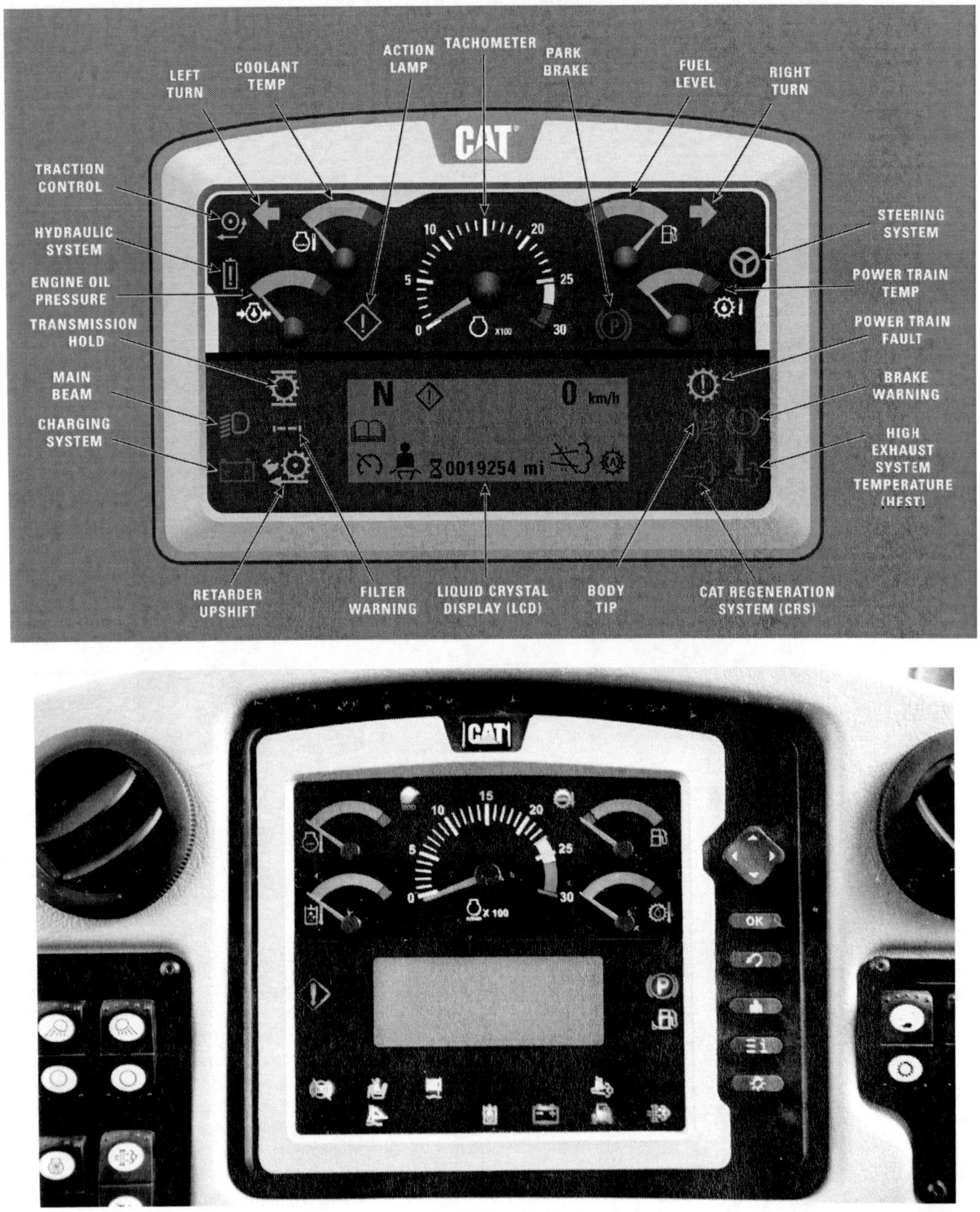

22104-12_F16.EPS

Figure 16 Operator control panels.

Teaching Tip/Notes

12 NCCER – *Heavy Equipment Operations Level One* 22104-12

3.4.1 Batteries and Cables

The electrical system of a heavy equipment machine is not all that different than the electrical system in a normal car or truck. The biggest difference is that the heavy equipment engines require more battery power to start, so the heavy equipment vehicles are equipped with multiple batteries. *Figure 17* shows a typical battery storage area similar to those found on heavy construction equipment.

> **WARNING!**
>
> Batteries contain acid. Most batteries are sealed, but may still leak. Anyone working with or near any kind of vehicle battery needs to be careful. Keep battery acid away from bare skin. Do not allow any kind of metal object (tool or whatever) to drop across the terminals of the batteries. If batteries must be changed out for new ones, make sure to review the machine's O&M manual for disconnecting and reconnecting batteries.
>
> If a machine's batteries are dead (containing little or no electrical charge), do not attempt to charge or jump the batteries without reviewing the company's jump-starting procedures, and the machine's O&M manual. Some companies do not allow the jumping of batteries.

When inspecting a machine's electrical system, start at its batteries. Make sure that the ground cables are secured to the machine's chassis ground. Also, make sure that the ground and positive cables are securely attached at the battery terminals. At the terminals, look for corrosion that could cause a bad connection between the cables and the battery terminals. Verify, as far as visually possible, that the battery cables appear to be undamaged.

Batteries must be recharged to continue working. An alternator, usually driven by a belt from the front of the engine, recharges the batteries (*Figure 18*). Verify that the alternator and all its terminals appear to be securely mounted and undamaged. When the engine is started, a current or amperage meter on the operator's console should indicate that the alternator is recharging the batteries. Until the engine is started, nothing more can be done to verify that the alternator is working.

3.4.2 Lights, Horn, and Alarms

Heavy equipment vehicles are often equipped with lights. Visually inspect any installed lights to make sure they are undamaged, clean, and securely fastened. Also, take a look at the physical positioning of the lights and make sure that they appear to be pointing in the proper direction. Verify that the wiring into the lights is securely fastened and undamaged. Do the same thing for the machine's horn and any alarms that may be installed on the machine.

3.4.3 Sensors

As noted earlier, modern construction machinery relies heavily on computers and their related devices. A vehicle's computer relies on sensing devices to tell it things about the vehicle, such as water temperature and hydraulic oil temperature. On construction equipment, sensors also tell the computer how the devices attached to the machine are positioned (raised, lowered, or at some

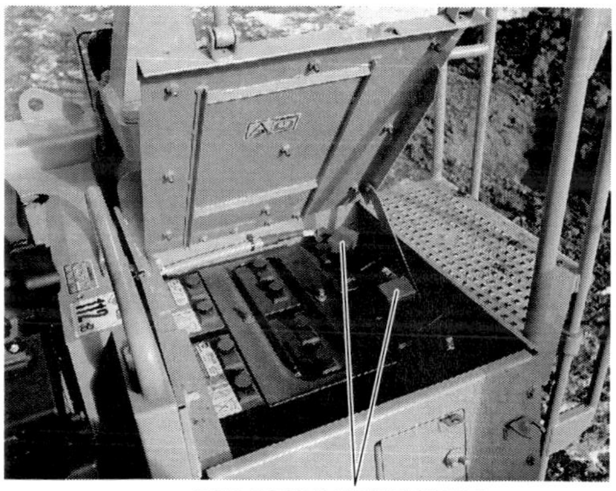

Figure 17 Batteries on heavy construction equipment.

Figure 18 An alternator.

Audiovisual

Show Slide 20 (Figure 17). Point out that some machines have multiple batteries.

Safety !

Explain the hazards associated with battery acid and jump starting batteries.

Audiovisual

Show Slide 21 (Figure 18). Describe the function of the alternator.

Classroom

Discuss the inspection requirements for lights, horn, and alarms.

Describe how sensors are used to monitor equipment functions.

Teaching Tip/Notes

Audiovisual

Show Slide 22 (Figure 19).

Classroom

Discuss common attachments used on dozers and graders.

Audiovisual

Show Slide 23 (Figure 20).

Show Slide 24 (Figure 21). Explain the purpose of a ripper.

Show Slides 25 and 26 (Figure 22 and Figure 23).

angle). *Figure 19* shows sensors that are used on modern graders.

When inspecting sensors, always verify that the sensor itself is securely attached and properly positioned. Verify that the cabling connected to it is securely fastened, and that the cable connector and cable are undamaged.

3.5.0 Attachments

The term *attachment* can apply to a lot of things associated with heavy equipment machinery. For bulldozers, their primary attachment is the **blade** and the arms and cylinders supporting it (*Figure 20*). Bulldozers may also be equipped with rippers that are usually attached to the rear end of the bulldozer (*Figure 21*).

Motor graders also have cutting blades, but they are called **moldboards** (*Figure 22*), and graders may also be equipped with ripper attachments.

For machines such as loaders, backhoes, and excavators, the primary attachments are the **buckets** used to dig or move materials such as dirt, rock, or waste products. *Figure 23* shows the tips of a bucket used on a loader. The buckets used on backhoes and excavators have similar tips.

Before working with any of these machines, the operator needs to inspect the attachments being used to verify that they are properly attached, and not worn, broken, or otherwise damaged. If any part is found to be unusable, repair it or report it in accordance with company procedures. Do not operate the machine with missing or damaged parts.

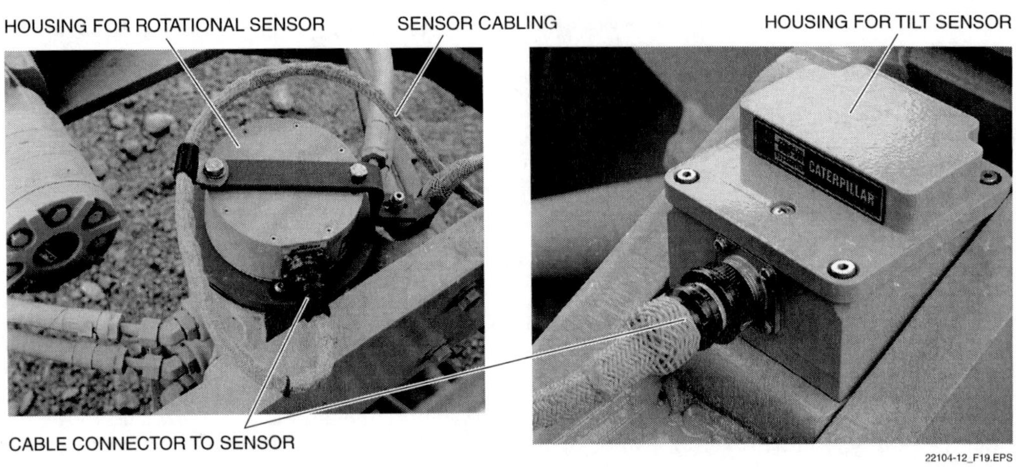

Figure 19 Sensor examples.

Figure 20 Bulldozer attachments.

Teaching Tip/Notes

Figure 21 Ripper.

3.6.0 Operator Cab

One of the last things an operator needs to check are the areas inside the cab of the machine being used, starting with the seat. *Figure 24* shows how the controls are mounted around the seat of a John Deere backhoe. The seat needs to be adjusted to a comfortable position for operator weight, pedal reach, and comfort. The seat belt needs to be checked to make sure it is not damaged.

All seats are adjustable, either manually or through pneumatic or hydraulic controls. While most seats are mounted in a fixed location, the seat of a backhoe can be reversed so that the operator can either see out the front of the machine where the bucket is mounted, or out the back of the machine where the backhoe attachment is mounted.

> **NOTE**
> The seat belt installation needs to be checked every three years. Five years from the manufacturing date, the seat belt needs to be replaced.

The complexity of the machine being run determines the complexity of the equipment inside these cabs. Machines like utility tractors, backhoes, loaders, forklifts, skid steers, and small trenchers normally have fairly simple controls. Machines such as dozers, motor graders, excavators, and scrapers most likely have more complex controls because they are used to cut and fill grades to specific levels. Some grade-cutting or filling machines have computer controls that are tied into a global positioning system (GPS) that tells the machine how deep to cut or fill. The angle of a dozer blade, or the moldboard on a grader, can be set or changed through computer controls mounted inside the operator cab of the machine. The person who will be operating such machines must make sure to check out all the computer controls inside the cab. The only way to really check the controls is to start the machine and test-operate each control to verify that it functions correctly. *Figure 25* shows views from inside the cab of a motor grader.

Figure 22 Moldboard of a motor grader.

Classroom

Point out that cabs differ from one machine to another and that controls will differ as well.

Audiovisual

Show Slides 27 and 28 (Figures 24 and 25). Note the differences between the cabs and controls of the backhoe and motor grader.

Classroom

Explain that some equipment, such as dozers and graders, are often equipped with and controlled by GPS.

Teaching Tip/Notes

22104-12 Basic Operational Techniques Module Four 15

Classroom

Discuss the different types of dump trucks and how their systems differ.

Point out the importance of documenting daily inspections.

Audiovisual

Show Slide 29 (Figure 26).

Homework

Have trainees review Sections 4.0.0–4.4.0.

Materials

Ensure that you have everything required for teaching this session.

Figure 23 Bucket tips.

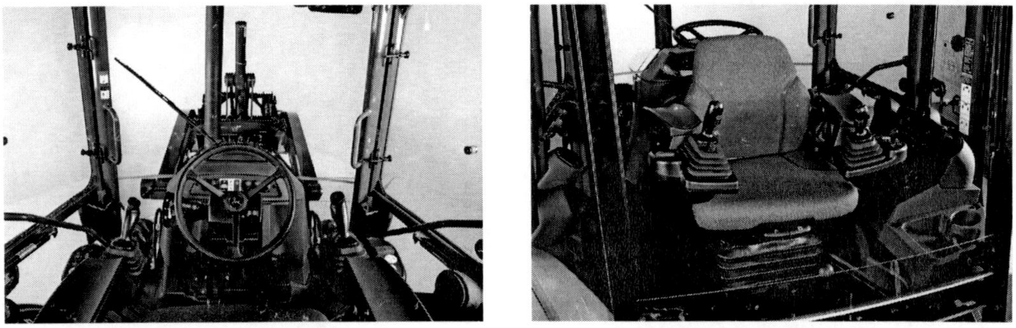

Figure 24 Backhoe operator seat.

3.7.0 Prestart Activities for Trucks

While all the previous prestart activities have been primarily focused on the heavy equipment machines used on a construction site, most of those prestart inspections also apply to the dump trucks used on such jobs. *Figure 26* shows several dump trucks used in construction.

Whether a dump truck is designed to travel only on public roads or only across off-road paths, they all have engines, transmissions, and drive axles just like the machines previously covered. They also use hydraulic systems for steering and to lift and lower their beds as they unload whatever they may have been carrying. The operator cab controls of most dump trucks get more complicated on the larger articulated off-road trucks and the rigid-frame dump trucks. Operators must still review the O&M manual for each truck to become familiar with the controls of the truck.

Good operations always include planning and paperwork. Complete any daily inspection forms. Note fluids added, service needed, and hours of operation. Some companies schedule service based on the equipment's hours of operation. Make sure that the equipment is serviced regularly.

Teaching Tip/Notes

22104-12_F25.EPS

Figure 25 Controls inside a Caterpillar motor grader cab.

4.0.0 Startup Activities

Starting equipment involves more than just turning the key. Certain actions must be taken before starting the engine. After startup, the engine needs to warm up for a short period before beginning operations. Refer to the O&M manual for specific amount of warmup time.

4.1.0 Indicator Checks

Before starting the engine, turn the key to the On position and check the instrument panel, but do not start the engine. Make sure that all gauges and indicators function in accordance with the operator's manual. The instrument panel should light for a few seconds. If an indicator or gauge does not light up, check the bulb. Repeat the test. If the indicator still fails to light, have the machine serviced.

4.2.0 Starting Aids

Gasoline and diesel engines are started differently. Most gasoline engines start the same way as the gasoline engine in a street machine. These engines rarely need any kind of prestart aid. If the temperature is extremely cold, engine block heaters and ether starting fluid can be used. The startup and warmup of an engine must be done in accordance with the manufacturer's instructions.

WARNING!

Ether must never be used in combination with another starting aid. The manufacturer's instructions must be carefully followed.

ON-ROAD DUMP TRUCK

ARTICULATED DUMP TRUCK FOR UNDERGROUND MINING

RIGID-FRAME OFF-ROAD DUMP TRUCK

22104-12_F26.EPS

Figure 26 Heavy equipment dump trucks.

Classroom

Describe the startup process. Point out the importance of warming up the engine.

Note that gasoline engines generally do not need starting aids, but ether is sometimes used in extremely cold weather.

Safety !

Emphasize that ether should never be mixed with another starting aid.

Teaching Tip/Notes

Classroom

Explain the use of glow plugs.

Audiovisual

Show Slide 30 (Table 1). Discuss glow plug timing.

Classroom

Explain how block heaters work. Point out that they should be unplugged before the engine is started.

Describe the machine startup sequence. Explain the purpose of the choke.

Point out that there is a limit to how long the starter should be run before giving it a rest.

! Safety

Remind the trainees to always check that the area is clear of personnel before moving a machine.

Classroom

Review typical vehicle warmup practices.

Homework

Have trainees review Sections 5.0.0–6.0.0.

Materials

Ensure that you have everything required for teaching this session.

Classroom

Explain that the purpose of this lesson is to familiarize the trainees with the operating characteristics of different machines.

4.2.1 Glow Plugs

Prestart aids are often used with diesel engines. **Glow plugs** are frequently used to warm the engine prior to starting. Many diesel engines have glow plugs. These heating elements are used to heat up the combustion chamber and aid in igniting fuel in a cold engine. A current supplied to each glow plug causes it to produce heat. When the switch is activated, the plugs glow. On some equipment, the glow plugs can be seen through a sight hole. When they stop glowing, the engine is warm enough to start. Other models have a timing chart for how long the glow plugs should be activated. The activation time is based on the temperature surrounding the engine. An example of a timing chart is shown in *Table 1*. The chart is in the operator's manual or mounted on the dash.

4.2.2 Block Heaters

Engine block heaters are used in colder climates. An engine block heater is an electrical heating element that is installed in the engine block water jacket. In cold weather, it keeps the engine water warm. This keeps the engine temperature warm enough to allow easy engine starts. Typically, the heater is plugged into a standard electrical outlet. Unplug the engine block heater before moving the machine. Properly secure the plug so it does not become fouled with ice and snow.

4.3.0 Starting

Verify that all controls are properly set. Most vehicles must be started in neutral or park. This includes both the drive and the operational controls. Also check that the parking brakes, if provided, are properly set.

> **WARNING!**
> Before starting an engine or moving the machine, make sure that the area is cleared of personnel. Check to make sure that no one is underneath, around, or on the machine. Alert nearby personnel in accordance with site procedures.

Start the machine only after completing any prestarting aids and fastening the seat belt. Move the gearshift to neutral. Set the engine control speed and choke as directed in the operator's manual. Sound the horn to alert people standing nearby that the machine is being started. Turn the key to the start position and/or depress the start button.

Table 1 Timing Chart

STARTING AID CHART	
STARTING TEMPERATURE	GLOW PLUG HEAT TIME
Above 60°F (16°C)	No
60°F (16°C) to 32°F (0°C)	1 Minute
32°F (0°C) To 0°F (−18°C)	2 Minutes
Below 0°F (−18°C)	3 Minutes

22104-12_T01.EPS

> **CAUTION**
> Do not push the starter button for more than 20 seconds. This can damage the starter. If the engine does not start after 20 seconds, wait two minutes before attempting to start again. Recheck all instrument settings and controls. Many controls automatically lock out the starter if not set properly.

4.4.0 Warmup

It is important to let the machine warm up before operating. Operating the machine without warming it up can damage it. The warmup period is directly proportional to the temperature of the air surrounding the machine. The colder the temperature, the longer the warmup period is necessary. Refer to the O&M manual for any specific warm-up time.

Run the machine for the time listed in the O&M manual. When the engine is running, check that all of the gauges and instruments are working properly.

If the hydraulic oil is cold, the hydraulic system will be sluggish. Circulate warm hydraulic oil throughout the machine by cycling the hydraulic controls. With a backhoe, for example, lift the bucket to hood height and lower it. Extend and retract the boom. Continue to cycle the system until it moves freely. Once the machine is properly warmed up, operations can begin.

5.0.0 Equipment Operations

The way heavy equipment is constructed and the way it steers makes a difference in how it is driven. The size of the machine must also be considered when it is operated. Tracked vehicles must be operated differently than wheeled vehicles.

Teaching Tip/Notes

5.1.0 Construction and Steering

Most wheeled machines with straight frames have either a front or rear axle with wheels that can be turned to make the machine change directions. A normal dump truck driven on public roads has a straight frame and is steered by its front wheels. Most wheeled utility tractors, backhoes, skid steers, and trenchers are steered the same way. Other heavy equipment machines are hinged in the middle to make the machine more flexible to maneuver. *Figure 27* shows a typical loader that is hinged in its middle.

Driving a straight-framed machine is pretty much the same as driving a personal vehicle. The biggest difference is the fact that the heavy equipment machines are usually much larger than the personal vehicles. Driving a hinged machine is not that much different than driving a straight-framed machine. The biggest difference is seen from the operator's seat as the whole front section of the machine turns when the steering wheel is turned.

Driving a wheeled machine verses driving a tracked machine is much different simply because of how the left and right tracks of the tracked machine are controlled. Most wheeled machines are steered by a steering wheel, and the vehicle must be rolling forward or backward to make any kind of turn. Tracked vehicles on the other hand are steered by levers or **joysticks** that control the speed of the individual tracks. A tracked vehicle can be essentially sitting still and make a 90-degree turn, as one track is held still and the other track is put into motion to pull the vehicle around a tight arc. A tracked vehicle can also be turned in an even smaller space as one track is reversed while the opposite track is made to move forward.

Figure 27 Example of a hinged machine.

5.2.0 Basic Operations

Regardless of whether the machine being operated is a wheeled vehicle or a tracked vehicle, operators must learn how to do the same basic maneuvers. The basic operations of any heavy equipment machine involves moving it forward slowly in a straight line, and then steering it left and right of that straight line before bringing the machine to a complete stop. In most cases, the machine needs to be operated in its lowest gear during these initial maneuvers. After these maneuvers can be performed confidently at a slow speed, the speed can be increased (as allowed by the training instructor or the company policies). As the operator becomes more comfortable maneuvering the machine at the increased speed, the speed can be increased to what is considered to be normal operating speed. The last forward maneuver is to drive the machine forward and then make a hard 90-degree turn to the right or the left before bringing the machine back onto a straight line leading 90 degrees away from the original straight line. Next, the same maneuver is repeated, but this time it needs to be turned in the opposite direction by 90 degrees and brought along a straight line before being stopped.

After an operator is comfortable driving a machine forward and steering it from side to side of the forward motion, the next step is to back the machine. Learning to back the machine needs to be done the same way the forward motions were done. Start by backing the machine in a straight line. Backing any machine is more difficult than driving it forward, but backing is a maneuver that all operators must be comfortable doing. After the operator can safely back the machine, the next step is to back it left and right at a slight angle off the straight line. After the slight backing maneuvers are completed, the last thing to do is to back the machine and make it turn a sharp 90-degree turn.

Operators often must operate their machines in fairly close quarters. To get comfortable doing that, a trainer may have some obstacles set up for the machine operator to work through. The idea is to verify that the operator can maneuver the machine close to an obstacle while not touching it. The more an operator uses a machine, the more comfortable the operator should become. Personnel operating heavy equipment must learn the feel of the machine as it is being asked to do certain things. That sense of touch is most critical in operating any machine, but especially when operating large machines. Heavy equipment operator training is usually started on level ground, but must later be moved to uneven surfaces. Opera-

Classroom

Describe the differences between articulating and rigid body machines.

Discuss differences in operation between wheeled and tracked machines.

Audiovisual

Show Slide 31 (Figure 27).

Classroom

Describe the maneuvers that the trainees will have to perform: moving forward in a straight line; steering left and right; backing up straight and at an angle.

Point out that these maneuvers will be performed at a very low speed until the trainees are more experienced.

Explain the importance of just getting a feel for the machine before trying any maneuvers.

Teaching Tip/Notes

Classroom

Explain that the operator manual supplied with a machine should always be consulted for specific operating instructions.

Describe the various makes and models of tractors used in construction.

Discuss the operation of a machine with a manual transmission.

! Safety

Describe the conditions that can cause a tractor to roll sideways or flip over backwards. Explain how to avoid these situations.

Explain the hazards associated with a PTO.

Audiovisual

Show Slide 32 (Figure 28). Explain how to set up a backhoe.

Classroom

Point out that a backhoe is basically a tractor with a backhoe attachment.

Describe the operation of a backhoe.

tors need to also learn a sense of balance when operating their machines because they often have to operate over hilly or otherwise uneven surfaces. The operating instructions for each machine will usually lists the maximum speed at which the machine can be operated. The instructions should also indicate a point at which the machine tips over. The operator must be aware of what slope angles the machine can be safely operated on. Utility tractors for example should not be driven across slopes of more than 25 degrees.

5.3.0 Machine-Specific Operations

Each machine operated by a heavy equipment operator has its own characteristics. The O&M manual for each machine lists any specific instructions on how to operate the machine.

5.3.1 Utility Tractors

Utility tractors, both large and small, operate about the same. They are straight-framed vehicles that are steered by their front wheels. Some are two-wheel drive (by the rear wheels) and others are four-wheel or all-wheel drive. Steering utility tractors is about the same as steering a personal vehicle.

Tractors with manual transmissions have a clutch that the operator must learn to carefully engage to make the machine move. The operator must also learn how to shift gears after the tractor is rolling. For operators unfamiliar with manual transmissions and clutch operations, the initial gear engagement and the gear shifting may take a little getting used to. Tractors with hydrostatic or other types of automatic transmissions are usually easier to learn.

By design, the front end of a utility tractor is much lighter than its rear end. Under normal operating conditions, that is not a problem. But a tractor can be made to flip over backwards under certain conditions. When an attachment is installed on the back end of a tractor, the additional weight makes the tractor more unbalanced. If a tractor is driven up a grade, with or without an attachment, it can flip backwards. Operators should avoid driving up hills or steep slopes. They should not cross slopes of more than 25 degrees.

If a tractor is driven into a ditch, always back out of ditches, gullies, and up steep hills. Utility tractors are used in off-road environments. If a tractor becomes mired or the rear wheels become locked, the front end will begin to lift off the ground. If that happens, immediately disengage the clutch to stop the drive to the rear wheels.

Utility tractors are designed to have a wide variety of attachments installed on either their front ends or the rear ends. Almost all utility tractors have **power takeoff (PTO)** drives at their rear ends, but some also have a PTO drive at the front end. Operators must be extremely careful when installing any attachment. They must stay clear of pinch points, and always stay clear of activated PTOs.

Utility tractors of any size have some type of hydraulic system to lift and lower the attachment(s) installed on the tractor. Operators must become familiar with the hydraulic system on the tractor being used. Hydraulic systems have a tendency to leak. Operators need to always watch for leaks in the hydraulic lines used on a tractor. If an attachment is lifted and in a holding position, watch for signs that the hydraulic system is slowly allowing the attachment to lower.

> **WARNING!**
> Never assume that a load held up by hydraulic system components will stay up. Never get caught under a raised attachment (and its load). Always lower the load before dismounting the equipment.

5.3.2 Backhoes

Backhoes are a lot like the utility tractors in that most are based on a tractor body. Wheeled backhoes steer and drive like a utility tractor. The bucket attachment mounted on the front of a backhoe is similar to a bucket assembly that can be mounted on the front of a utility tractor. The rules about driving a backhoe up or down hills or slopes is the same as those that apply to utility tractors.

The biggest difference between a utility tractor and a backhoe is the backhoe attachment mounted on the back of a backhoe machine. The backhoe attachment has outriggers or load **stabilizers** that the operator uses to level and stabilize the machine as the backhoe attachment is being used. The front bucket is also lowered to raise the backhoe's front end while the backhoe attachment is being used. *Figure 28* shows a backhoe with all four wheels off the ground as the outriggers and bucket support the backhoe.

Any time an operator raises a backhoe by its front bucket and its outriggers, the risk of tipping over increases. The operator must be constantly aware of how the outriggers and front bucket are being used to support and level the backhoe.

After the backhoe is positioned to dig, the operator begins the digging process. The operator extends the machine's boom and bucket out and

Teaching Tip/Notes

Figure 28 Backhoe supported by front bucket and outriggers.

lowers the bucket teeth into the ground and then draws the bucket back toward the backhoe. The dirt scooped up in the bucket is moved to the left or right of the dug trench and dumped. Operators must be constantly aware of where the backhoe boom and bucket are being moved. They must watch out for utility lines above them and in or on the ground near the work area. As the digging progresses, the distance limit to which the bucket can safely extend is reached. At that point, the backhoe must be moved. To move the backhoe, the outriggers and front bucket must be raised so that the backhoe wheels can be used to relocate the backhoe. After the backhoe is relocated, the front bucket and the outriggers are lowered again to raise and level the backhoe. At that point, the digging can continue.

5.3.3 Skid Steers

Skid steers can be equipped with either wheels or tracks (*Figure 29*). Both types are steered by levers that allow the wheels (or tracks) on one side to move slower or faster than those on the opposite side. Operating a skid steer is pretty much the same, regardless of traction type (wheeled or track).

A wide variety of attachments can be installed on a skid steer. Hydraulic controls inside the operator cab allow the operator to lift or lower the attachments. For those attachments with forks or a bucket, the operator can also tilt the forks or bucket up or down as needed. Since buckets are used to scoop up materials, such as dirt or gravel, they can also be tilted forward enough to dump the materials they carry.

Figure 29 Different types of skid steers.

Some skid steers can be equipped with rotating tools such as trenching devices, post hole diggers, or street sweepers (*Figure 30*). One or more hydraulic motors on the skid steer generates the rotating power to the specialized attachments.

5.3.4 Loaders

The term *loader* is used on a wide variety of heavy equipment machines. Track loaders are essentially dozers with controllable buckets mounted on their front ends. Wheeled loaders are similar, except for the fact that they move on wheels and not tracks. Multi-terrain loaders and compact track loaders are simply skid steer loaders with tracks and a variety of attachments. Knuckleboom loaders (*Figure 31*) are specialty loaders more often found in the wood-cutting and wood-hauling businesses. They are specialized loaders designed to grasp, lift, and move objects such as logs.

Audiovisual

Show Slide 33 (Figure 29). Explain how to steer a skid steer.

Show Slide 34 (Figure 30). Describe the attachments used with a skid steer.

Show Slide 35 (Figure 31). Explain what a knuckleboom loader is.

Teaching Tip/Notes

Audiovisual

Show Slide 36 (Figure 32). Describe the different types of controls used on loaders.

SKID STEER WITH POST HOLE DIGGER

SKID STEER WITH SWEEPER BROOM

22104-12_F30.EPS

Figure 30 Specialized skid steer attachments.

22104-12_F31.EPS

Figure 31 Knuckleboom loader in wood yard.

Loaders are steered by either a steering wheel or by levers (*Figure 32*), depending upon whether they are wheeled loaders or tracked loaders. They also have foot controls. Loader attachments are controlled by levers that operate the hydraulic systems used to drive the attachments.

Always travel with the mounted attachment in the stored position. For loaders with a bucket attachment, travel with the bucket lowered for better stability and visibility. When approaching an obstacle that must be crossed, approach it at an angle if driving a wheeled loader. If traveling on a tracked loader, approach the obstacle straight on. When traveling a loader, or working with one, the operator must stay aware of all things both behind and in front of the loader. The basic operations of a loader involves going forward and backward, and the lifting and tilting of the attachments.

OPERATOR AREA OF A CATERPILLAR WHEELED LOADER

OPERATOR AREA OF A CATERPILLAR TRACKED LOADER

22104-12_F32.EPS

Figure 32 Loader controls.

Teaching Tip/Notes

5.3.5 Dozers

Most people relate the term *dozer* to the tracked bulldozers, but they also exist in wheeled versions (*Figure 33*). Dozers are usually equipped with a blade attachment on their front ends. The blade is used primarily for pushing materials. Tracked dozers are steered by levers or joysticks that in turn control the movement of the left and right tracks. Wheeled dozers may also be steered by levers, but more likely are steered by conventional steering wheels. Dozer blades are raised, lowered, or tilted by hydraulic system components controlled by levers in the operator's cab.

To move the machine forward, raise the blade high enough to clear obstructions. Push down on the service brake, and release the parking brake. Move the gear shift to the forward position. Move the **throttle** control forward to set the desired speed. Release the brake pedal. Move the governor to the desired engine speed. The following are a few tips for forward operations:

WHEEL DOZER

TRACK DOZER

22104-12_F33.EPS

Figure 33 Dozers.

- Reduce engine speed when maneuvering tight spaces or when breaking over a rise.
- Set the proper gear before starting down grade in order to control the speed. Do not change gears while going downhill unless the machine is equipped with a hydrostatic transmission.
- A good practice is to use the same speed going downhill that would be used to go uphill.
- Do not allow the engine to over-rev while traveling downhill. Maintain control of the machine by tapping the service brake to reduce speed.

> **NOTE**
> Drive the machine in the forward direction for best visibility and control.

It is possible to change a dozer's speed and direction at full engine speed on machines equipped with a power shift or hydrostatic transmission. However, to ensure operator comfort and maximize machine service life, brake before changing speed or direction. Slow down by using the hand throttle or by depressing the decelerator. Push the brake pedal to stop the machine. Move the transmission control to neutral and shift to the desired direction and speed. Release the brake pedal. Increase engine speed by adjusting the hand throttle or releasing the decelerator.

Although dozer machine controls vary, the turning principles are the same. To turn a tracked dozer, disengage or brake the track on one side, while keeping power to the other side. This causes the machine to turn in the direction of the slower side. Returning both tracks to the same speed stops the turn. Wheeled dozers are articulated and the front wheels control the turning process.

On older tracked machines, levers and pedals are used for turning. An operator must learn to coordinate applying power to one track and brak-

On Site

FNR Lever

Some bulldozers and other heavy equipment that are equipped with hydrostatic transmissions have a lever or control to select the direction of movement. Typically these are known as FNR controls, for forward, neutral, or reverse. They are separate from a standard gear selection control. The operator moves this lever to make the machine go in the right direction. Only move this lever when the machine is at a complete stop.

Classroom

Point out that there are both track and wheeled dozers and discuss the differences.

Audiovisual

Show Slide 37 (Figure 33).

Classroom

Describe procedures for moving a dozer.

Describe safe operating practices for dozers.

Describe how to maneuver a track dozer with foot pedals and joystick.

Teaching Tip/Notes

Classroom

Explain how to operate the dozer blade using the lift, angle, tilt, and pitch controls.

Point out that blade pitch is generally not changed.

Explain that there are both wheeled and track excavators, but the track excavators are more common.

Audiovisual

Show Slide 38 (Figure 34). Discuss the special attachments used with excavators.

Classroom

Discuss the differences in controls between towed and powered scrapers.

Audiovisual

Show Slide 39 (Figure 35).

ing the other to accomplish the turn. The quickness of the turn is proportional to the amount of power and braking applied to each side. For a gradual turn, disengage power to one track while powering the other. For a sharper turn, completely disengage and brake one track while powering the other track.

Newer machines have joystick controls for engine control and movement. A joystick can be designed for steering or combined with transmission controls. The newer joysticks include steering, engine speed, and transmission controls. On these machines, moving the joystick forward or backward controls forward or rearward movement. Moving the joystick sideways causes the machine to turn. For example, moving the joystick to the left causes a left turn. With the joystick in neutral for forward and reverse, moving it sideways causes the dozer to pivot in place.

CAUTION

Bulldozers are some of the largest and most powerful pieces of equipment on the job site. You must be very careful so that you have time to recognize any potential hazards and try to avoid them. Make sure you have a clear view in all directions.

Before operating a bulldozer, the operator must understand how to operate the blade and its controls. The blade position can be changed in lift, angle, tilt, and **pitch.** Changing the position of the blade allows the bulldozer to perform different grading operations. Refer to the O&M manual for each dozer for the location and operation of the blade controls.

The lift control lowers or raises the blade. Lowering the blade allows the operator to change the amount of bite or depth to which the blade will dig into the material. Raising the blade permits the operator to travel, shape slopes, or create stockpiles. The lift lever can also be set to **float.** The blade adjusts freely to the contour of the ground. The float position is commonly used in reverse to smooth the surface.

The angle control adjusts the blade in relation to the direction of travel. When moving the load, the blade should be perpendicular to the line of travel. For filling a ditch, the blade should be angled to permit the load to be pushed off to the side.

The tilt control changes the angle of the blade relative to the ground. This permits the blade to cut deeper on one side than on the other. This is useful for performing side hill work where the blade tends to hang lower on the downhill side. It is also useful for crowning roads and grading slopes and curves. Practice moving the blade into different positions.

The pitch of the blade is the slope of the blade from top to bottom. The greater the slope, the more the blade tends to dig in. On most dozers, blade pitch must be changed manually. Because of its difficulty to control, pitch is only changed for unusual circumstances.

5.3.6 Excavators

Excavators come in all sizes, but they all do basically the same kind of work when equipped with buckets. Buckets are used for digging. Excavators, like skid steers and backhoe machines, may also be equipped with special attachments that allow them to do work activities other than digging. *Figure 34* shows excavators equipped with attachments other than buckets.

Excavators can be mounted on either tracks or wheels. Their movements across the ground are controlled by either levers/pedals (for tracks) or a steering wheel (for wheeled undercarriages). The operator controls the excavator's movement along the ground as it is moved into a work position.

After the excavator is properly positioned for work, the operator uses the joysticks or levers and other controls inside the cab to control the excavator's boom and attachments. When excavators are equipped with specialized attachments such as the hammer or grapple, the operator must also learn how to operate the new attachment. When such special attachments are used, refer to the additional operator manual for instructions on operating the attachment.

5.3.7 Scrapers

There are two types of scrapers. A towed scraper is towed behind a dozer or a heavy utility tractor. It simply follows whatever machine is towing it. The other type of scraper is called a wheeled tractor scraper. These tractor scrapers may be powered by a single engine located in the front or the rear of the scraper, or they may be powered by engines located at each end of the scraper. *Figure 35* shows both a towed scraper and a wheeled tractor scraper. This particular tractor scraper has two engines.

Hydraulic systems are used on both types of scrapers to control their scraping and dumping functions. On a towed scraper, those hydraulic controls are controlled from whatever machine is being used to tow the scraper. On the wheeled tractor scrapers, the scraping and dumping functions are controlled from the operator cab of the

Teaching Tip/Notes

TRACKED EXCAVATOR EQUIPPED
WITH HAMMER ATTACHMENT

WHEELED EXCAVATOR EQUIPPED
WITH GRAPPLE ATTACHMENT

22104-12_F34.EPS

Figure 34 Excavators equipped with other attachments.

scraper. The wheeled tractor scraper is steered by a steering wheel inside the operator cab.

5.3.8 Motor Graders

The motor grader is a rubber-tired, hydraulically operated, single-engine machine used to shape and finish materials in earth moving, construction, and maintenance. It is one of the most used pieces of equipment on a construction job. Its main purpose is to mix, place, and smooth material that is on the ground or on other surfaces.

Figure 36A shows a modern, medium-size motor grader grading a level surface. This particular model has a rigid frame. The large articulated motor grader in *Figure 36B* has a frame that is

TOWED SCRAPER

WHEELED TRACTOR SCRAPER

22104-12_F35.EPS

Figure 35 Towed and wheeled tractor scrapers.

hinged in the middle. This particular machine is designed for mining operations.

There are seven primary operations that motor graders can perform using their blade. Other functions, such as **scarifying,** ripping, or heavy snow plowing can be done using attachments to the grader.

- Rough grading
- Mixing rows of material
- Making rows of material for others to pick up (windrowing)
- Leveling new material
- Finish grading
- Ditch cutting or cleaning
- Light snow plowing

A motor grader can have many individual parts; however, all these parts can be grouped into the following four major components:

- The frame
- The power unit, including the engine, transmission, and differential
- The steering and other controls
- The circle, drawbar, moldboard, and blade

Classroom

Discuss the different types of work that can be performed using a motor grader.

Audiovisual

Show Slide 40 (Figure 36). Point out the differences between rigid body and articulating motors graders.

Classroom

Review the key components of a motor grader.

Teaching Tip/Notes

22104-12 Basic Operational Techniques Module Four 25

Classroom

Describe the types of controls found on motor graders.

Explain that many motor graders are laser- or GPS-guided.

GOING GREEN

Hybrid Excavator

As part of an ongoing effort to reduce emissions and improve fuel efficiency, heavy equipment manufacturers are looking for new ways to power their equipment. This Komatsu hybrid excavator uses an electric motor to drive the upper structure. The energy created when the structure brakes and comes to a stop is converted to electricity, which is stored in an electronic device known as an ultra-capacitor. The stored energy is then released and used to drive the swing and to assist the vehicle's engine. Fuel savings can be 20 to 40 percent and the carbon footprint is reduced by a like amount.

22104-12_SA01.EPS

Both rigid-frame and articulated motor graders have each of these components. Attachments that can be used on both types of motor graders are mounted at the rear, at the front, or underneath the frame behind the front wheels.

Motor grader controls consist of foot pedals, levers, switches, and a steering wheel. The levers that operate the moldboard and blade, circle, articulation, and wheel lean are placed in front of, and to either side of, the steering pedestal. The levers that control the engine and transmission are placed on the right side of the operator's seat. The instrument panel with switches and gauges is usually placed on the steering pedestal. The layout of the controls for the moldboard and wheel positioning will vary, depending on the control mechanisms and the manufacturer's design.

The location of the controls and their names may vary from one manufacturer to another, but the basic operation of every motor grader is generally the same. Newer motor graders may be equipped with laser-guided controls in addition to the manual controls listed here. The functions of the controls are as follows:

- *Accelerator pedal* – Increases engine rpm and increases the speed of travel. When the accelerator is released, the engine speed returns to the hand-throttle setting.
- *Articulation control lever* – Moves the rear portion of the motor grader from the center to the right or to the left of center.
- *Blade shift lever* – Moves the moldboard and blade to the right or left.
- *Blade tilt lever* – Tilts the moldboard and blade rearward or forward.
- *Brake pedal* – Stops or slows the machine. Some models have right and left pedals that assist in turning.

Teaching Tip/Notes

Figure 36 Rigid-frame motor grader with ROPS cab.

- *Circle shift lever* – Moves the circle to the right or left.
- *Circle rotation lever* – Moves the circle clockwise or counterclockwise.
- *Deceleration pedal* – Decreases speed quickly.
- *Differential lock* – Provides additional traction when engaged. When engaged, the speed of the right and left wheels will be equal. When disengaged, the speed of the wheels will be different.
- *Engine speed control (hand throttle)* – Governs the speed of the engine.
- *Forward/reverse control lever* – Moves the motor grader forward or backward. The midpoint of the lever travel is neutral.
- *Gear range lever* – This is the transmission control. It selects the gear range for the ground speed of the grader.
- *Parking brake lever* – Secures the motor grader when it is stopped or parked.
- *Right and left blade lift levers* – Used to raise and lower the right or left end of the blade. Blade lift levers are usually located to the outside of the control lever panel. They may be operated one at a time to raise one end of the blade, or both levers may be used together to raise and lower the entire blade.
- *Saddle pin release button or pedal* – Hydraulically releases the saddle pin for changing the position of the blade.
- *Scarifier/accessory control lever* – Raises or lowers the scarifier or other accessory.
- *Wheel lean lever* – Tilts the front wheels to the right or left.

Teaching Tip/Notes

22104-12 Basic Operational Techniques Module Four 27

Classroom

Describe how compacting equipment is used on a job site.

Audiovisual

Show Slide 41 (Figure 37). Describe how compactors are used to compact soil.

Show Slide 42 (Figure 38). Describe how compactors are used in highway construction.

Classroom

Discuss the controls used on compactors. Compare them to the complex controls found on motor graders.

On Site

Grader Evolution

The basic design of the motor grader has not changed much from the earliest days when graders were pulled by horses or steam-powered tractors. Today, power is supplied by a large diesel engine with the controls being either hydraulic or electronic instead of mechanical.

NOTE

The saddle pin release button must be engaged or disengaged when changing the blade position.

Operators learning to operate a motor grader need to start with the simple movements of the machine. After those movements can be easily controlled, the next step is to control the blade enough to do basic grading work.

5.3.9 Compacting Equipment

The term *compact* means to squeeze or compress together to increase the density of the material being compacted. On construction jobs, machines called soil compactors are used to pack the fresh soil until it is firm enough to support a road or a building. Soil compactors look like a loader equipped with a large drum-like roller with raised points that pack the soil like thousands of small feet. After the graded soil is compacted to the desired density, gravel may be added on top of the packed soil to build a stronger foundation. Another type of compactor using a smooth packing drum may be used to compact and smooth the gravel. *Figure 37* shows compactors routinely used to compress the soil and gravel into a usable foundation.

If the construction job involves the laying of asphalt, another type of compactor will most likely be employed. These compactors roll back and forth to compress the asphalt until it is at the desired density. Some compactors have vibrating rollers, which increase the compaction. A motor mounted inside the roller causes the vibration. Asphalt compactors may be constructed with solid metal drums on both ends, or they may have a solid metal drum on one end and multiple rubber tires on the other end. *Figure 38* shows compactors working on asphalt.

COMPACTOR PACKING SOIL

COMPACTOR PACKING GRAVEL

22104-12_F37.EPS

Figure 37 Soil and gravel compactors.

Compacting machines are built to be heavy. Operators running these compactors must be aware of their machine's weight and how much braking is required to stop. Compactor operators must also stay away from soft spots that could allow the machine to bog down or even sink. As for steering a compactor, it is usually steered by a steering wheel located in the operator cab. Other than the steering wheel, the only other controls are for the throttle and brakes.

Most compactors are articulated, which allows the front section to pivot left or right of the rear section. Personnel operating these compactors are expected to drive a smooth line, especially when working with materials such as asphalt. The finished surface of asphalt must be extremely smooth.

Teaching Tip/Notes

5.3.10 Trenchers

Trenchers come in all sizes from the simple walk-behind or ride-along models that are often rented for backyard use to the large ones that may be used to dig trenches for large piping sections. *Figure 39* shows a skid steer equipped with a trenching attachment.

Figure 38 Asphalt compactors.

Figure 39 Trencher attachment on a skid steer.

Operating a trencher is a lot like operating a backhoe or excavator in that the trenching machine must be maneuvered into a position that allows the trenching equipment to dig at the right spot. As a trencher is being traveled, the trenching bar and chain assembly is raised so that it strikes nothing on the ground. An operator maneuvering a trencher must be careful to not swing the trenching assembly into anything because the sharp chain teeth can damage most of the things they encounter. The chain drive must also be in neutral whenever the machine is being moved from one spot to another.

WARNING!

Before starting work on a trench, make sure any buried utilities in the area have been located and marked.

The trenching equipment is driven and controlled by a hydraulic system. The trenching assembly can be raised and lowered. Some trenchers can also pivot the trenching assembly left or right of center. When activated, the hydraulically driven chain drive moves the chain forward across the top of the chain bar toward the outermost end of the bar. The chain teeth are positioned to dig into the ground as they pivot around the end of the chain bar on their way back toward the machine. With the chain activated, the trenching assembly bar is lowered slowly toward the ground until the chain teeth begin digging. A slight amount of downward pressure must be applied to keep the chain digging deeper and deeper into the ground. After the trenching chain has dug the trench as deep as planned, the assembly is raised out of the

Audiovisual

Show Slide 43 (Figure 39). Describe the different types of trenching machines.

Classroom

Describe the proper method for clearing roots and other obstructions form the trencher chain.

Safety !

Stress the importance of making sure that any buried utilities have been located and marked before digging.

Discuss the importance of keeping the trenching attachment clear of people and objects.

Teaching Tip/Notes

Classroom

Discuss the different types of dump trucks and their uses.

Point out that some trucks have a secondary braking system.

Audiovisual

Show Slide 44 (Figure 40).

ground and the machine is moved enough for another section of trench to be dug. Some trenchers may be able to continue digging as the machine is moved, but the operator must be careful to not put the trenching assembly into a bind while the machine is being moved. After the final section of trench is dug, the assembly is raised and the chain drive is stopped. At that point, the machine can be moved away from the trench and parked. Always check the trencher after completing a job and look for any loose or broken teeth. Trenching chains often cut through roots. Roots tend to get stuck in the chain links and sprockets. Shut the machine off and clear any roots, rocks, or other types of debris from the trenching assembly.

5.3.11 Dump Trucks

Trucks classified as dump trucks are designed to transport materials, such as dirt, rocks, and other loose materials. All dump trucks use a heavy duty hydraulic system that lifts and lowers the trucks' beds in order to dump their loads. Some off-road trucks have a hydraulic ram, located at the front of the truck's bed, to push the load out the tailgate of the bed. *Figure 40* shows a few construction trucks dumping.

All dump trucks have steering wheels, gear shifters, brake pedals, throttle pedals, instrument panels, and mirrors. The hydraulic system controls used on the truck are inside the operator's cab. Because of their size, some of the larger trucks have camera systems that allow the driver to see what is behind the vehicle. Such camera systems are extremely important when dumping a load.

When loaded, dump trucks are extremely heavy, which means that they cannot be stopped quickly. Operators must learn to judge the stopping distance of a loaded truck verses an unloaded truck. Some trucks have secondary braking systems that help keep a loaded truck under control. Before operating any truck, review

Figure 40 Examples of off-road dump trucks.

Teaching Tip/Notes

the operator manual to ensure that all systems are clearly understood.

> **WARNING!**
> Check for overhead power lines and other obstructions before raising the bed.

Operators must also be aware that when a loaded truck bed is being raised for dumping, it changes the truck's center of gravity. Changing the center of gravity means that the truck may be easier to tip over. Operators need to make sure that their truck is on level, solid ground before attempting to lift the bed to dump a load. After a load has been dumped, the bed must be slowly brought back down onto the truck frame.

Most dump trucks have four basic control positions: raise, hold, float, and lower. The raise position causes the dump body (bed) to lift into a dumping position. When the raise control is released, the hydraulic system shifts into the hold mode, which means that the dump body neither raises nor lowers. When the operator selects the float position, the hydraulic system causes the dump body to seek its own level. When the lower position is selected, the dump body begins to lower. The dump body continues to lower until either it has shifted into the hold mode or it settles onto the frame of the truck. When hauling a load, the dump body must be lowered and placed into the float position.

From a safety point of view, truck operators must never travel with the dump body raised. They must make sure that nobody is near the area where they will be dumping their load. If any work must be performed under a raised dump body, make sure that the dump body is adequately supported.

> **WARNING!**
> Do not raise the dump body while the truck is on a slope. The truck may tip over.

Dump trucks used on a construction site are usually loaded by a loader or an excavator. The truck operator is responsible for driving the truck into the designated loading position. Since the truck operator cannot see inside the bed of the truck, the truck driver must have some form of communication with the loader operator to know when the truck is loaded.

5.3.12 Forklifts

Forklifts are selected for the environment in which they are being used. The forklifts used in a typical warehouse cannot function on the unfinished and rough surfaces of a construction site. Warehouse forklifts are also designed to work in close quarters while most construction-site forklifts are operated in open spaces. To minimize emissions, warehouse forklifts are usually powered by engines that run off rechargeable electric batteries, or by engines that burn propane gas. Construction-site forklifts use gasoline or diesel engines. *Figure 41* shows examples of a warehouse forklift and two rough-terrain forklifts.

Warehouse forklifts are built lower to the ground and operate on smaller hard tires with smooth tread. Rough-terrain forklifts are built higher off the ground and operate on larger treaded tires. Forklifts have either two-wheel drives or four-wheel drives. Most forklifts have a straight solid frame, but others have articulated frames that allow them to be maneuvered more easily in close quarters. Most forklifts can be steered from one end or the other, but some can be steered from both ends. All forklifts are required to have protective bars around and over the operator's seat area.

The lifting forks on all forklifts are driven and controlled by hydraulic system components. The forks on most forklifts simply move up and down, and tilt slightly forward or backward. The forks on most lifts stay in a fixed horizontal position on the forklift's mast until manually reset to another position, but a few forklifts have controls that allow the forks to be spread, closed, or shifted left or right from the operator area. On some specialized forklifts, the fork assembly may be made to tilt slightly left or right, or even rotate.

On smaller forklifts, the operator often stands inside the main body of the forklift and operates it with a steering wheel and levers. On most forklifts, the operator sits inside a protected area

> **On Site**
>
> **Higher and Higher**
>
> Some masts can extend to make higher lifts. They can either be two-stage or three-stage. This refers to the number of telescoping channels built in the mast. A two-stage mast has one telescoping channel. A three-stage mast has two telescoping channels. The telescoping channels provide greater lift height.

Safety !

Stress the importance of checking for overhead power lines before raising the truck bed.

Explain that raising the bed changes the center of gravity, which potentially creates a tipping hazard.

Classroom

Discuss the different control positions: raise, hold, float, and lower.

Point out that operating efficiency dictates that the driver brings the truck to the loading equipment.

Audiovisual

Show Slide 45 (Figure 41). Discuss the different types of forklifts and their uses.

Classroom

Discuss the different tasks that forklifts are used for on a construction site.

Teaching Tip/Notes

Classroom

Point out that each forklift has a maximum lifting capacity that should be reflected on a load chart on the machine.

Explain that loads must be carried on the forks, not suspended from them.

Describe the typical shutdown procedure for heavy equipment and compare it to that of a personal vehicle.

Explain that the shutdown procedure for any machine will be specified in its operator manual.

WAREHOUSE FORKLIFT

ROUGH-TERRAIN FORKLIFT

FIXED-MAST ROUGH-TERRAIN FORKLIFT

22104-12_F41.EPS

Figure 41 Examples of forklifts.

on top of the forklift and operates it from there. Other than the steering wheel, the basic controls on most forklifts include levers to raise and lower the lifting forks, and levers to tilt the forklift's mast back or forward to pick up or set off a load.

The most important factor to consider when using a forklift is lifting capacity. Each forklift is designed with an intended capacity, which must not be exceeded. Exceeding the capacity jeopardizes the equipment, the operator, and anyone nearby. Each manufacturer supplies a capacity chart for each forklift. Read and follow the capacity chart.

To operate a forklift safely, the operator must understand the capability and limitations of the machine. Do not operate beyond the capacity of the machine. Plan the work. Know the area in which the work will be done. Good planning can prevent many accidents.

Do not lift materials that are too heavy. Even if a load can be initially lifted, the machine can fail during transportation. Such a failure can damage the materials and the machine or cause injury. Carry the load as low as possible.

Pick up and carry loads using the forks. Suspending a load is very dangerous. Only use approved attachments or devices. Do not modify the equipment or remove shields or guards.

Follow the rules of the road. Know the rules covering traffic at the work site. Learn the meaning of all signs, flags, and signals. Operators must make sure that they communicate with others as needed. Do not carry passengers on the forks or in the cab.

6.0.0 Shutdown Activities

When a driver is through driving a personal vehicle, the vehicle is stopped and placed into Park (if automatic) before the ignition switch is turned to the Off position to stop the engine. There is no cooldown period needed. If the personal vehicle is being parked on an incline, most operator manuals recommend that the vehicle's parking brake be set before the transmission is placed into Park.

When heavy equipment machines are shut down, similar actions are taken. But there may be additional actions needed for machines due to attachments installed on the heavy equipment. This section looks at the common shutdown activities associated with heavy equipment machinery. Always refer to the O&M manual of a given machine for specific shutdown activities.

Teaching Tip/Notes

32 NCCER – *Heavy Equipment Operations Level One* 22104-12

The shutdown of any heavy equipment machine should follow a specific procedure applicable to that machine. Review the operator's manual for any shutdown specifics. Proper shutdown reduces engine wear and prevents possible damage to the machine.

When a job is not completed but the machine needs to be shut down, make sure it is safe to leave the machine in setup position. If not, then move it to a level spot away from the work area. When shutting down a machine over the work area, make sure that it is secure. Add stability by placing the bucket or other attachments on the ground.

Park the machine on a level spot. Lower the bucket or other attachments to the ground. Depress the clutch and place all levers in neutral. Place the transmission in neutral or park, and engage the parking brake (if provided).

Before turning the engine off, let it cool down. Decrease engine speed to a fast idle and allow it to run with no load for two to five minutes, or for the time listed in the O&M manual. Move the engine speed to slow idle after the engine has cooled. Turn the key to the Off position, and remove it from the switch. Operate the control levers to release the hydraulic pressure. Engage the brake lock and any other security devices and chock the wheels. If parking on an incline, position the vehicle at a right angle to the slope, and block the wheels.

> **NOTE**
> When shutting down a machine, remember to observe security precautions. If possible, park the machine away from main roads. Lock the instrument panel or cab if the machine is so equipped. Unauthorized personnel may attempt to steal or operate the machine if it is not properly secured. Follow company policy regarding the ignition key.

Dismount the machine using the proper hand and foot hold. Maintain three points of contact. Inspect the machine at the end of each day. Perform a walk-around similar to the morning daily inspection. Clean off mud and any other debris. Note hours of service, any damage, and maintenance required. Have the machine serviced if maintenance or repairs are needed. If the machine needs to be refueled, refuel it right away to prevent condensation from forming in the tank.

SUMMARY

Operating heavy equipment can be very dangerous. A heavy equipment operator must be aware of hazards and follow procedures that minimize danger. These procedures include daily inspection of the equipment, mounting, basic start-up, movement, and shutdown.

A prestart inspection must be performed before operating any piece of heavy equipment. A checklist is often provided in the operator's manual, the safety manual, or by the company. The daily inspection also includes checking and topping off fluids and lubricating joints. It also includes becoming familiar with the day's operations and job-site restrictions.

Always mount equipment at the appropriate location using three points of contact. Adjust the seat and mirrors and check the safety equipment before starting the machine. Consider the temperature and follow procedures for starting and warming up cold engines.

Operate equipment within the rated capacity, and follow all safety warnings on the equipment. Only leave the equipment set up on the job site if it is safe to do so. For the most part, equipment should be parked on a safe, level spot at the end of the work shift. The engine should be allowed to idle down before shutting it off. Lower attachments to the ground and cycle controls to relieve hydraulic pressure.

Follow these basic procedures when operating heavy equipment. These procedures will help keep you safe and the equipment in good working order.

Classroom

Emphasize the importance of lowering buckets, forks, or attachments before shutting off the machine.

Demonstration

Show trainees the basic prestart inspection, startup, operational movement, and shutdown procedures for selected heavy equipment machines.

PT/Laboratory

Under your supervision, have trainees perform basic prestart inspection, startup, operational movement, and shutdown for selected heavy equipment machines. This task corresponds to Performance Task 1.

Classroom

Summarize the major concepts presented in this module.

Teaching Tip/Notes

Classroom

Have the trainees complete the Review Questions, and go over the answers prior to administering the Module Examination.

Have the trainees complete the Trade Terms Quiz, and go over the answers prior to administering the Module Examination.

? Examination

Administer the Module Examination. Record the results on Training Report Form 200, and submit the results to the Training Program Sponsor.

✓ Performance Testing

Administer the Performance Test and fill out the Performance Profile Sheets for each trainee. If desired, trainee proficiency noted during laboratory sessions may be used to complete the Performance Test. Record the results on Training Report Form 200 and submit the results to the Training Program Sponsor.

Review Questions

1. When dismounting a machine, the operator should _____.
   a. face the machine making three points of contact
   b. climb down using any route available
   c. face away from the machine
   d. always dismount from the rear of the machine

2. OSHA and MSHA require that virtually all heavy equipment be equipped with approved seatbelts and _____.
   a. rollover protection structures
   b. rollover alarms
   c. a backup alarm
   d. brush guards

3. O&M manuals may also be called operator's manuals.
   a. True
   b. False

4. Tire inflation pressures are often posted _____.
   a. under the engine hood
   b. somewhere near each tire
   c. above the front windshield
   d. on the operator instrument panel

5. The metal tracks used on some machines are made up of what devices attached to a track chain?
   a. Rollers
   b. Chain links
   c. Track shoes
   d. Molded cleats

6. If bearings must be greased, before attempting to add grease to them _____.
   a. remove the vent plugs
   b. loosen the grease fittings
   c. tighten the grease fittings
   d. make sure that the grease fittings are clean

7. On a fuel-injected engine, fuel filters usually mounted _____.
   a. near the fuel tank
   b. inside the fuel tank
   c. near the fuel injector
   d. inside the air filter housing

8. Before removing a radiator cap, you should _____.
   a. put a bucket under the radiator
   b. allow the engine to cool
   c. start the engine and let it warm up
   d. use the dipstick to check coolant level

9. The level of oil in a drive axle is usually checked by _____.
   a. looking at a sight glass
   b. opening an inspection port
   c. checking the oil temperature
   d. checking the oil on the dipstick

10. The heart of a hydraulic system is its reservoir and the pump that moves the hydraulic fluid.
    a. True
    b. False

Teaching Tip/Notes

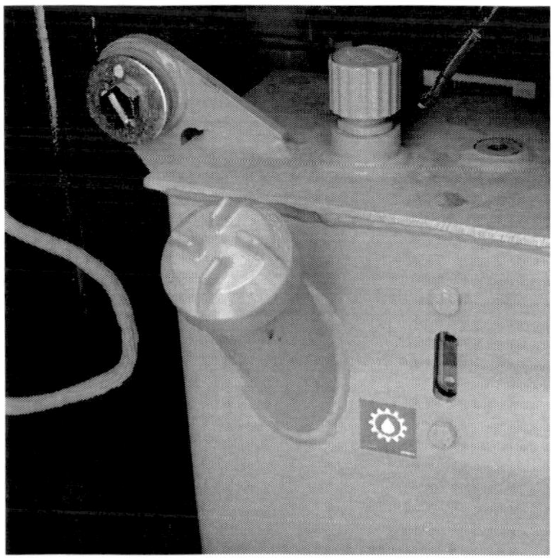

22104-12_RQ01.EPS

Figure 1

11. *Figure 1* shows a(n) _____.
    a. fuel tank
    b. storage compartment
    c. hydraulic reservoir
    d. engine oil sump

12. How much recommended glow plug heat time is needed for an engine when the temperature surrounding the engine is between 32°F and 60°F?
    a. None
    b. 1 minute
    c. 2 minutes
    d. 3 minutes

13. What method is used to keep a machine's engine water warm in cold weather when the engine is turned off?
    a. Glow plugs
    b. Block heater(s)
    c. Electric lights
    d. An external oil pump

14. Most heavy equipment vehicles must be started in neutral or park.
    a. True
    b. False

15. What dozer control is used to change the angle of the blade relative to the ground?
    a. Lift
    b. Angle
    c. Tilt
    d. Pitch

Teaching Tip/Notes

Trade Terms Quiz

Fill in the blank with the correct term that you learned from your study of this module.

1. Loosening the top surface of a material using a set of metal shanks or teeth is ________.
2. The angle of the blade and/or moldboard in relation to a vertical plane is called the ________.
3. The ________ is a control setting where the bucket, forks, or blade will follow the ground contour.
4. The ________ is the lever that regulates the supply of fuel to an engine.
5. A(n) ________ is a control mechanism that pivots about a fixed point in four directions and that is used to control the motion of an object.
6. The hydraulic arms on backhoes and other equipment that can be positioned on the right and left sides of a machine to give it additional stability are called ________.
7. Heating elements used to preheat the combustion chamber and aid in igniting fuel in a cold engine are called ________.
8. The metal structure which is attached to the equipment's cab and designed to protect the operator from equipment rollover is called a(n) ________.
9. A mechanism attached to a motor vehicle engine that supplies power to a non-vehicular device, such as a pump or a pneumatic hammer, is called a(n) ________.
10. The U-shaped, closed-end scoop attached to a loader or backhoe is called a ________.
11. The ________ is the primary attachment on a bulldozer that is used to push the material in front of the equipment.
12. The ________ is the cutting plate on a motor grader.

Trade Terms

Blade
Bucket
Float
Glow plugs
Joystick
Moldboard
Pitch
Power takeoff (PTO)
Rollover protection structure (ROPS)
Scarifying
Stabilizers
Throttle

Teaching Tip/Notes

Trade Terms Introduced in This Module

Blade: The primary attachment on a bulldozer used to push material in front of the equipment. It is typically a concave metal plate.

Bucket: A U-shaped, closed-end scoop attached to a front-end loader or backhoe.

Float: A control setting where the bucket, forks, or blade will follow the ground contour.

Glow plugs: Heating elements used to preheat the combustion chamber and aid in igniting fuel in a cold engine.

Joystick: A control mechanism that pivots about a fixed point in four directions. It is used to control the motion of an object.

Moldboard: The cutting blade on a motor grader.

Pitch: The angle of the blade and/or moldboard in relation to a vertical plane.

Power takeoff (PTO): A mechanism attached to a motor vehicle engine that supplies power to a non-vehicular device, such as a pump or pneumatic hammer.

Rollover protective structure (ROPS): A metal structure attached to the equipment's cab designed to protect the operator should the equipment roll over.

Scarifying: To loosen the top surface of material using a set of metal shanks (teeth).

Stabilizers: Hydraulic arms on backhoes and other equipment that can be positioned on the right and left sides of the machine to give it additional stability.

Throttle: A lever that regulates the supply of fuel to an engine.

Teaching Tip/Notes

Additional Resources

This module presents thorough resources for task training. The following resource material is suggested for further study.

The Earthmover Encyclopedia, 2007. St. Paul, MN: MBI Publishing.

Construction Safety. 1996. Jimmie Hinze. Englewood Cliffs, NJ: Prentice Hall.

Field Safety Participant Guide, 2003. NCCER. Upper Saddle River, NJ: Prentice Hall.

Handbook of OSHA Construction Safety and Health, Second Edition, 2006. Charles D. Reece and James V. Eidson. Boca Raton, FL: CRC Press.

HazCom for Construction. DVD. DuPont Sustainable Solutions – Training Solutions. Virginia Beach, VA.

Manual on Uniform Traffic Control Devices for Streets and Highways. 2009 edition. Washington, DC: US Department of Transportation, Federal Highway Administration.

OSHA Standards 29 CFR, Part 1926 Safety and Health Regulations for Construction. Washington, DC: Occupational Safety and Health Administration (OSHA), US Department of Labor, US Government Printing Office.

Figure Credits

Reprinted courtesy of Caterpillar Inc., Module opener, Figures 2–23 and 25–40

Deere & Company, Figure 24

Komatsu America Corp. SA01

Landoll Corporation, Figure 41 (top photo)

Sellick Equipment Limited, Figure 41 (middle and bottom photo)

Teaching Tip/Notes

ANSWERS TO REVIEW QUESTIONS

Answer		Section Reference
1.	a	2.2.0
2.	a	2.2.0
3.	a	3.0.0
4.	b	3.1.1
5.	c	3.1.2
6.	d	3.1.3
7.	c	3.2.1
8.	b	3.2.2
9.	b	3.2.5
10.	a	3.3.1
11.	c	3.3.1
12.	b	4.2.1; Table 1
13.	b	4.2.2
14.	a	4.3.0
15.	c	5.3.5

ANSWERS TO TRADE TERMS QUIZ

1. Scarifying
2. Pitch
3. Float
4. Throttle
5. Joystick
6. Stabilizers
7. Glow plugs
8. Rollover protection structure (ROPS)
9. Power takeoff (PTO)
10. Bucket
11. Blade
12. Moldboard

NCCER CURRICULA — USER UPDATE

NCCER makes every effort to keep its textbooks up-to-date and free of technical errors. We appreciate your help in this process. If you find an error, a typographical mistake, or an inaccuracy in NCCER's curricula, please fill out this form (or a photocopy), or complete the online form at **www.nccer.org/olf**. Be sure to include the exact module ID number, page number, a detailed description, and your recommended correction. Your input will be brought to the attention of the Authoring Team. Thank you for your assistance.

Instructors – If you have an idea for improving this textbook, or have found that additional materials were necessary to teach this module effectively, please let us know so that we may present your suggestions to the Authoring Team.

NCCER Product Development and Revision

13614 Progress Blvd., Alachua, FL 32615

Email: curriculum@nccer.org

Online: www.nccer.org/olf

❑ Trainee Guide ❑ AIG ❑ Exam ❑ PowerPoints Other ______________

Craft / Level: Copyright Date:

Module ID Number / Title:

Section Number(s):

Description:

____

____

____

____

Recommended Correction:

____

____

____

____

Your Name:

Address:

____

Email: Phone:

22105-12

Utility Tractors

Module Five

Materials

reminds you to gather materials needed for classes, labs, and testing.

Homework

suggests that you assign reading for the next session.

Classroom

prompts you to discuss a topic, stress a point, explain a concept, or provide examples.

Audiovisual

indicates an appropriate time to show a slide or transparency.

! Safety

used to emphasize safety issues.

Demonstration

directs you to show trainees how to perform tasks.

Laboratory

used when trainees are to practice performing tasks.

✓ Performance Testing

recommends administering a performance test, or a portion thereof.

PT / Laboratory

indicates that lab practice corresponds with a performance task. If desired, you can note the proficiency of the trainees during the lab to satisfy performance testing requirements.

Teaching Tip

indicates tips for practice exercises, Quick Quizzes, field trips, or other activities designed to encourage class participation.

? Examination

recommends administering the written module exam.

NCCER Standardized Craft Training Program

NCCER provides a standardized national program of accredited craft training. Key features of the program include instructor certification, competency-based training, and performance testing. The program provides trainees, instructors, and companies with a standard form of recognition through the National Registry. The program is described in full in the *Guidelines for Accreditation*, published by NCCER. For more information on standardized craft training, contact NCCER by writing to 13614 Progress Blvd., Alachua FL 32615; calling 1-888-NCCER20; or emailing **info@nccer.org**. More information is available at **www.nccer.org**.

How to Use this Annotated Instructor's Guide

Each page presents two sections of information. The larger section displays each page exactly as it appears in the Trainee Module. The narrow column ties suggested trainee and instructor actions to each page and provides icons (detailed on the left) to call your attention to material, safety, audiovisual, or testing requirements. The bottom of each page includes space for your notes, or when available, teaching tips that suggest activities or class exercises that will help your students learn the material.

Preparation

Before teaching this module, you should review the Objectives, Performance Tasks, Materials and Equipment List, and Module Outline. Be sure to allow ample time to prepare your own training or lesson plan and gather all required materials and equipment.

Instructors purchasing a single module receive the printed exam and performance tasks with the AIG module. Perfect-bound versions of this title no longer include the printed exam booklet and performance task sheets. Log on to the Instructor Resource Center at **www.NCCERIRC.com** to download these materials. Your access code is on the card bound into the front of this AIG. This code will allow you to access exams and performance tasks for all levels of this craft. Be sure to keep all testing materials in a secure place at all times.

Utility Tractors
Annotated Instructor's Guide

Module 22105-12

Module Overview

This module describes the use of tractors in construction. It covers basic tractor operation, controls, attachments, maintenance, and safety guidelines.

Objectives

Upon completion of this module, the trainee will be able to do the following:

1. Identify the operating controls of a typical utility tractor.
2. Describe the different types of transmissions used on utility tractors.
3. Explain the safety measures necessary to operate utility tractors and hydraulic systems.
4. Describe the proper methods for operating a utility tractor on slopes and hills.
5. Explain the proper method for adjusting a drawbar.
6. Perform prestart inspection and maintenance procedures.
7. Start, warm up, and shut down a gasoline-powered and diesel-powered tractor engine.
8. Perform basic maneuvering with a tractor.
9. Attach implements to a drawbar, three-point hitch, or power takeoff.
10. Connect hydraulic-powered attachments to the tractor.

Performance Tasks

Under the supervision of the instructor, the trainee should be able to do the following:

1. Perform prestart inspection and maintenance procedures.
2. Properly start, warm up, and shut down a gasoline-powered and diesel-powered engine tractor.
3. Perform basic maneuvering with a tractor.
4. Attach implements to a drawbar and three-point hitch.
5. Attach and detach implements to a power takeoff.

Materials and Equipment

Markers/chalk
Pencils and paper
Whiteboard/chalkboard
Heavy Equipment Operations Level One
PowerPoint® Presentation Slides
(ISBN 978-0-13-292167-1)
Multimedia projector and screen
Computer
Appropriate personal protective equipment
Tractor
Attachments:
Drawbar/hitch types
PTO types
Tractor operator manual
Operator maintenance checklists
Manufacturer literature on tractor transmissions
AEM safety manual
Trade Terms Quiz*
Module Examinations**
Performance Profile Sheets**

* Located at the back of the Trainee Guide module

** Single-module AIG purchases include the printed exam and performance task sheet. If you have purchased the perfect-bound version of this title, download these materials from the IRC using your access code.

Safety Considerations

Ensure that the trainees are equipped with appropriate personal protective equipment and know how to use it properly. Review safety guidelines associated with working on or around heavy equipment. Emphasize the importance of proper housekeeping.

Additional Resources

This module presents thorough resources for task training. The following resource material is suggested for further study.

Site Layout Levels 1 and 2, Latest edition. Alachua, FL: NCCER.

Teaching Time for This Module

An outline for use in developing your lesson plan is presented below. Note that each Roman numeral in the outline equates to one session of instruction. Each session has a suggested time period of 2½ hours. This includes 10 minutes at the beginning of each session for administrative tasks and one 10-minute break during the session. Approximately 17½ hours are suggested to cover *Utility Tractors*. You will need to adjust the time required for hands-on activity and testing based on your class size and resources. Because laboratories often correspond to Performance Tasks, the proficiency of the trainees may be noted during these exercises for Performance Testing purposes.

Topic	Planned Time
Session I. Introduction; Identification of Equipment	
A. Introduction	____________
B. Identification of Equipment	____________
1. Basic Tractor Component Operation	____________
2. Instruments and Controls	____________
3. Instruments	____________
4. Controls	____________
Session II. Safety Guidelines	
A. Safety Guidelines	____________
1. Safety Rules for Operating Tractors	____________
2. Safety Rules for Effective Tractor Operation	____________
3. Safety Rules for Operating a Tractor on Slopes or Hills	____________
4. Safety Rules for Operating the Hydraulic System	____________
5. Safety Rules for Operating the Power Takeoff	____________
6. Safety Rules for Tractor Tires	____________
Session III. Hitches and Attachments	
A. Hitches and Attachments	____________
1. Types of Hitches and Attachments	____________
2. Drawbar Adjustments	____________
3. Connecting an Attachment to the Drawbar	____________
4. Ballasting a Tractor	____________

Session IV. Basic Preventive Maintenance and Operation

A. Basic Preventive Maintenance and Operation

1. Servicing a Tractor
2. Start, Warm Up, and Shut Down of a Gasoline- or Diesel-Powered Tractor
3. Safely Stopping a Tractor
4. Basic Tractor Operations
5. Connecting an Attachment to a Three-Point Hitch
6. Attaching a PTO
7. Operating a PTO
8. Connecting and Disconnecting Attachment Hydraulic Hoses

Sessions V and VI. PT/Laboratory

A. PT/Laboratory

1. Have trainees perform prestart inspection and maintenance procedures. This task corresponds to Performance Task 1.
2. Have trainees start, warm up, and shut down a gasoline-powered and diesel-powered engine tractor. This task corresponds to Performance Task 2.
3. Have trainees perform basic maneuvering with a utility tractor. This task corresponds to Performance Task 3.
4. Have trainees attach implements to a drawbar and three-point hitch. This task corresponds to Performance Task 4.
5. Have trainees attach and detach implements to a power takeoff. This task corresponds to Performance Task 5.

Session VII. Review and Testing

A. Review

B. Module Examination

1. Trainees must score 70% or higher to receive recognition from NCCER.
2. Record the testing results on Training Report Form 200 and submit the results to the Training Program Sponsor.

C. Performance Testing

1. Trainees must perform each task to the satisfaction of the instructor to receive recognition from NCCER. If applicable, proficiency noted during laboratory exercises can be used to satisfy the Performance Testing requirements.
2. Record the testing results on Training Report Form 200 and submit the results to the Training Program Sponsor.

22105-12

Utility Tractors

Module Five

Teaching Tip/Notes

Homework

Assign the reading of this module.

Trainees with successful module completions may be eligible for credentialing through NCCER's National Registry. To learn more, go to **www.nccer.org** or contact us at **1.888.622.3720**. Our website has information on the latest product releases and training, as well as online versions of our *Cornerstone* newsletter and Pearson's product catalog.
Your feedback is welcome. You may email your comments to **curriculum@nccer.org**, send general comments and inquiries to **info@nccer.org**, or fill in the User Update form at the back of this module.

V.1 4/12

Teaching Tip/Notes

22105-12 UTILITY TRACTORS

Objectives

When you have completed this module, you will be able to do the following:

1. Identify the operating controls of a typical utility tractor.
2. Describe the different types of transmissions used on utility tractors.
3. Explain the safety measures necessary to operate utility tractors and hydraulic systems.
4. Describe the proper methods for operating a utility tractor on slopes or hills.
5. Explain the proper method for adjusting a drawbar.
6. Perform prestart inspection and maintenance procedures.
7. Start, warm up, and shut down a gasoline-powered and a diesel-powered tractor engine.
8. Perform basic maneuvering with a tractor.
9. Attach implements to a drawbar, three-point hitch, or power takeoff.
10. Connect hydraulic-powered attachments to the tractor.

Performance Tasks

Under the supervision of your instructor, you should be able to do the following:

1. Perform prestart inspection and maintenance procedures.
2. Properly start, warm up, and shut down a gas-powered and diesel-powered engine tractor.
3. Perform basic maneuvering with a tractor.
4. Attach implements to a drawbar and three-point hitch.
5. Attach and detach implements to a power takeoff.

Trade Terms

Bead
Draft
Hydrostatic
Power takeoff (PTO)
Retard
Rockshaft unit
Seat
Spline
Travel
Turbocharger

Industry Recognized Credentials

If you're training through an NCCER-accredited sponsor you may be eligible for credentials from NCCER's Registry. The ID number for this module is 22105-12. Note that this module may have been used in other NCCER curricula and may apply to other level completions. Contact NCCER's Registry at 888.622.3720 or go to nccer.org for more information.

Materials

Ensure that you have everything required to teach the course. Check the Materials and Equipment list at the front of this module.

Classroom

Explain that terms shown in bold (blue) are defined in the Glossary at the back of this module.

Audiovisual

Show Slide 1, Objectives, and Slide 2, Performance Tasks. Review the goals of the module, and explain what will be expected of the trainees.

Teaching Tip/Notes

When you call on someone to answer a question, the rest of the class relaxes or even tunes out because they expect that the question and answer will take place only between you and the trainee you called on. Instead, use this technique to involve more trainees in answering questions and to keep them on their toes:

1. Ask trainees to define a term or explain a concept.
2. After one trainee has answered, ask a trainee seated nearby if the answer is right. Then ask whether a trainee in the back of the room agrees.
3. Ask trainees to explain why they think an answer is right or wrong.
4. Use the session to clear up incorrect ideas and encourage trainees to learn from their mistakes.

Contents

Topics to be presented in this module include:

Teaching Tip/Notes

Teaching Tip/Notes

Figures and Tables

Teaching Tip/Notes

1.0.0 INTRODUCTION

Tractors are manufactured by many different companies and come in various sizes and configurations. Many machines used on a construction site, such as scrapers, are pulled by tractors. This module describes a class of tractors called utility tractors, which range from less than 100 horsepower (hp) to well over 300 hp. Some have only rear-wheel drive, but many are equipped with optional mechanical front-wheel drive (MFWD). Such tractors can be used on construction sites for light or medium work, including staging of materials and equipment, finish grading, fence installation, and moving construction trailers. The John Deere 5 and 8 Series of utility tractors are good examples of such tractors. The basic versions of these tractors are equipped with diesel engines, mechanical three-point hitches, drawbars, and independent mechanical **power takeoffs (PTO).** They may be equipped with a variety of transmissions. They are also available with additional options that include a rollover protection structure (ROPS) operator cab, electro-hydraulic (EH) three-point hitch, EH PTO, and EH selectable control valves (SCVs). *Figure 1* shows two different views of slightly different John Deere tractors.

22105-12_F01.EPS

Figure 1 Examples of John Deere utility tractors.

2.0.0 IDENTIFICATION OF EQUIPMENT

While the components, controls, and indicators of tractors vary with the manufacturer, model, and the options available, they are all similar. *Figure 2* shows the general location of components on a typical John Deere utility tractor.

> **NOTE**
> The callout numbers in the following text refer to *Figure 2*. Before operating any tractor, operators must obtain, read, and understand the operator's manual for the tractor they will be operating.

2.1.0 Basic Tractor Component Operation

A diesel, gasoline, or, in some cases, a propane (LP) engine (1) provides power for these tractors. Air is drawn through the air cleaner (2) into the cylinders. Fuel from a fuel tank (3) is injected. On some tractors, especially diesels, a **turbocharger** (4) compresses air for better combustion and results in more power delivered to the crankshaft. The cooling system (5) dissipates heat from the engine and, in some cases, from one or more hydraulic systems. The crankshaft transmits torque to either a clutch (6) that provides power to a mechanical transmission (7) or to a hydraulic (**hydrostatic**) transmission. The clutch stops and starts power flow to a mechanical transmission. A hydrostatic transmission's power flow is controlled by the rate and direction of oil flow within the transmission. The differential (8) transmits power from the transmission, through final drives, to the rear axles (9). The power takeoff (PTO) shaft (10) and mechanical front-wheel drive axles (11) are driven through separate EH clutches from the main power train. Power from the engine is also coupled to a variable-displacement pump (12), providing hydraulic power for other systems. The variable-rate hydraulic system provides hydraulic fluid power through a filter (13) for various operations. These include the EH clutches, the rockshaft and SCV controls (14) to the rear SCVs (15), mid-mount SCVs (16) if so equipped, and the **rockshaft unit** (17) for the three-point hitch (18).

Classroom

Describe how tractors are used on construction sites.

Audiovisual

Show Slide 3 (Figure 1). Explain that several companies manufacture tractors of this class.

Show Slides 4 and 5 (Figure 2). Point out and name the major components of a tractor and explain the function of each.

Teaching Tip/Notes

22105-12 Utility Tractors Module Five 1

Classroom

Describe the operation of sliding gear, constant mesh, hydraulic assist, and hydrostatic drive transmissions.

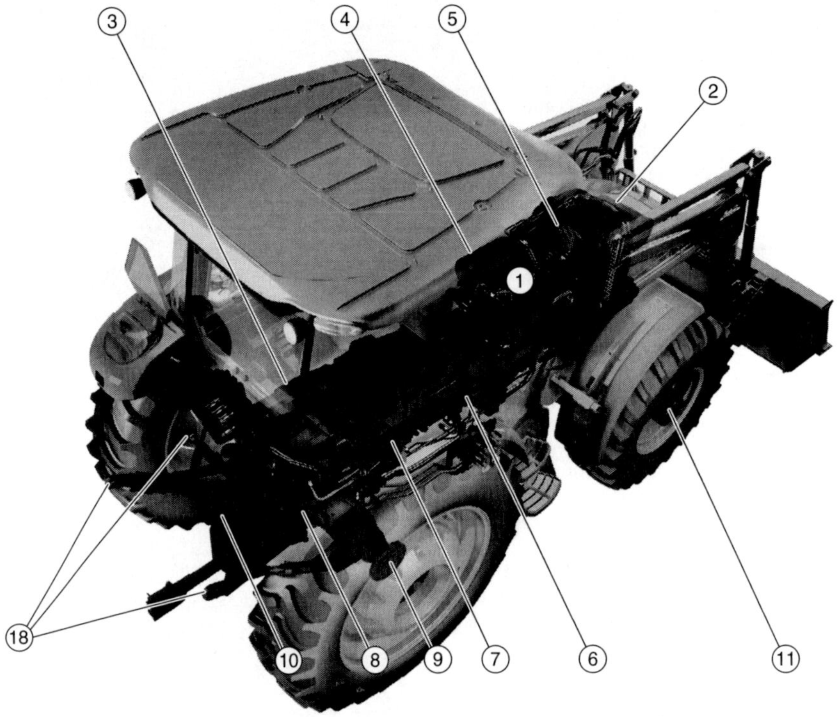

Figure 2 Components of a John Deere utility tractor. (1 of 2)

2.1.1 *Transmissions*

A variety of transmissions are used on tractors. They provide a number of forward speeds and one or more reverse speeds. The most common transmission in utility tractors is the synchromesh transmission. Transmissions can also be used to drive a PTO shaft to operate equipment attached to the tractor. Mechanical shift transmissions, located after a clutch and before the differential, consist of two major types—sliding gear and constant mesh. Some are equipped with a range selector as well as a gear selector. These allow the operator to select one of several speed ranges while the tractor is stopped, and then to select finer gear speeds within that range. Some tractors have a high-low shifter that is like a range selector, but can be shifted on the go. Mechanical transmissions, along with hydraulic assist transmissions and more expensive hydrostatic (hydraulic) drive transmissions, are described as follows:

- *Sliding gear* – Sliding gear transmissions have two or more internal parallel shafts. When moved by a shift lever, sliding spur gears are arranged on the shafts to mesh with each other to provide speed and direction changes. Movement of the tractor must be stopped to shift gears in this type of transmission.
- *Constant mesh* – The two types of constant mesh transmissions are collar or shuttle shift and synchromesh transmissions. In the collar or shuttle shift version, the selected free-running

Teaching Tip/Notes

Obtain manufacturers' literature that describes the types of transmissions used on tractors. This information should be available in brochures from local distributors or manufacturers. Other sources, such as tractorsmart.com and howstuffworks.com, are available online.

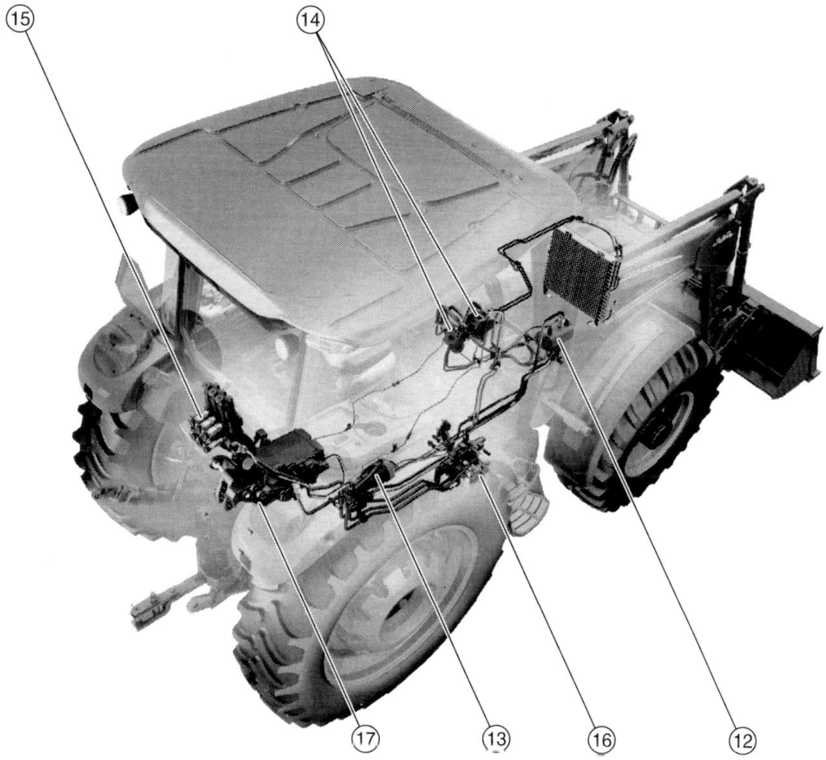

12 – VARIABLE DISPLACEMENT PUMP
13 – HYDRAULIC FLUID FILTER
14 – ROCKSHAFT AND SCV CONTROLS
15 – REAR SCV
16 – MIDMOUNT SCV (IF EQUIPPED)
17 – ROCKSHAFT UNIT

22105-12_F02B.EPS

Figure 2 Components of a John Deere utility tractor. (2 of 2)

gears are locked to their shafts by a sliding collar or shuttle. These versions of the transmissions usually require that the tractor be stopped to shift gears. Sliding gear and collar or shuttle shift transmissions are sometimes called standard transmissions. In synchromesh transmissions, gears can be selected without clashing by synchronizing devices in the transmission that match the speeds of mating gears before they engage. With these types of transmissions, the shifting can be accomplished on the go with the use of the clutch.

- *Hydraulic assist* – A hydraulic assist transmission is a train of constantly meshed gears that can be shifted on the go, without a clutch, to prevent interruption of the power flow through the transmission. When the operator selects a range and gear speed, hydraulic fluid activates hydraulic disc clutches and/or brakes for the appropriate gear combinations that are required for the selected speed. This type of transmission, although built for heavier duty, operates in a manner similar to 4-, 5-, or 6-speed automatic transmissions in cars, except that the shifts must be manually selected.
- *Hydrostatic drive* – A hydrostatic-drive transmission allows an infinite adjustment of tractor speed within a range without any gear shifting. However, it is relatively expensive and inefficient compared to a geared transmission. These types of transmissions use internal hydraulic fluid instead of gear trains to transmit power to the differential. Energy from the engine is transferred to a pump and then transferred by the fluid to a motor in essentially a closed circuit. The output shaft of the motor is coupled to

Teaching Tip/Notes

Audiovisual

Show Slide 6 (Castered MFWD Axle). Explain the benefit of the castered MFWD axle.

Show Slide 7 (Figure 3). Explain the operation of a hydrostatic transmission. Point out that it provides an infinite adjustment of speed with a range.

On Site

Castered MFWD Axle

Some manufacturers of MFWD tractors use some form of a castered front axle to provide a tight turning angle. When the wheels are turned from a straight line of travel, the rearward tilt of the king pins on both sides of the axle causes the top of both front wheels to tilt with a 12-degree caster toward the direction of the turn to reduce sideslip. The caster also allows the inside wheel to tuck under the frame slightly to achieve a tighter turning angle.

22105-12_SA01.EPS

the differential. While the fluid in the multiple paths between pump and motor moves back and forth slightly, it is considered a static flow pressure system. The pressure of the fluid is what transfers energy. *Figure 3* is a very simplified diagram of the inside of a hydrostatic transmission with the pump and motor cylinder blocks separated. In use, the blocks are coupled so that each of the two fluid ports, each containing multiple pistons, mates with a similar port containing pistons in the motor block. The incompressible fluid moving back and forth through the ports transfers energy from the pump to the motor. Energy transfer occurs when the variable-angle swashplate rotates around the fixed position pump block, causing the pistons in a port to move into the block in a rotating pattern, forcing oil against mating pistons in the port of the fixed motor block. This, in turn, forces mating pistons out of the fixed motor block and against a rotatable fixed-angle swashplate attached to the output shaft. Depending on the pressure exerted, this action causes the fixed-angle swashplate to rotate and turn the output shaft. The angle of the variable swashplate can be changed to increase or decrease the displacement of the pistons to adjust the speed of the output shaft and the direction of rotation. In an actual hydrostatic transmission, the fluid is also circulated through an external radiator for cooling, then through a reservoir and filter, and back to the transmission.

Teaching Tip/Notes

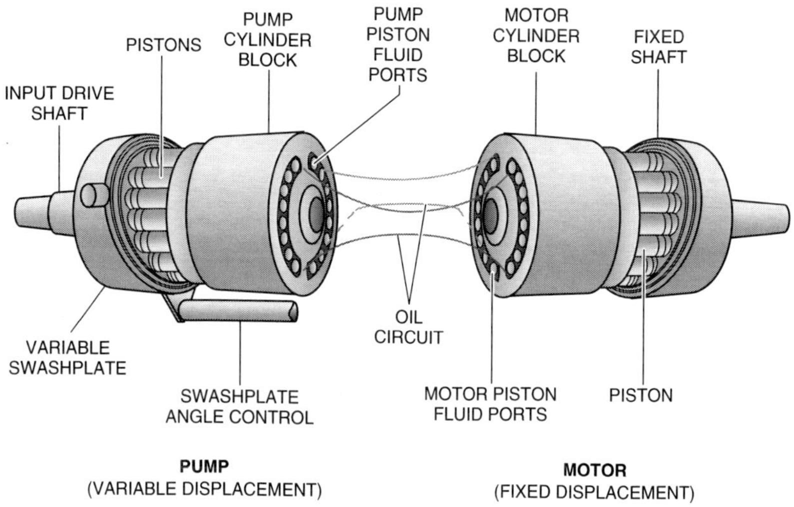

Figure 3 Simplified hydrostatic-drive transmission diagram.

2.1.2 Clutches

Various types of disk, overrunning, and cone clutches are used on tractors to engage and disengage input power for transmissions and auxiliary power trains.

A disk clutch is commonly used to connect a tractor engine to a mechanical transmission. Depressing a foot pedal on the operator platform disengages the engine power from a transmission. Releasing the pedal engages the power. Two types of disk clutches are used: wet or dry. Both operate in a similar manner, except that a wet disk clutch is immersed in an oil bath or spray for cooling. The oil is circulated through an external radiator to remove the heat.

A basic disk clutch consists of three main components: a drive plate, a clutch plate, and a pressure plate assembly. The drive plate is usually the flywheel of the engine, which revolves all the time the engine is running. The clutch plate is a metal plate coated on both sides with a friction material, or it may contain a separate clutch disk made of friction material. The clutch plate is mounted by **splines** on the output shaft of the clutch. The clutch disk is the driven plate and is free to slide back and forth on the output shaft splines, but not free to rotate on the shaft. The spring-loaded pressure plate assembly, on the other side of the clutch disk, is mounted to the flywheel with standoff bolts and rotates with the flywheel. As the clutch is engaged, the springs of the pressure plate squeeze the friction surfaces of the clutch disk between the flywheel surface and the pressure plate surface. This gradually locks the clutch disk to the rotating flywheel and pressure plate, causing the clutch disk to turn and rotate the output shaft.

Heavy-duty clutches contain more than one clutch disk splined to the output shaft and separated by steel plates that slide back and forth on the standoff bolts between the flywheel and pressure plate assembly. The standoff bolts cause the steel separator plates to rotate with the flywheel and pressure plate assembly. This provides multiple friction surfaces between the clutch disks to reduce heating and extend the friction surface life of the clutch disks under heavy loads. The pressure plate is released and retracted by a clutch release bearing fork (throw-out bearing) on the output shaft that is activated by mechanical linkages or a hydraulic system activated by the clutch pedal. EH clutches used for accessories or MFWD operate similarly, but are engaged and disengaged by electrical switch action instead of pedals.

An overrunning clutch automatically engages in one direction, but freewheels in the opposite direction. This is accomplished with a number of roller bearings and ramps on the input shaft inside the output shaft. When the input turns in the proper direction, the bearings roll up their ramps and jam between the ramps and output shaft. This locks the output shaft to the rotating input shaft. If the input shaft rotation slows or stops, the bearings are forced down the ramps, allowing the output shaft to freewheel until it begins to rotate slower than the input shaft. At that time, the bear-

Classroom

Explain the types of clutches used on tractors.

Describe a disk clutch and explain how it functions.

Describe heavy-duty and overrunning clutches and explain how they function.

Teaching Tip/Notes

Classroom

Explain that a cone clutch is a type of friction clutch.

Audiovisual

Show Slide 8 (Figure 4). Point out the locations of controls and indicators.

Show Slide 9 (Figure 5). Discuss the individual items on the instrument panel.

Classroom

Describe the conditions that cause the charging system indicator to light.

ings are again jammed against the output shaft by the ramps of the input shaft, causing the output shaft to rotate with the input shaft. These devices are most often used inside various transmissions.

Cone clutches are another form of a friction clutch. They are simpler and more compact than an equivalent basic disk clutch because the friction surfaces are cone-shaped instead of flat. Cone clutches are composed of two pieces. The drive piece fastened to the input shaft has an internal cone with a machined surface. The driven piece is a cone that fits inside the drive piece cone. It is mounted, and slides back and forth, on the splines of an output shaft. The mating surface of the driven cone is covered with a friction material. When the driven cone is forced into the drive cone by a throw-out bearing, the cones gradually lock up and the output shaft turns with the input shaft. Cone clutches are commonly used to control gear trains that supply auxiliary power, including PTOs.

2.2.0 Instruments and Controls

The variety of instruments and controls for a tractor must be understood before operating the tractor. Study the operating manual for the tractor being used to understand the specific controls and instruments, including engine operating temperatures, charging system readings, oil and hydraulic system pressures, and other operating characteristics. The following sections and figures describe the common instruments and controls found on most tractors. Most of the pictures in this book are based on John Deere tractors. *Figure 4* is a view of a tractor operator's cab.ww

2.3.0 Instruments

A tractor's instrument panel, located on the front control console, must be closely observed during operation to avoid serious damage to the tractor. Instrumentation varies among different makes and models of tractors, but generally includes the instruments covered in the following sections. A typical instrument panel is shown in *Figure 5*.

> **NOTE**
> The parenthetical callout letters in the following sections refer to *Figure 5*.

2.3.1 Charging System Warning Indicator

This warning indicator (L) lights if the charging system is out of a normal range because the battery is discharging due to an excessive load or a malfunctioning alternator. Some tractors are equipped with an ammeter or voltmeter in addition to, or instead of, a warning indicator. These types of meters indirectly register the rate of electrical current charge or discharge by monitoring the electrical system voltage variations caused by loads and alternator input. After starting, the meter should show a moderate positive charging value, which tapers off to a lower value during operation.

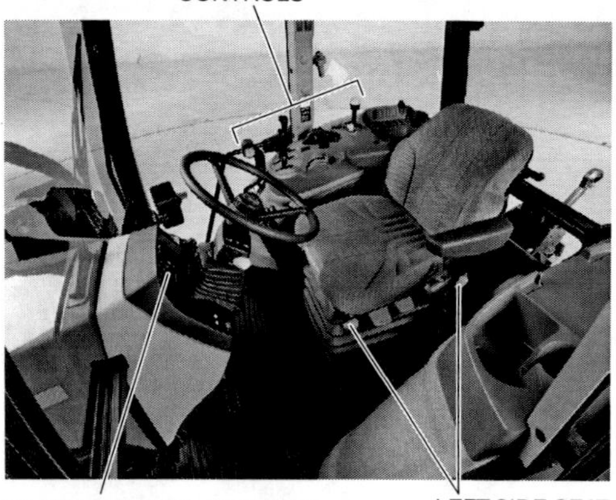

Figure 4 Overhead view of a John Deere tractor's operator cab.

> **NOTE**
> If the indicator light and/or a meter shows a high negative discharge rate when the tractor is operating above one-third throttle, the tractor should be stopped and inspected to determine the cause.

2.3.2 Engine Oil Pressure Warning Indicator

The engine oil pressure warning indicator (N) lights if the engine oil pressure falls below an acceptable level. Some tractors are equipped with a warning indicator and/or an oil pressure gauge. The gauge shows the actual engine oil pressure and, usually, the normal operating range. Normal engine oil pressures vary widely due to the type of engine and manufacturer. Consult the tractor's operating manual for the normal oil pressure range.

Teaching Tip/Notes

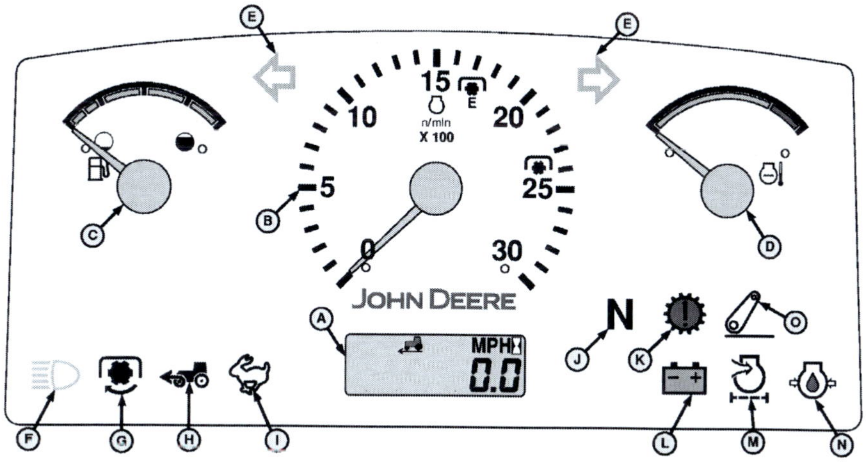

A - Speedometer/Hour Meter
B - Tachometer
C - Fuel Level Gauge
D - Coolant Temperature Gauge
E - Hazard/Warning Turn Signal Indicators
F - High Beam Indicator
G - PTO Engaged Indicator
H - MFWD Engaged Indicator (if Equipped)
I - Hi/Lo Indicator (if Equipped)
J - Neutral Indicator (if Equipped)
K - Transmission (with Reverser) Information Indicator (if Equipped)
L - Charging System Indicator
M - Air Cleaner Restriction Indicator
N - Engine Oil Pressure Indicator
O - EH Hitch Indicator (if Equipped)

22105-12_F05.EPS

Figure 5 Indicators on a typical tractor instrument panel.

CAUTION

To avoid serious engine damage, always stop the tractor engine immediately if the oil pressure falls below an acceptable value or an oil pressure warning light comes on during operation.

2.3.3 Fuel Level Gauge

The fuel level gauge (C) indicates the amount of fuel in the tractor's fuel tank. On diesel engine tractors, the gauge may contain a low fuel warning zone, or a low fuel warning light may be present. Avoid running out of fuel on diesel engine tractors because the fuel lines and injectors must be bled of air before the engine can be restarted.

2.3.4 Coolant Temperature Gauge

The coolant temperature gauge (D) shows the temperature of the coolant flowing through the engine. Because operating temperatures vary between tractor makes and models, always refer to the tractor's operating manual to determine the correct operating range. Most gauges are calibrated so that the range or needle is centered on the gauge. If the engine overheats, warning lights may also be activated.

CAUTION

Do not place a tractor under heavy load until the correct operating temperature is attained. Operating under heavy load with a cold engine increases fuel and oil consumption and shortens engine life. If the engine overheats under load with no visible loss of coolant, stop the tractor and take it out of gear, but do not stop the engine. Let the engine fast idle for several minutes until the temperature returns to the operating range. Then shift to a lower gear or reduce the load to prevent overheating. If the engine is shut off while it is overheated, temperatures rise even more and can result in burned valves, heat distortion, and lubrication failure. However, if the engine overheats from loss of coolant because of a broken hose or leaking radiator, shut the engine off immediately to prevent any more serious damage to the engine.

Classroom

Point out that the engine must be shut off immediately if the engine oil pressure light comes on.

Explain that the fuel level gauge shows how much fuel is in the tank.

Point out that the tractor should not be moved until the engine temperature has reached a specified level.

Explain what to do in engine-overheat and loss-of-coolant situations.

Teaching Tip/Notes

22105-12 Utility Tractors Module Five 7

Classroom

Describe how the tachometer reading relates to PTO operation.

Safety

Explain that PTO-driven equipment must never be operated above its rated speed.

Classroom

Explain that the speedometer/hour meter shows both speed and number of hours of operation.

Describe other types of indicators normally found on the instrument panel of a tractor.

Audiovisual

Show Slide 10 (Figure 6). Describe the information displayed on a performance monitor.

On Site

Weight Transfer Pressure Gauge

Some tractors equipped with a weight transfer hitch sometimes have a weight transfer and rockshaft pressure gauge. This gauge indicates the relative amount of weight transferred to the rear wheels of the tractor. The gauge normally has two zones. One zone indicates the weight transfer and the other indicates hydraulic pressure at the rockshaft. During operation, the rockshaft control lever setting is adjusted so that the gauge moves about slightly when the tractor is in motion.

2.3.5 Tachometer

The tachometer (B) indicates engine speed in revolutions per minute (rpm). Most tachometers are marked in hundreds on the meter face. For instance, 15 on the meter face would represent 1,500 rpm. The normal operating range for tractors varies between makes and models. Check the operating manual for the correct range. Most tachometers also have special marks for operation of a PTO. For tractors equipped with a 540/540E PTO, the marks coincide with an economy (E) engine speed (1,700 rpm) for 540 rpm PTO operations using light loads. For tractors equipped with a standard 540 rpm or EH 540 rpm PTO, with or without a 1,000 rpm stub shaft, the normal full-power engine speed (2,400 rpm) must be used.

WARNING!

Never operate PTO-driven equipment above the rated speed listed in the tractor's operator manual. Doing so can cause catastrophic failure of the equipment connected to the PTO, which may result in injury or death to the tractor operator or nearby persons due to flying parts.

2.3.6 Speedometer/Hour Meter

The speedometer/hour meter (A) indicates the forward speed of the tractor in miles per hour (mph) and the number of hours of operation. The speedometer is helpful when the effectiveness of an attachment depends on the speed of the tractor.

2.3.7 Additional Status and Warning Indicators

Besides the warning indicators described in the previous sections, many tractors also have other status indicators and system warning indicators, depending on the options that are fitted to the tractor. These lights call attention to the operating status of tractor systems such as the power train, PTO, or other hydraulic systems. Some may require immediate engine shutdown to prevent damage to the tractor. These indicators can include the following:

- Hazard warning/turn signal (E)
- Headlight high beam (F)
- PTO engaged (G)
- MFWD engaged (H)
- Hi/Lo (I)—Lights when hi position of hi/lo shift is engaged.
- Neutral (J)—Lights when gearshift and electro-hydraulic reverser levers are in neutral position. Flashes when gearshift is in neutral and reverser is in forward position. Goes out when gearshift is in park position.
- Transmission (with reverser) information indicator (K)—Lights if malfunction occurs in transmission; terminate operation when lit.
- Air cleaner restriction (M)—Indicates a clogged prefilter or air cleaner.
- EH hitch (O)—Indicates a malfunction in the hitch control system; terminate operation when a malfunction is indicated.

Other tractors may also have the following warning indicators that require terminating operation:

- Transmission wet-clutch low oil
- PTO clutch low oil

2.3.8 Performance Monitor

Some large tractors are equipped with a computerized performance monitoring system (*Figure 6*). These devices are capable of displaying, among other things, digital values for ground speed, productivity, area covered, wheel slip, PTO speed, engine speed, distance, and hours since last service. These monitors usually use a bar graph to continuously display wheel slip, along with an excessive slip alarm.

Teaching Tip/Notes

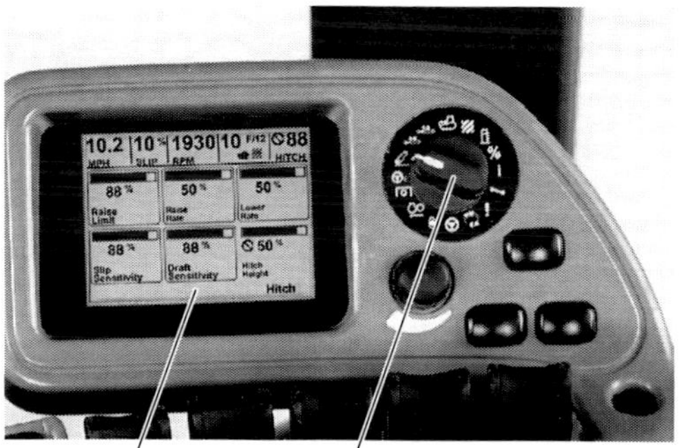

Figure 6 Typical computerized performance monitoring system.

Other computerized systems are also available for agricultural uses on large tractors. Among them are the following types of systems:

- *Global positioning system (GPS) guidance* – Programmable GPS guidance systems can perform hands-off steering of the tractor to maintain repetitive straight-line operation over long distances at night, in fog, or in other conditions.
- *Attachment management systems* – These systems allow two or more programmable sequences. Each sequence can perform 12 functions or more with a single operator command. The programming allows precise, repetitive control of the attachment and tractor for such things as attachment depth, raise/lower, float, SCV extend/retract, and speed shifts.

2.4.0 Controls

Controls and their location can vary between makes and models of tractors. To aid in identifying controls and their functions, some manufacturers color code them. The following is one example of a color code:

- *Yellow* – PTO or auxiliary power
- *Red* – Gear shift levers or tractor movement
- *Black* – Hydraulics or miscellaneous functions

Controls for most tractors are located on the seat, on and under the front control console, on the side and rear control console, and on the overhead air conditioning control console. Most of the common controls are described in the following sections.

2.4.1 Seat and Steering Wheel Adjustment

Proper seat and steering wheel adjustments are important to safe tractor operations. After entering the cab area using the steps on the side of the tractor, the operator should adjust the seat and steering wheel position and then fasten the seat belt before operating the tractor. *Figure 7* shows a typical seat used on most John Deere tractors.

The adjustments are usually located left and right of the seat, but some may be located below the front edge of the seat bottom itself. Most seats can be moved up or down and forward or backward. Some provide an adjustable shock absorber function using springs or a hydraulic mechanism. The seat should be adjusted so that the operator's legs are almost straight when the clutch or brake pedals are fully depressed and the operator's back is flat against the back of the seat.

The steering wheel (*Figure 8*) on some tractors can be adjusted up or down and in or out as desired after the seat is correctly positioned. Because seat and steering wheel adjustment devices vary widely for various makes and models, refer to the operator manual for specific instructions.

> **NOTE**
> Never adjust the steering wheel or seat when the tractor is in operation.

Figure 7 Mechanical suspension seat adjustments.

Classroom

Point out that some manufacturers color-code their controls.

Stress the importance of properly adjusting the seat and discuss how the seat should be adjusted.

Audiovisual

Show Slide 11 (Figure 7). Point out the locations of seat adjustment controls.

Show Slide 12 (Figure 8). Discuss the proper relationship of the steering wheel location to that of the seat.

Teaching Tip/Notes

Audiovisual

Show Slide 13 (Table 1). Explain the relationship between gear selection and ground speed.

Classroom

Describe how a high-low shifter works.

Explain that a reverser lever allows the tractor to change direction without using the clutch or changing gears.

Describe the purpose of the front-wheel drive selector and the limitations on its use with MFWD.

Audiovisual

Show Slide 14 (Figure 9).

Safety

Point out the tipping hazard associated with MFWD.

Audiovisual

Show Slide 15 (Figure 10). Discuss the clutch function. Point out that the clutch may need to be cycled at startup.

Figure 8 Position of adjustable steering wheel to the operator seat.

2.4.2 Transmission Shift Lever Operation

On tractors equipped with standard, synchronized, or hydraulic-assist mechanical transmissions, there are five basic transmission shift options:

- *Range selector* – The range selector lever, if present, is used to select a range of operating speeds for the transmission where the individual speed is then selected using the gear selector. Usually the lever is separate from the gear selector, but on some tractors the same lever is used for both functions. In some cases, a creeper gear selection for the range selector is an option that allows very slow tractor speeds in forward and reverse gears. For most tractors, the tractor must be stopped and the clutch depressed before the range selector can be shifted. *Table 1* shows the typical range (letter) and gear selection (number) ground speeds for one series of John Deere tractor fitted with standard tires.
- *Gear selector* – The gear selector on all tractors is used to shift into a selected forward speed or shift into reverse. On most tractors, the gear selector must be in the park position before the tractor can be started. The gear selector can be shifted using the clutch when the tractor is in motion.
- *High-low shifter* – Some tractors have a high-low shifter that may be actuated electrically or mechanically. When used, it basically shifts the tractor speed a half-gear up or down from the currently selected gear without clutching while the tractor is in motion.
- *Reverser* – The reverser lever, if present, allows changing the direction of the tractor without clutching and changing gears. It may have two or three positions: forward, reverse, and on some tractors, neutral. Reverse is engaged by pulling the lever downward. Some tractors are equipped with reverser gear ratios that are higher than the respective forward speeds. As a result, use the reverser at lower speeds or throttle down before engaging it. Reversers are often used in tight places that require a number of back and forth motions. For most tractors, the reverser lever must be in the neutral position to start the tractor.
- *Front-wheel drive selector* – On tractors with power assist front-wheel drive, the lever or switch is used to engage and disengage the front-wheel drive. For tractors with hydrostatic front-wheel drives, the selection can be made with the tractor in motion. On a tractor with an MFWD that is not equipped with a hydraulic device for engagement or disengagement, tractor movement must be stopped before engagement or disengagement of the MFWD. If the MFWD has an EH device for engagement or disengagement (*Figure 9*), it can be engaged or disengaged while the tractor is in motion. Some of these switches are equipped with an automatic position or brake assist position that disengages the MFWD when the tractor is turned or if a brake is applied. The MFWD is re-engaged automatically under certain conditions.

WARNING!

MFWD greatly increases traction. Use extra caution on slopes because MFWD maintains more traction on steeper slopes than just two-wheel drive, increasing the possibility of the tractor tipping over backwards.

2.4.3 Clutch Pedal Operation

The clutch pedal (*Figure 10*) engages and disengages power flow from the engine to a mechanical transmission. The clutch is located on the left side of the operator platform and is depressed to disengage and released to engage. The twin pedals on the right side are brake pedals.

Teaching Tip/Notes

Table 1 Estimated Ground Speeds for Range and Gear Selections

	Forward		Reverse
Range-Gear	1,600/2,400 rpm km/h (mph)	Range-Gear	1,600/2,400 rpm km/h (mph)
A-1	1.16/1.73 (0.72/1.07)	A-1	1.26/1.89 (0.78/1.17)
A-2	1.57/2.36 (0.98/1.47)	A-2	1.72/2.58 (1.07/1.60)
A-3	2.15/3.22 (1.34/2.00)	A-3	2.34/3.52 (1.45/2.19)
A-4	2.88/4.32 (1.79/2.63)	A-4	3.14/4.71 (1.95/2.93)
B-1	3.33/5.00 (2.07/3.11)	B-1	3.64/5.45 (2.26/3.39)
B-2	4.54/6.80 (2.82/4.23)	B-2	4.95/7.42 (3.08/4.61)
B-3	6.19/9.29 (3.85/5.77)	B-3	6.76/10.13 (4.20/6.29)
B-4	8.30/12.44 (5.16/7.73)	B-4	9.05/13.57 (5.62/8.43)
C-1	9.64/14.45 (5.99/8.98)	C-1	10.51/15.77 (6.53/9.80)
C-2	13.12/19.67 (8.15/12.22)	C-2	14.31/21.46 (8.98/13.33)
C-3	17.90/26.86 (11.12/16.69)	C-3	19.53/29.30 (12.14/18.21)
C-4	23.98/35.97 (14.90/22.35)	C-4	26.16/39.24 (16.26/24.38)

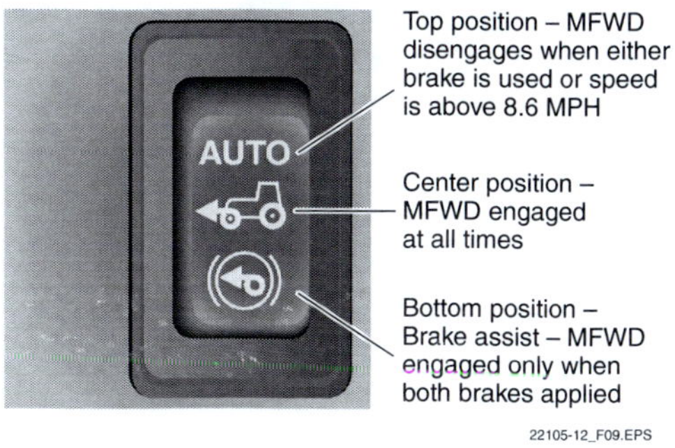

Figure 9 An EH MFWD three-position selector switch.

Tractors with a hydraulic assist transmission have an inching pedal in place of the clutch pedal. The inching pedal is used to help connect attachments, for making emergency stops, or when slow starts are necessary. On tractors equipped with full power shift or a hydrostatic transmission, no clutch or inching pedal is present. On most tractors, the clutch must be cycled once at tractor startup to disengage an engagement override valve. Then the clutch must be depressed again to shift the range selector lever to the desired range, if necessary, and to shift the gearshift lever. To change the range selector lever during operation, always depress the clutch and stop all tractor movement before shifting the range selector lever.

2.4.4 Brake Pedal Operation

Tractors with two-wheel drive, including MFWD tractors, have two brake pedals on the right side of the operator platform (*Figure 10*). The left pedal controls the left rear wheel brake and the right pedal controls the right rear wheel brake. The brake pedals should be depressed simultaneously with the right foot to stop/park the tractor, or individually when making tight turns. For tight right hand turns, the right hand brake pedal can be depressed while steering the tractor to the right. The left hand brake pedal can be used in the same manner when steering to the left. The pedals must be locked together for safe braking when the tractor is driven at speeds over 15 mph. Full-time four-wheel drive tractors have only one brake pedal.

WARNING!

Before operating a tractor on the road, lock the brake pedals together. Use of only one brake can cause the tractor to overturn or turn sharply in one direction or the other. Use reduced speed if a towed load weighs more than the tractor and is not equipped with brakes. Reduce speed and downshift before traveling downhill. Use both brakes.

CAUTION

Make sure to check that MFWD is disengaged or disengages when a turn is attempted.

2.4.5 Differential Lock Operation

Some tractors have an electro-hydraulic differential lock that automatically engages when the rear wheels begin to slip, and disengages when the traction equalizes, a brake is applied, or a turn is attempted. Other tractors have a mechanical engagement lever or a foot pedal that engages the

Classroom

Point out that tractors with two-wheel drive have two brake pedals.

Explain how the brake pedals can be used to aid in steering the tractor.

Safety !

Point out the hazard of using only one brake pedal when traveling on the road.

Classroom

Describe how the differential lock engages when the wheels start to slip.

Teaching Tip/Notes

22105-12 Utility Tractors Module Five 11

Classroom

Discuss the difference between the automatic and manual differential locks.

Audiovisual

Show Slide 16 (Figure 11). Point out the locations of lights on the tractor.

Show Slide 17 (Table 2). Discuss the different switch positions and their functions.

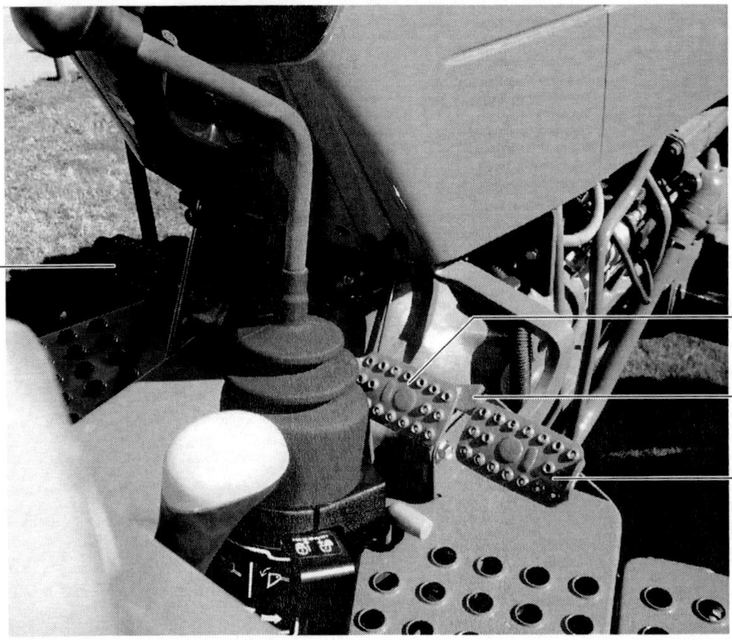

22105-12_F10.EPS

Figure 10 Location of tractor clutch and brake pedals.

lock. Tractors with mechanical locks must not be turned while the lock is engaged. To use a mechanical type of lock, the tractor wheels must be stopped or turning at the same speed before engaging the lock. The lock is engaged by pressing down on the differential lock foot pedal. The pedal remains down and engaged until the traction equalizes or a brake pedal is depressed. Then it disengages itself by spring action. If the wheels repeatedly slip, the pedal can be held down to maintain traction.

CAUTION

Never operate a tractor at high speeds or attempt to turn the tractor with a mechanical differential lock engaged. To prevent drive train damage, do not engage a mechanical differential lock when one rear wheel is turning faster than the other wheel.

2.4.6 Tractor Light Operation

Tractors are required to be equipped with legally suitable lights (*Figure 11*) for any on-road use. In addition, many have front and rear work lights for off-road use.

The lights are controlled from a light switch on the front control console. Most light switches have a number of positions for various operational conditions as shown in *Table 2*. The warning position is for on-road use during daylight hours in clear conditions. The work light position is for off-road use only. The transport positions are for on-road use at night or when visibility is limited. Switching between transports 1 and 2 is used to change the headlights back and forth between the high beams and low beams when meeting other vehicles on the road. When the high beams are on, the high-beam indicator on the instrument panel should also be on. When the warning or transport positions are used, both of the turn signal indicators on the instrument panel will flash, along with amber hazard lights. The hazard-flashing function is overridden while the turn signal is in use.

Teaching Tip/Notes

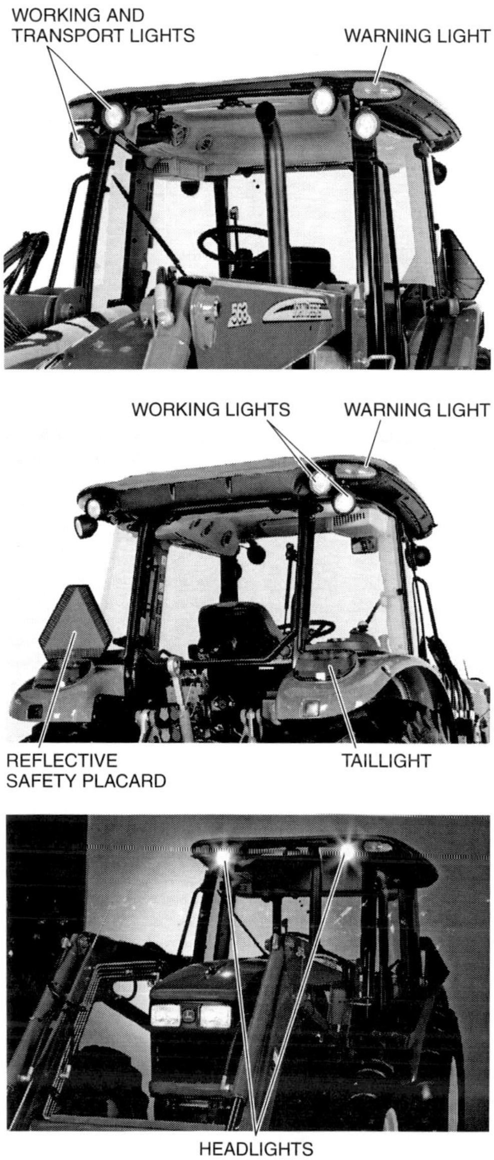

Figure 11 Example of tractor work and road lights.

> **WARNING!**
>
> Because tractors are slow-moving vehicles, always use the warning, or transport, switch positions when traveling on public roads. Never use the work light position because the very bright, clear lights can blind or confuse other drivers as they approach from the front or rear of the tractor.

2.4.7 Cab Heating and Air Conditioning Operation

Tractors with an optional enclosed operator's cab usually have a heating and air conditioning unit as standard equipment in the cab. The heating and air conditioning controls are usually on the operator's right side, or overhead. Because of the wide variation of these units among tractor manufacturers and models, consult the operating manual for the tractor for specific instructions. Keeping the cab pressurized and the blower speed on high will keep dust out of the cab.

2.4.8 Power Takeoff Operation

A PTO supplies mechanical power to certain attachments connected to the rear of the tractor at the three-point hitch or drawbar hitch. The PTO is a splined shaft at the rear of the tractor above the drawbar hitch (*Figure 12*).

Power is coupled to the attachment via a PTO propeller shaft on the attachment that is coupled by a universal joint to the PTO splined shaft at the rear of the tractor. Tractors may have one of the following types of PTOs:

- *Transmission-driven* – When a PTO selector lever is engaged, the tractor driveline transmission clutch operates the transmission and PTO at the same time.
- *Continuous-running* – This type of PTO is controlled by a double-acting tractor driveline transmission clutch. With the clutch fully depressed, both the driveline and the PTO are disengaged. With the clutch released halfway, only the PTO is engaged. With the clutch completely released, both the driveline and PTO are engaged.
- *Independent* – Most tractors use either a mechanical or an electro-hydraulic clutch PTO that operates separately from the driveline transmission and clutch.

Tractors with the optional EH PTO are normally equipped to operate only at 540 rpm. However, an optional 540/1,000 rpm stub shaft conversion kit

Classroom

Explain that heating and air conditioning systems, if provided, vary widely on all makes and models.

Audiovisual

Show Slide 18 (Figure 12). Point out the location of the PTO.

Classroom

Describe the differences in the three types of PTO mechanisms.

Explain how to change rpm on the EH PTO.

Teaching Tip/Notes

Audiovisual

Show Slide 19 (Figure 13). Describe how to swap out a PTO shaft.

Safety

Point out that the existing shaft could be hot if recently used.

Audiovisual

Show Slide 20 (Figure 14).

Table 2 Light Activation in Various Switch Positions

Switch Position	Warning Lights Amber	Tail Lights Red	Work Lights Rear Facing	Work Lights Front Facing	Headlights Front Grille	Auxiliary Front Work Lights (Optional Cab)
Off	Off	Off	Off	Off	Off	Off
Warning	On flashing	Off	Off	Off	Off	Off
Work light	Off	Off	On	On	On - high beams	On
Transport 1	On flashing	On steady	Off	Off	On - high beams	Off
Transport 2	On flashing	On steady	Off	Off	On - low beams	Off

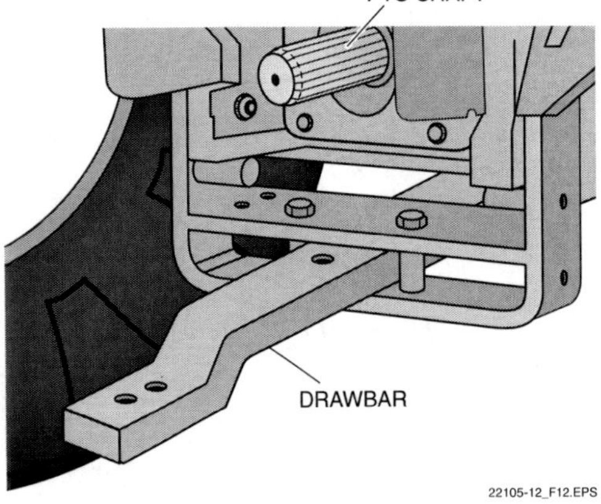

Figure 12 Tractor PTO and drawbar.

is available to allow 1,000 rpm operation. The exposed end of the shaft, when the stub shaft is inserted, has either 6 splines for 540-rpm operation or 21 splines for 1,000-rpm operation. To change the shaft to the alternate speed, first make sure the tractor engine is stopped, the gear selector is in park, and any prior attachment is disconnected from the tractor.

> **WARNING!**
> If a prior attachment powered by the PTO was attached and being used, the shaft may be hot from operation. Wear gloves and allow the stub shaft to cool before changing.

> **NOTE**
> Make sure any PTO-driven attachment attached to the tractor is rated for the power of the tractor and the speed at which the PTO will be operated. Always refer to the attachment manufacturer's specifications or operating manual for this information.

At the PTO housing, align the snap ring with the flats on the stub shaft, remove the snap ring, and pull the shaft out of the PTO housing (*Figure 13*). After cleaning and lubricating with grease, make sure the borehole in the stub shaft is clean. Then, insert the stub shaft into the PTO housing with the desired set of splines exposed. Align the exposed snap ring groove on the shaft with the snap ring groove in the PTO housing, and reinstall the snap ring over the flats on the shaft into both snap ring grooves.

Make sure the desired PTO-driven attachment is properly installed to the tractor drawbar or three-point hitch and the PTO shaft of the attachment is properly connected to the splined shaft on the tractor with all shields in place. *Figure 14* shows a tractor with a PTO-driven attachment connected to the tractor's three-point hitch and its PTO.

When the PTO-driven attachment is properly installed, and while the tractor engine is stopped, follow the instructions in the tractor's operator manual for setting the PTO controls on the operator console of the tractor. After the controls are set, the tractor can be started and the PTO can be engaged. The tractor's engine speed may not exceed the maximum specified for a given PTO

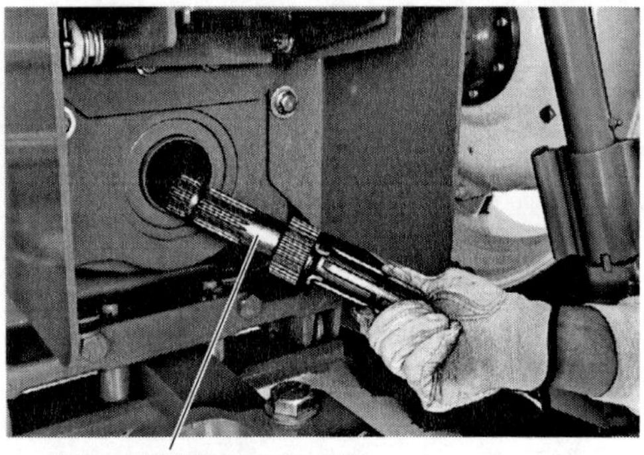

Figure 13 Swapping out a PTO shaft.

Teaching Tip/Notes

Figure 14 PTO-driven attachment on a tractor.

operation. Therefore, the tractor's speed can only be adjusted using transmission range and gear selections.

2.4.9 Hitch Control Operation

A three-point hitch is used to pull as well as raise or lower certain types of attachments attached to a tractor. The attachment procedures for a three-point hitch are outlined later in this module. The components of a tractor with a three-point hitch are shown in *Figure 15*.

> **NOTE**
> The parenthetical letters in the following paragraph are for callouts in *Figure 15*.

In *Figure 15*, the attachment is connected to the **draft** links (E), which impart the pulling power of the tractor and establish the attachment's operating height above or below ground level (height/depth or draft). To accomplish this, the attachment's topmast must be connected to the center link (D). The triangular attachment configuration, consisting of the two draft links and the center link, is normally adjusted so that the attachment is level behind the tractor. During tractor operation, the rockshaft lift arms (A), powered by hydraulic cylinders and connected to lift links (B) are used to raise or lower the draft links (E) which, in turn, raise or lower the attachment. For some applications, the center link is used to sense the load on the attachment and automatically adjust the attachment's operating height for varying terrain conditions, depending on the rockshaft control settings. The sway bars (C) provide lateral stability to the draft links (E). Other manufacturer's tractors may use other methods of providing lateral stability, including the use of chains in place of the bars.

> **NOTE**
> Since this training manual cannot cover the operational details of the hitch lifts on all tractors, operators must follow the instructions available in the operator manual associated with the tractor being used.

Audiovisual

Show Slide 21 (Figure 15). Use the callouts on the slide to identify the components of a three-point hitch.

Classroom

Explain the importance of following the manufacturer's instruction for a particular tractor.

Teaching Tip/Notes

Safety

Identify the pinch points on a tractor hitch.

Classroom

Explain how a typical tractor hitch is controlled.

Discuss the relationship between the rockshaft and the hitch control lever.

Explain how to use external raise/lower switches on an EH hitch.

Safety

Discuss safety considerations when operating the external raise/lower switches on an EH hitch.

Classroom

Describe the functions of the height limit control and the depth stop wheel.

Describe the functions of the rate-of-drop control, transport lock position, and EH load/height controls on an EH hitch.

Figure 15 Three-point hitch components.

> **WARNING!** A tractor hitch lifts and controls the attachments connected to the tractor. The movements of the lifting devices can create pinch points that can cause serious personal injury. Review all operational warnings listed in the operator manual for the tractor and attachment being used.

For the hitch controls to work, the tractor's engine must be running. The rockshaft arms that move the hitch are controlled by hitch controls located near the operator's seat. The hitch control lever is used to raise and lower the hitch and to set working height or depth of the attachment. The lever is pushed forward to lower and rearward to raise the attachment. Most tractors also have a raise/lower switch that can be used independently of the hitch control lever to raise or lower the hitch between the settings of the hitch control lever and height-limit control. This helps when turning at the end of an operation. The settings of these controls and others are explained in the following paragraphs:

- *External raise/lower switches* – With the transmission gear selector lever in park, pressing and holding the raise or lower switches (usually mounted on the left fender of a tractor) allows for rockshaft control outside of the cab. The rockshaft raises or lowers the hitch at a very slow rate. Any height limit and working height (depth) settings are ignored. The switches are disabled when the hitch control lever in the cab is in the transport lock position.

> **WARNING!** To prevent personal injury or damage caused by tractor movement, make sure the tractor transmission is in park position before using the external raise or lower switches. Stay clear of any pinch points when using the switches, and make sure that the attachment does not interfere with the tractor's fenders.

> **NOTE** Once any external raise/lower switch is pressed, the rockshaft is prevented from moving accidentally. To make the rockshaft ready for operation again, the hitch control lever in the cab must be moved to a position that corresponds to the position of the draft links. Then, raise/lower switch in the cab must be actuated.

- *Height limit control* – Turning the height limit control counterclockwise reduces maximum hitch height; clockwise increases maximum height. For the initial setting with an attachment connected, the control should be turned clockwise to the maximum position. The hitch should be raised with the hitch control lever to a desired maximum height setting without interference to the tractor. Then, the control should be turned counterclockwise until the hitch just begins to lower.
- *Depth stop wheel* – Turning the depth stop wheel control sets a mechanical stop at the desired working height (depth) of the attachment. After setting the working height (depth) of the attachment with the hitch control lever, press and rotate the depth stop wheel to move the mechanical stop up to the forward edge of the hitch control lever. Then, if the hitch control lever is moved rearward to raise the hitch for some reason, it can be moved forward to the stop, which will return the hitch to the preset working height (depth). If it becomes necessary to lower the hitch below the preset height (depth) for some reason, the hitch control lever can be lifted and pushed past the stop.
- *Rate-of-drop control* – Turning the rate-of-drop control counterclockwise slows the rate of drop of the hitch; clockwise increases the rate of drop. For any one setting, heavy attachments

Teaching Tip/Notes

will drop faster and light attachments will drop slower. By varying the setting, a satisfactory rate of drop of two seconds or more can be achieved for the weight of any attachment rated for use with the size and power of the tractor.

> **WARNING!** Excessive drop speed may cause personal injury or damage to the tractor and attachment. Fully lowering the attachment should require at least two seconds.

- *Transport lock position* – Pulling the hitch control lever back to the highest setting and placing the lever in the lock slot raises the hitch to its maximum height limit control setting and disables all raise/lower switches. This prevents the hitch from lowering during transport. The hitch automatically rises when starting the tractor if the hitch control lever is in the transport lock position.

> **WARNING!** To prevent personal injury when starting the tractor, make sure no one is near the hitch or any attachment attached to the hitch. To prevent personal injury or equipment damage during transport, make sure the hitch control lever is in the transport lock position.

- *Load/height (depth) control* – The load/height (depth) control can be rotated to a number of positions for automatic adjustment of the working draft (height/depth) of the attachment due to load resistance and terrain contour variations. Turning the control counterclockwise reduces draft response; clockwise increases draft response. The higher the draft control setting number, the more the attachment is raised and lowered due to load resistance variations. The draft settings of the control are used primarily for subsurface attachments, such as a ripper. They are also used to maintain tractor speed and load, and to maintain, as much as possible, the working depth of the attachment that was set by the height control lever. Rotation of the control fully counterclockwise to the position control detent holds an attachment, such as a scraper, at a previously set fixed-height position, regardless of the attachment loading. This setting, along with a minimum setting of the height control lever (fully forward), is also used for float control of a three-point mounted attachment with gauge wheels or skid pads so that the attachment freely follows any terrain contours.

> **WARNING!** When connecting or disconnecting an attachment to a tractor's hitch system, turn the load/depth control fully counterclockwise to the position control detent before attachment or detachment. Use only the hitch control lever or, if provided, the finder-mounted raise/lower switches when connecting or disconnecting an attachment. Do not use the raise/lower switch on the side control console hitch controls.

2.4.10 Selectable Control Valve (SCV) Operation

Most newer tractors are equipped with one or more SCVs to operate attachments with hydraulic motors and single- or double-action hydraulic cylinders. They can range from valves with basic hydraulic cylinder extend/retract functions to optional valves with enhanced operational capabilities. *Figure 16* shows a tractor equipped with a set of SCV couplers. Tractors may have multiple sets of SCV couplers, depending upon the number of functions needed on the attachments they are pulling. A set of SCV control levers are located near the operator's seat, but some tractors may also have an additional set of control levers mounted near the SCV couplers at the rear of the tractor.

Most SCV control levers, including the enhanced SCV control levers, operate the same. Each control lever usually has four operating positions:

SCV COUPLER TO EXTEND ATTACHMENT CYLINDER(S)

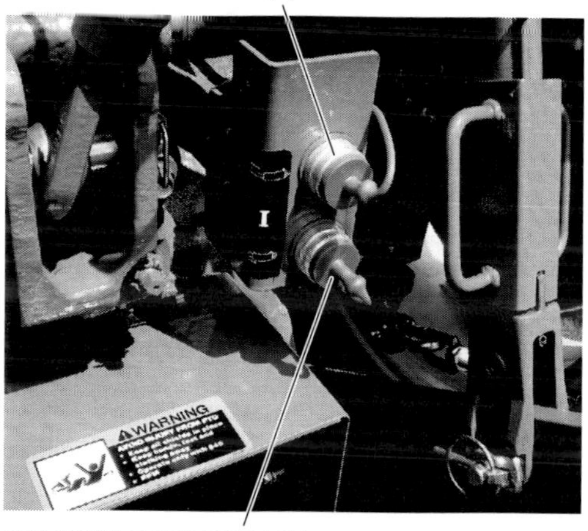

SCV COUPLER TO RETRACT ATTACHMENT CYLINDER(S)

22105-12_F16.EPS

Figure 16 Tractor with SCV couplers.

Safety !

Discuss safety considerations when operating the rate-of-drop control, transport lock position, and EH load/height controls on an EH hitch.

Audiovisual

Show Slide 22 (Figure 16). Describe the function of selectable control valves.

Classroom

Discuss the four operating positions of an SCV control lever.

Teaching Tip/Notes

Classroom

Explain that the float detent mode allows an attachment to follow ground contours.

Discuss the four operating positions of an ignition switch: accessory, stop, run, and start.

Explain that there are different types of ignition switches.

Homework

Have trainees review Sections 3.0.0–3.6.0.

Materials

Ensure that you have everything required for teaching this session.

- *Neutral* – The control is spring loaded to the neutral position, which is in the middle of the movement range for the lever. In the neutral position, hydraulic oil pressure is equalized for both couplers of the SCV, which hydraulically locks a cylinder in a fixed position.
- *Extend* – When the spring-loaded control lever is moved rearward from neutral, hydraulic oil is directed out of the top extend coupler of the SCV to the hydraulic cylinder at a rate that is proportional to the distance the lever is moved. This allows for varying the rate of cylinder extension. For double-acting cylinders, the oil in the cylinder used for retract is forced out of the cylinder through the depressurized bottom retract coupler. The top extend coupler is used for both single-action and double-action cylinders and as the fluid return line for hydraulic motors. When used for hydraulic cylinder operation, the lever should not be held in the extend position if the cylinder reaches its end of **travel**. Overheating of the hydraulic oil can occur.
- *Retract* – When the lever is moved forward from neutral, hydraulic oil is directed out of the bottom retract coupler of the SCV to a double-acting hydraulic cylinder at a rate that is proportional to the distance the lever is moved. This allows varying the rate of cylinder retraction. For double-acting cylinders, the oil in the extend cylinder is forced out of the cylinder through the depressurized top extend coupler. The bottom retract coupler is used only for double-acting cylinders and, along with an adjustable flow rate valve, as the pressure source for hydraulic motors. For single-acting cylinders, the bottom retract coupler is blocked by an internal check valve and the top extend coupler is depressurized when the lever is moved to a retract position. This allows the gravity of the load on the single-acting cylinder to push oil back through the top extend coupler in proportion to the distance the lever is moved forward. When used for double-acting cylinder operation, the lever should not be held in the retract position if the cylinder reaches its end of travel. Overheating of the hydraulic oil can occur.
- *Float detent* – When the lever is pushed all the way forward through retract to a detent position, both couplers are depressurized and the hydraulic cylinder operates in a float mode that allows the cylinder to extend and retract freely. This mode is used to allow an attachment, such as a loader, to follow ground contours or to stop operation of a hydraulic motor. The control lever must never be placed in the float detent position when any attachment operation involves loads suspended in the air; doing so will cause the loads to immediately fall to the ground.

2.4.11 Ignition Switch Operation

The types and functions of ignition switches can vary widely between makes and models of tractors. Some only activate the starter and ignition system for gasoline engines. Others may activate fuel pumps, fuel valves, and starting aids for diesel engines as well as the starter. In some cases, diesel fuel valves and starting aids are engaged by other manual controls. For specific instructions, refer to the operator manual for the tractor being used. The multipurpose ignition switch used on most tractors is located on the front control console. It normally has the following four positions:

- *Accessory* – Pushing the key in and turning it to the accessory position applies power to an accessory power outlet located in the rear console or to devices such as radios that may be connected to accessory power. Other than a battery gauge or indicator, none of the other electrical controls is activated.
- *Stop* – Turning the key to the stop position stops the low-pressure primer fuel pump that feeds the mechanically driven high-pressure injector pump and closes the electrical fuel valve that stops the engine. It also shuts off all other electrical power in the tractor including accessory power.
- *Run* – Turning the key to the run position starts the low-pressure primer fuel pump, opens the electrical fuel valve that enables starting and running the engine, and initiates a momentary instrument panel indicator bulb check. If air temperature is below a given temperature such as 5°C (40°F), the run position is used to activate cold weather starting devices, such as glow plugs or intake air warmers. This is accomplished by pushing the key in, holding it in for a period specified by the tractor manufacturer, and then turning to the start position. The key returns to the run position when it is released at engine startup.
- *Start* – The key is turned to the start position to crank the engine and high-pressure injector pump, which starts the engine. This position is spring loaded to return to the run position when the key is released. To reduce battery load during starting, the ignition switch of some tractors may be configured to shut off power to accessories and lights when the key is in the start position.

Teaching Tip/Notes

> **CAUTION**
> The average time allowed to crank most tractor engines is around 20 seconds. Attempting to crank longer than that can damage the starter. Always allow two minutes of cooling time before trying to start the engine again.

3.0.0 Safety Guidelines

Safety should be the number one priority of the operator. Working safely involves knowing the proper operation of the machine being used, knowing job-site rules and regulations, and being aware of nearby workers.

3.1.0 Safety Rules for Operating Tractors

Keep the following safety rules in mind when operating tractors:

- Read the operator's manual thoroughly and know the machine. If you do not understand something, ask someone, such as a dealer or factory representative, to explain it.
- Study the company's safety manual.
- Wear close-fitting clothing.
- Wear ear protection if needed.
- Provide for ventilation.
- Become familiar with controls. Be sure to know how to stop the tractor and engine.
- Do not let another person on or near the tractor when it is in operation. With rotating attachments, shut down the tractor if a person or animal approaches.

> **NOTE**
> Some tractor cabs have instructor seats, which must be equipped with a seat belt.

- Remove or tie down loose tools and equipment.
- Keep seat belt fastened.
- Stay seated during operation.
- Use all lights and safety warning devices.
- Check brakes.
- Disengage PTO when not operating an attachment.
- Do not smoke while working on the tractor, handling fuel or solvents, or checking or filling batteries. Hydrogen gas buildup from the battery may ignite.
- Never use attachments that are not matched to tractor size, weight, power, PTO speed, and drawbar or three-point hitch capacity.
- Make sure all safety devices, guards, shields, and safety decals or signs are properly installed as specified by the manufacturer's manual. Tractors should have ROPS equipment or a ROPS cab. PTO shields must be in place for the tractor and attachment shafts.
- Observe and follow all safety signs or decals, and use any steps or handholds to mount tractor.
- Make sure front and rear frame counterweights, wheel weights, and wheel ballasts are used as recommended by the manufacturer for the attachment being used. Never add extra counterweights to compensate for an overload.

3.2.0 Safety Rules for Effective Tractor Operation

Use the following tips to operate the tractor efficiently:

- Perform prestart inspection and maintenance procedures each day before starting the tractor.
- Adjust the seat and steering wheel, and fasten seat belt before starting the tractor.
- When starting the tractor, follow the manufacturer recommended starting procedure.
- After first starting a tractor, quickly check that the controls function properly before moving the tractor.
- Before moving a tractor, make sure that adequate clearance exists in all directions.
- When first moving the tractor, make sure that the steering works both left and right and the brakes function properly.
- Make sure that any attachments used do not exceed the load rating of the tractor. Any attachments using a PTO must match the PTO speeds available on the tractor. Attachments using the SCV capabilities of a tractor must not exceed the tractor rating for this service.
- Keep all parts of your body inside the operator's compartment when operating the tractor.
- To prevent the tractor from tipping over backwards, do not lock both rear wheels with the brakes at the same time while the clutch is engaged and the tractor is in a low range and gear with the engine at working speed. If both rear wheels become mired (stuck) or frozen and do not move and rotate, do not attempt to move forward. Always use reverse gear to free mired or frozen wheels. Also, make sure that any towed load is attached below the bottom of the tractor's rear axle.
- Keep speed down.
- Avoid sudden starts, stops, and turns.
- Slow down before a turn. Some tractors will tip over at approximately 8 mph when turning on level ground.

Classroom

Review the listed safety points.

Explain how following safety rules and procedures can increase operating efficiency.

Review the list of safety rules for effective tractor operation.

Teaching Tip/Notes

Provide a copy of the Association of Equipment Manufacturers (AEM) tractor safety manual or similar manual for the trainees to examine.

Ask the trainees to list ways that performing required operator preventive maintenance can improve operating safety.

Classroom

Describe the particular hazards of operating a tractor on a slope or hill.

Review the list of rules for operating a tractor on a slope or hill.

Describe the hazards associated with hydraulic systems.

Review the list of safety guidelines for working with hydraulic systems.

Point out that hydraulic fluid is under high pressure and can penetrate the skin.

Explain that pressure must always be relieved before disconnecting hoses or otherwise opening the hydraulic system.

- Take care when entering the highway. Consider the condition of the road, the tractor, the traffic, and the weather. As speed doubles, the chances of a tractor tipping over becomes four times as great. Always use extreme caution.
- Signal all turns and stops.
- Travel completely on the shoulder whenever possible. Avoid driving half on the road and half on the shoulder.
- Depending on the soil conditions, stay at least 4 to 10 feet from the edges of deep ditches, embankments, and excavations.
- Do not touch electrical wires with any part of the tractor.
- When traveling forward and turning with a towed attachment connected to the rear of the tractor, make sure that it does not come in contact and ride up the tractor tire that is on the inside of the turn. In reverse, make sure that the tire does not ride up on and crush the attachment. If turning with an attachment connected to the three-point hitch, make sure that it will not strike anything on the outside of the turn.
- If possible, always park in a non-operating or designated area and on level ground. Check the ground to make sure it is firm. If parking on a grade, position the tractor at a right angle to the slope and block the wheels.
- Before dismounting a tractor, disengage the PTO, lower any attachments to the ground, and retract any hydraulic cylinders, especially single-acting cylinders, as much as possible. Then, place the tractor in park (or neutral) and/or set the brakes. Shut off the engine and cycle all the SCV control levers through all positions to relieve hydraulic pressure, then return the levers to the neutral position. If the tractor is to be left unattended, remove the ignition key before dismounting. To help prevent water condensation, make sure the fuel tank is full if the tractor will be parked overnight.
- Follow the company's preventive maintenance program.
- When transporting a tractor, make sure it is properly secured, with the brakes locked. Cover the rain cap to protect the turbocharger. If the rain cap is uncovered, wind will turn the blades and overheat the turbocharger unit.

3.3.0 Safety Rules for Operating a Tractor on Slopes or Hills

Follow these guidelines when operating on slopes or hills:

- Avoid working on slopes and hills. If possible, avoid driving up hills.
- Always reduce speed on slopes or hills.
- If forced to move forward out of a ditch, gully, or up a steep slope, use the lowest gear possible and engage the clutch very slowly. Be prepared to disengage the clutch immediately if the front wheels rise off the ground.
- Avoid driving forward out of deep ditches or gullies. Rearward tip-over can occur in less than a second. Always back out if possible.
- Go up steep slopes in reverse.
- If work on a slope is unavoidable, go up and down the slope, perpendicular to the toe of the slope.
- Do not go downhill with heavy loads.
- Keep the tractor in gear when going downhill. Use the same gear used for traveling uphill.
- Keep all raised equipment close to the ground when working on slopes.
- Make sure the heavy end of the tractor faces uphill.
- Do not cross slopes of more than 25 degrees. One rule of thumb is that most tractors will not tip over sideways on a slope during straight line travel if a vertical line from the side-to-side center of the top of a cab or ROPS falls inside the downhill rear tire.
- When crossing steep slopes, avoid turning uphill. If it is unavoidable, slow down and make a wide turn. Also, avoid any depressions with the downhill tires and any bumps, large rocks, stumps, ledges, or raised areas with the uphill tires.

3.4.0 Safety Rules for Operating the Hydraulic System

Follow these guidelines when working with hydraulic systems:

- Use caution when working with the hydraulic system.
- Wear gloves and eye protection when connecting, disconnecting, or checking any high-pressure hoses. High-pressure fluid from pinhole leaks is usually not visible and can easily penetrate exposed skin. If an accident occurs, fluid injected into the skin must be surgically removed within a few hours or an infection may result. When checking for pinhole leaks in a pressurized hydraulic system, use a flat piece of wood as a search tool. Also, look for a stain below the suspected line, or a ring of dirt on the line.
- Before disconnecting high-pressure hydraulic hoses, always relieve pressure as much as possible by first lowering attachments and loads. Then, retract any cylinders, especially single-

Teaching Tip/Notes

There are numerous resources on tractor safety available online. As a homework assignment, consider having the trainees research and report on tractor accidents and their avoidance. Include what manufacturers have done and are doing to improve tractor safety.

acting cylinders, as much as possible. Shut off the tractor engine and cycle the SCV control levers through all positions to relieve remaining pressure, and then return them to the neutral position. Before connecting hoses, make sure the tractor engine is stopped and the appropriate SCV control levers are in the float detent position if connecting hydraulic motors, or in the neutral position if connecting hydraulic cylinders.
- When adjusting hydraulic motor speeds, never use the SCV control lever; always use an adjustable flow control valve with the control lever in full retract position. Always adjust the motor to its rated speed or lower. Never place an SCV control lever in neutral with a hydraulic motor connected to its couplers.
- Never place the SCV control levers in the float detent with a load suspended from an attachment. The load will immediately fall to the ground. Always lower the load first.
- Use of single-acting cylinders lowers the hydraulic fluid level in the hydraulic system reservoir. Always check that the hydraulic fluid level does not fall below the minimum level with all single-acting cylinders in the fully extended position.
- Wear gloves and use a blotting material, such as cardboard or rags, when checking hydraulic lines for leaks.
- Stand clear of equipment when working with hydraulic lines and tools.
- Keep the hydraulic system clean. Always use dust caps on quick-couple disconnects when the hydraulic system is not in use. Wipe dirt off of hoses and connectors before attempting to connect the hoses.
- Use the proper fittings for hoses and attachments.
- Never depend on the hydraulic system to support machine parts. Block or jack up rotary wings; if they are raised to a vertical position, secure them with chains or fastening rods.
- Avoid prolonged contact with hydraulic fluid. Hydraulic fluid is a skin irritant.

3.5.0 Safety Rules for Operating the Power Takeoff

Tractors have 540 rpm PTO shafts, 1,000 rpm PTO shafts, or both. Make sure the attachment is compatible with the PTO. Some tractors have interchangeable stub shafts and are equipped with a selector lever to provide the different speeds. With a selector lever, always be sure the selector is set for appropriate rpms. Otherwise, engine

On Site

PTO Safety

Every year, serious and often fatal accidents occur when a worker gets too close to an engaged PTO. Working around a moving PTO is very dangerous because it can easily catch loose clothing or hair. The PTO rotates very rapidly and has very high torque. A person can get wrapped around a PTO so quickly that there is no time to react.

damage and serious personal injury can occur. Always keep the following safety guidelines in mind when operating the PTO:

- Keep PTO guards and shields in place, even when the PTO is not running.
- Always disconnect the PTO when not in use. Failure to disconnect the PTO when not in use accounts for many operator injuries.
- Never engage the PTO while the machine engine is shut off.
- Keep hands, feet, and clothing away from the PTO parts.
- Never step or cross over a PTO part.
- Be sure that clothing fits snugly.
- Never operate the PTO shafts at extreme angles.
- Never ride, or let others ride, on the drawbar of the tractor.
- Be sure the PTO spinner shields (*Figure 17*) rotate freely at all times. Disengage all power and shut off the tractor engine before checking spinner shields.
- Always be sure the PTO stub shaft is properly secured to the machine's power shaft.

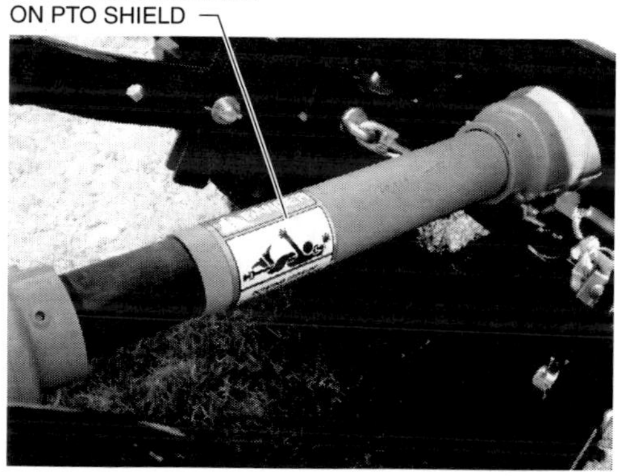

Figure 17 PTO warning decal.

Classroom

Point out that hydraulic fluid is a skin irritant.

Point out hazards associated with a power takeoff.

Review the list of safety guidelines for working with a power takeoff.

Audiovisual

Show Slide 23 (Figure 17). Point out the importance of making sure that spinner shields are present and that they rotate freely.

Teaching Tip/Notes

Ask the trainees to list different ways of relieving hydraulic pressure.

22105-12 Utility Tractors Module Five 21

Safety

Explain that tractor tires can blow out with explosive force.

Advise the trainees to stand to one side when inflating a tire.

Classroom

Describe how to determine what the tire pressure should be and how check it.

Homework

Have trainees review Sections 4.0.0–4.4.0.

Materials

Ensure that you have everything required for teaching this session.

Audiovisual

Show Slide 24 (Figure 18). Describe the types of attachments used on tractors in construction work.

Classroom

Explain that attachments, such as the bush hog and grader blade, are driven off the PTO.

Audiovisual

Show Slide 25 (Figure 19). Describe how the drawbar length is adjusted and how the drawbar is aligned.

- Do not service, lubricate, or perform other work without first disengaging the PTO and shutting off the tractor engine.

3.6.0 Safety Rules for Tractor Tires

An inflated tire can be very dangerous. Under pressure, tires can burst with explosive force. Always follow the manufacturer's instructions for tire inflation pressures and tire changing procedures along with the following rules:

- Check tires daily for damage or noticeably low pressure.
- Check inflation pressure with an accurate gauge to prevent overinflation. Overinflation reduces performance and increases the strain on both the tire and wheel rim, which can cause tire or rim failure.
- Inflation pressures of less than 12 psi should be monitored regularly because the increased risk of tire **bead** leaks can result in tire damage.
- If the tires wrinkle, buckle, or slip on the rim under high-traction conditions, the tire pressure can be temporarily increased by 4 psi to alleviate the problem, but traction will be reduced.
- Tire pressures of tractors operating on side slopes or hills should be increased by 4 psi to compensate for weight transfer from side-to-side or front-to-back and back-to-front.
- When inflating a tire, never stand directly in front of the tire. Stand to one side at the tire tread.
- Stand clear of the tire and rim if it is being lifted with a chain or cable sling during a tire changing procedure.
- Never weld or perform torch work on wheels with inflated tires.

4.0.0 Hitches and Attachments

Attachments for the tractor are normally connected at the rear. The attachment can either be pulled using a drawbar or carried using a three-point hitch. Different manufacturers have different setups for their hitches. Be sure to read the operator's manual before trying to connect an attachment. Sometimes special tools or connectors are required.

4.1.0 Types of Hitches and Attachments

Figure 18 shows various types of attachments used for tractors. Trailers or attachments with gauge wheels and their own manual or hydraulic adjustments are usually towed using a drawbar. Some construction attachments, such as grader blades, rotary cutters, and counterweights, are carried by the rear three-point hitch at the rear of the tractors.

Finish grading can be accomplished with the adjustable rear grader blade. Such blades can be used with full hydraulic or manual adjustment for tilt, angle, and offset as well as an optional tail wheel to aid in leveling. Grader blades come in widths from 83 to 119 inches.

Box scrapers are used for leveling, grading, material removal, and backfilling. The model shown in *Figure 18* can be equipped with scarifiers that rip up and loosen soil ahead of the scraper. Box scrapers for utility tractors come in capacities of 15 to 28 cubic feet and widths varying from 60 to 84 inches.

The rotary cutter, often called a bush hog, is driven from the tractor's PTO. Some augers are pushed down by hydraulic cylinders mounted on the tractor's lift arms. Such hydraulic cylinders are controlled through the tractor's SCV controls. Such augers are capable of drilling straight holes through solid rock, large rocks and dirt, or large tree roots and dirt/rocks with augers up to 18 inches in diameter and 48 inches long.

A counterweight can be attached to the three-point hitch when a tractor is equipped with a front end loader bucket, as shown in *Figure 18*.

4.2.0 Drawbar Adjustments

When attachments are to be towed behind a tractor and powered by its PTO, the drawbar must be adjusted to the proper position. The following are the various adjustments that can be made to the drawbar of a tractor:

- *Length* – The drawbar can be used in the short position or the extended position (*Figure 19*). The short position is for PTO equipment rotating at a lower rpm (around 540 rpm). The positioning of the hitching point is measured from the output end of the PTO shaft. For the lower rpm equipment, the hitching point is 14 inches from the end of the PTO shaft. The 16-inch extended position is usually used for PTO equipment rotating at 1,000 rpm. In both positions, the drawbar must be centered with the PTO shaft. On some models, length adjustment is made under the tractor body.
- *Height* – Maintain the proper drawbar height and length to prevent the tractor from upsetting and to protect the PTO shaft. Leave 13 to 20 inches ground clearance, measured from the top of the drawbar. Height can be adjusted by

Teaching Tip/Notes

Figure 18 Different tractor attachments.

22105-12 Utility Tractors Module Five 23

Teaching Tip/Notes

Audiovisual

Show Slide 26 (Figure 20). Explain that drawbar height can be changed by inverting the drawbar.

Show Slide 27 (Figure 21). Explain that the hitch can be pinned in an off-center position for offset operations.

Show Slide 28 (Figure 22). Explain how to connect an attachment to a drawbar.

Safety

Explain that connecting attachments can be dangerous. Discuss common hazards and how to avoid them.

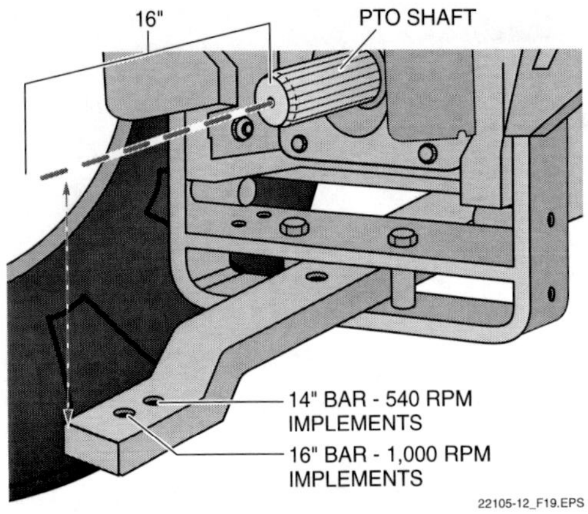

Figure 19 Drawbar length and alignment.

inverting the drawbar. *Figure 20* shows drawbars in the lowered and raised positions.

- *Lateral adjustment* – Pinning the hitch in one of the swinging positions is for offset operations. The center fixed position is used for transporting (*Figure 21*).

4.3.0 Connecting an Attachment to the Drawbar

Some attachments are pulled by their tongue. They must be fastened to the tractor's drawbar. Adjust the drawbar height from 13 to 20 inches clearance between ground level and drawbar hitch point (*Figure 22*).

Be sure to match the attachment to the tractor power. Be very careful performing this operation. Accidents are common. Avoid injuries from the following possible hazards:

- *Lifting* – Know the proper way to lift heavy objects.
- *Cutting* – Wear gloves when handling sharp objects.
- *Pinching* – Use caution when connecting and disconnecting joints; keep hands and fingers out of joints.
- *Crushing* – Keep feet out from under equipment.

Read the operator's manual thoroughly and perform one or more of the following actions:

- Turn the drawbar over.
- Adjust the drawbar bracket.
- Adjust the drawbar frame assembly where it bolts to the tractor frame.

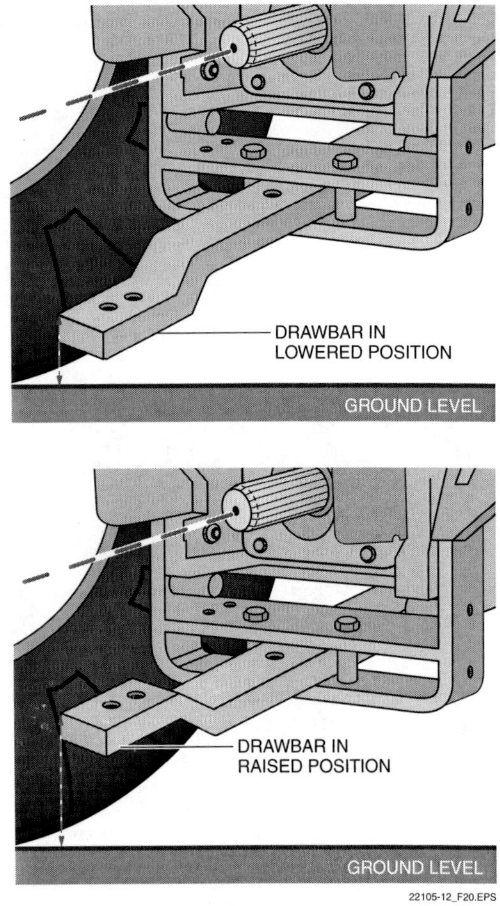

Figure 20 Drawbar position options.

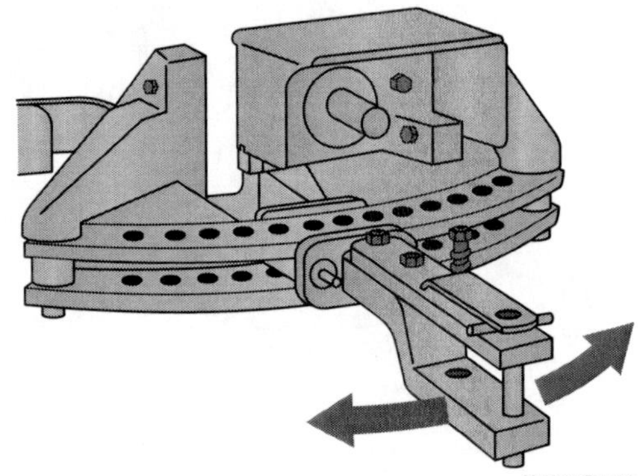

Figure 21 Lateral drawbar adjustment.

Teaching Tip/Notes

24 NCCER – *Heavy Equipment Operations Level One* 22105-12

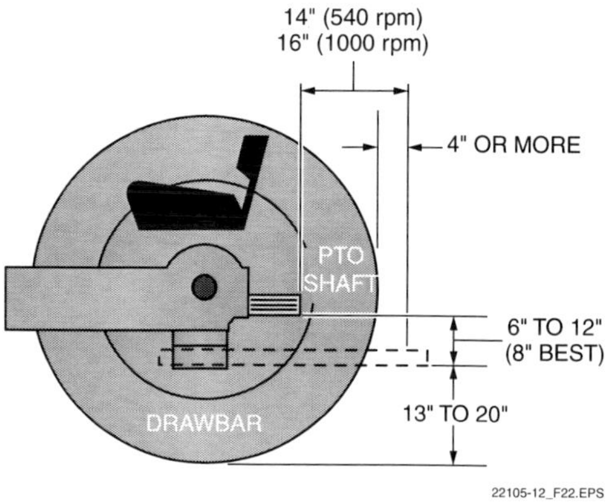

Figure 22 Connecting an attachment to the drawbar.

The final vertical distance between the drawbar hitch point and the PTO shaft should be about 8 inches.

If the hitch will be used with an attachment that must be offset, extend the hitch. If necessary, move it laterally, and secure it into position.

To use the hitch for a PTO attachment, adjust the hitch as follows:

- For power takeoff of 540 rpm, the hitch point must be 14 inches from the end of the PTO shaft.
- For power takeoff speed of 1,000 rpm, the hitch point must be 16 inches from the end of the PTO shaft. *Figure 23* shows the extended position hitch set-up.

If the hitch is used for a PTO attachment, center the drawbar hitch point directly under PTO shaft and fasten it securely. The vertical distance between the PTO shaft and the hitch point should be 6 to 12 inches, ideally 8 inches. Back the tractor into position so that the hole in the drawbar is in line with the hole in the attachment hitch. If the machine is on sloping ground, set and lock the tractor brakes. Then, set the hitch pin in place.

WARNING! Do not try to position the tractor by operating the transmission controls from the ground. You may get pinned between the tractor and the attachment.

If the hitch pin cannot be inserted by moving the attachment, get back on the tractor and realign the drawbar with the holes in the hitch. Be sure to

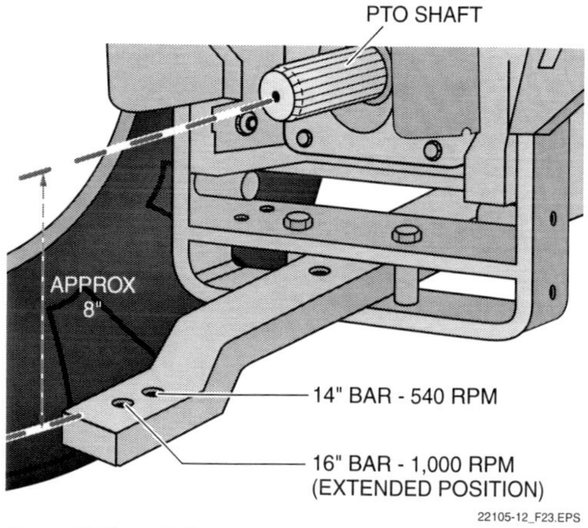

Figure 23 Extended position.

use a cotter pin or some other means of holding the hitch pin in place so that it does not work itself out. Secure the cotter pin or other device. Make the electrical or hydraulic connections if needed.

4.4.0 Ballasting a Tractor

Some tractors may require wheel weights to prevent wheel slip when towing heavy loads. MFWD tractors usually require front end weights or front wheel weights as well as rear tire weights to increase traction. For heavy attachments carried on a three-point hitch, weights may have to be added to the front of the tractor to act as counterweights for the attachment. This counterweight is necessary to prevent tractor tip-up when the attachment is lifted and transported, especially up hills. To see where and how to install ballasting weights on the tractor being used, refer to its operator manual. Make sure to securely fasten the weights.

5.0.0 Basic Preventive Maintenance and Operation

Preventive maintenance of the equipment is important. Each day before beginning operation of the tractor, perform a walk-around inspection and any prescribed daily maintenance checks.

5.1.0 Servicing a Tractor

Servicing a tractor involves checking the operating components for safe operation and providing any service materials that are needed. The follow-

Audiovisual

Show Slide 29 (Figure 23). Explain how to connect an attachment to an extended drawbar.

Classroom

Explain how to install a PTO-driven attachment on a hitch.

Safety !

Explain that the tractor must not be positioned by operating the transmission controls from the ground.

Classroom

Explain why wheel weights are used.

Point out that information on installing wheel weights should be obtained from the operator's manual.

Homework

Have trainees review Sections 5.0.0–5.8.2.

Materials

Ensure that you have everything required for teaching this session.

Classroom

Stress the importance of performing all operator maintenance in accordance with the established schedule.

Teaching Tip/Notes

Audiovisual

Show Slide 30 (Figure 24). Discuss the types of maintenance checks required on a daily basis.

Show Slide 31 (Figure 25). Describe how to properly check engine oil level. Stress the importance of using only specified oil.

Classroom

Point out that transmission oil is checked while the engine is hot.

Explain that daily inspections and checks must be performed before the tractor is started.

ing equipment and materials may be needed to service a tractor:

- Operator's manual
- Tool kit
- Air supply
- Grease gun
- Lubricants
- Cleaning solvents
- Rags
- Water
- Engine coolant
- Engine oil
- Hydraulic fluid
- Transmission fluid
- Tire pressure gauge
- Fuel

5.1.1 Typical Daily Inspection and Maintenance Procedures

The following basic service requirements are given as an example only. Follow the company's preventive maintenance program when servicing equipment. *Figure 24* shows a portion of a typical service interval chart for a tractor. Other manufacturer's recommendations may differ significantly from this example.

NOTE

The location of the filters and points where the liquids can be checked or replenished vary from one tractor to another. Refer to the operator manual for each machine to determine where these points are located, and at what levels the liquids should be. Also, check the operator manual to see what liquids can be used to replenish any oils or coolants.

In addition, attachments connected to a tractor usually have periodic service requirements that must be performed. Always refer to the tractor and attachment operator's manual for the lubrication chart and service for specific components.

- Visually inspect for leaks and broken, worn, or missing parts. Major repairs and replacements should be reported to the proper authority and completed by a trained mechanic.
- Check the engine oil level (*Figure 25*). Add or change engine oil if necessary.

CAUTION

Make sure that only an approved oil is added. Refer to the operator manual or company guidelines to determine the oil needed.

GOING GREEN

Fluid Disposal

To protect the environment, always contain and properly dispose of any leaking fluids or fluids removed from a tractor. If parts are being removed from a tractor, make sure to drain and contain any residual fluids in the part(s) being removed.

- Check the engine coolant level. Add coolant if needed. If the engine has a radiator cap, do not remove it unless the engine is cool; open the cap slowly to release pressure.
- Clean the air filter and replace as needed.
- Check the fuel level and drain the water and sediment from the fuel tank and filter when applicable. Fill the fuel tank as needed.
- Check the hydraulic fluid level when hot and add fluid when necessary. Also, check the filters, hose, lines, and fittings and look for leaks.
- Check the transmission fluid level when hot. Add fluid if needed.
- Clean and lubricate grease fittings according to operator's manual.
- Check the tire pressure, inflate to recommended pressure.
- Check condition of tires.
- If operating in extremely dusty or trashy conditions, check the grille, radiator, and air conditioning condenser, if so equipped, for dirt or trash buildup. Clean as necessary.

5.2.0 Startup, Warmup, and Shutdown of a Gasoline- or Diesel-Powered Tractor

The information in the following sections is general and varies between makes and models of tractors. Before attempting any prestart checks or startup, always read the operator's manual for the tractor to become familiar with specific controls and procedures.

5.2.1 Prestart Checks

Before starting or moving a tractor, perform any required daily inspection, maintenance, and safety checks. Also, make sure that the tractor's fuel tank is full. Check that all other persons are clear of the tractor. Then, observe the following:

- *Pinch points* – Note if any attachments connected to the tractor will interfere with the rear tires when turning the tractor.
- *Obstructions* – Note any obstructions nearby or in the working area that could interfere with the movement of the tractor.

Teaching Tip/Notes

Service Interval Chart-Daily or 10 Hours/50 Hours/First 100 Hours/First 300 Hours/300 Hours					
ITEM	Daily or 10 Hours	50 Hours	First 100 Hours	First 300 Hours	300 Hours
Check Engine Oil Level	•				
Drain Water and Sediment from Fuel Tank and Fuel Filter	•				
Check Engine Coolant Level		•			
Check Transmission-Hydraulic System Oil Level		•			
Check MFWD Axle Housing and Wheel Hub Oil Level		•			
Inspect Tires and Check Inflation Pressure		•			
Lubricate Adjustable Front (2WD) Axle Steering Spindles and Cylinder Ends		•			
Lubricate MFWD Steering Kingpins, Axle Pivot Pin and Rear Axle	•*	•			
Inspect Tractor for Loose Hardware		•			
Change Engine Oil and Filter			•		
Replace Transmission-Hydraulic Filter			•		
Check Clutch Pedal Free Play			•		
Change Engine Oil and Filter (5225 and 5325)				•	
Service Engine Air Cleaner		•*			•
Change Engine Oil and Filter					•
Lubricate Hitch Components					•
Check Neutral Start System					•
Check Clutch Pedal Free Play					•
Clean Cab Air Filters		•*			•
Adjust PTO Clutch Lever Linkage					•

*Service if operated in extremely dusty or wet conditions.

22105-12_F24.EPS

Figure 24 Portion of a typical service interval chart.

Teaching Tip/Notes

Safety

Explain that the tractor must be started while the operator is in the operator seat.

Classroom

Describe the procedure for starting up a tractor. Point out that the transmission must be in park or neutral with the parking brake on.

Describe the shut-down procedure for tractors.

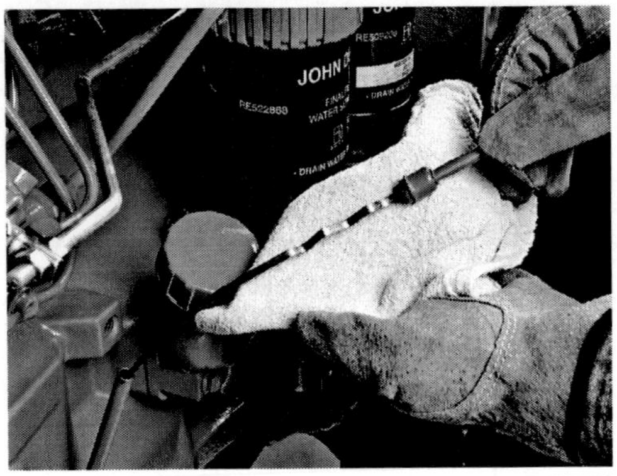

22105-12_F25.EPS

Figure 25 Checking tractor engine oil level.

5.2.2 Startup

WARNING! Always start a tractor from the operator's seat. It is dangerous to do otherwise.

First, mount the tractor while maintaining three points of contact. Adjust the seat and then fasten the seat belt if the tractor is equipped with a ROPS. Make sure that the transmission is in Park or the parking brake is set and the transmission is in Neutral. Check that the PTO, hitch, and any high/low shift and reverser controls are in the Off or Neutral position. If SCV controls are present, make sure they are in the neutral position; however, if a hydraulic motor of an attachment is connected to any SCV, make sure that control is set to the float position.

Once the controls are set, turn on the ignition and check that all warning/option indicator lights come on. Then set the throttle to the one-third or one-half position and, for a diesel, turn on the fuel supply if required. In cold weather, pull out the choke (if provided) or activate any cold weather starting aids. Next, press the clutch if required and activate the starter to crank the engine.

CAUTION If the engine is cranked for more than 20 seconds, allow the starter to cool for two minutes.

NOTE The engine starter may be activated by a button, lever, combination ignition switch and key, or a gear lever position. For a diesel engine, open the throttle beyond halfway when the engine is cranking.

Release the starting device when the engine starts. If a choke was used, push it in slowly as the engine warms up. Check that the warning lights go off and that any gauges register the correct readings while the engine is warming up. Release the clutch while the engine warms up. If provided, observe the temperature gauge to determine when the engine has reached the proper operating temperature.

If the tractor has power steering, turn the wheels back and forth occasionally to determine if the hydraulic fluid is warm enough to allow easy steering. In very cold weather, consult the operator's manual for any procedures to allow rapid warming of the hydraulic fluid.

5.2.3 Shutdown

NOTE Some tractors require that the clutch be depressed and released a second time while the engine is running to deactivate a hydraulic system lockout for power steering or other functions.

When ready to shut down a tractor, stop its movement. Shift the transmission to park, or set the parking brake and shift the transmission to neutral. If attachments are connected, lower them to the ground and retract any hydraulic cylinders as much as possible to protect the cylinder rods from weather exposure. If the attachments will be detached, it is especially important to retract single-action cylinders as much as possible to return hydraulic fluid to the tractor's hydraulic system reservoir.

Set the PTO, hitch, and any high/low shift and reverser controls to the Off or Neutral position. Then, **retard** the throttle. For diesel engines, turn off the fuel, if required. Next, turn off the ignition. Cycle all SCV controls through all positions to relieve hydraulic system pressure, and then return them to neutral. Turn all other switches off and, if the tractor will be left unattended, remove the ignition key. Then, dismount while maintaining

Teaching Tip/Notes

three-point contact. If the tractor will be parked overnight, make sure that the fuel tank is full to help prevent water condensation.

5.3.0 Safely Stopping a Tractor

WARNING!

Start preparing to stop far in advance and look for possible hazards. Avoid sudden stops.

First, depress the clutch (if present) or reduce forward speed and gradually apply both brakes. If the tractor will be stopped for a period of time with the engine running, shift the transmission to Park or set the parking brake and shift the transmission to Neutral. Then, set the throttle position to about one third for gasoline engines or one half for diesel engines for cooling purposes. If the tractor engine will be shut down, refer to the previous section.

5.4.0 Basic Tractor Operations

CAUTION

For tractors equipped with a speed range selector, the desired speed range can only be selected when the tractor is stopped.

Start and warm up the engine. Then depress the clutch (if present) and select the speed range (if present). Next, with the brakes applied, release the parking brake (if present) and shift the transmission as follows:

NOTE

Hand-shifted standard or synchronized transmissions may have lever(s) that work within various slots and notches of a forward or reverse quadrant, or lever(s) that are positioned according to a chart that shows gear positions.

- *Standard (non-synchronized) transmission* – Select the gear to be used for operation and place the tractor in gear. Start out in this gear. Operate the tractor at medium to fast idle and slowly release the clutch pedal. Move the throttle to the desired operating speed after the clutch is fully released. Never attempt to shift a standard transmission while the tractor is in motion; it must be stopped to shift gears.
- *Synchronized transmission* – Set the engine at slow or moderate rpm and place the transmission in the lowest gear. Slowly release the clutch. To shift gears while the tractor is in motion, depress the clutch pedal and shift the gearshift lever, gear by gear, to the desired ground speed while increasing the throttle to the desired engine rpm. The gears may take a few seconds to engage. Keep a steady pressure on the lever until it slides into position, then slowly release the clutch.
- *Power-assisted transmission* – In this transmission, the inching pedal replaces the clutch pedal. It is not used for shifting gears but for creeping speeds when connecting attachments, for making slow starts, and for emergency stops. It is possible to skip gears while shifting, but this results in a jerky motion. For safety, it is best to shift the tractor gear by gear. Place the gearshift lever into low gear with the engine running at a slow or moderate rpm. Shift the gearshift lever, gear by gear, to the desired ground speed while increasing the throttle to the desired operating rpm.
- *Hydrostatic transmission* – Set the throttle to obtain the desired engine operating rpm. Then, move the speed control lever gradually forward or rearward until the desired ground speed is reached.

CAUTION

Do not make tight turns with a towed attachment. Doing so when moving forward can cause the attachment to catch on and then ride up the rear tire that is on the inside of the turn. In reverse, the attachment can jackknife and a rear wheel could crush and ride up over the attachment. In both cases, injury and/or damage may result.

As the tractor moves forward or rearward, left or right turns are made by turning the steering wheel and, if necessary, applying a corresponding rear wheel brake to obtain a tighter turn. *Figure 26* shows the operations to make a tight right turn. Left turns are made in the opposite manner.

If equipped with a high-low shift, the speed of the tractor can be changed up or down by one-half gear without clutching while the tractor is moving. If equipped with a reverser, the tractor can be instantly moved backward while in a forward gear or forward while in a reverse gear without clutching, provided the selected tractor gear speed is for ground speeds under about 8 mph. If equipped, MFWD can be engaged for ex-

Classroom

Describe the procedure for safely stopping a tractor.

Discuss the importance of changing speed range selections only when the tractor is stopped.

Explain how to shift a standard transmission.

Describe how to shift synchronized and power-assisted transmissions.

Safety !

Point out that tight turns should never be made while towing an attachment due to the risk of damaging a tire or the attachment itself.

Audiovisual

Show Slide 32 (Figure 26).

Teaching Tip/Notes

Audiovisual

Show Slide 33 (Figure 27). Point out that bushings must be used when connecting a Category I attachment to a Category II tractor hitch.

Show Slide 34 (Table 3).

Show Slide 35 (Figure 28). Review the procedure for connecting an attachment to a three-point hitch.

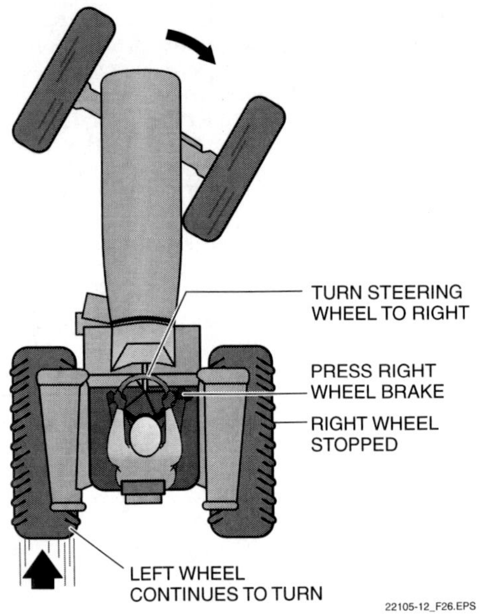

Figure 26 Turning to the right.

tra traction; however, it must be disengaged when making turns.

5.5.0 Connecting an Attachment to a Three-Point Hitch

A three-point hitch is the standard hydraulic linkage for tractor attachments. Three-point hitches consist of two lower links and an adjustable top link.

5.5.1 Preparing the Attachment

> **CAUTION**
> When connecting Category I attachments to a tractor, any sway bars may need adjustment to prevent binding and limiting full raise of the hitch.

As shown in *Figure 27,* Category I (A), a three-point attachment hitch is narrower and is used for smaller attachments than Category II (B) attachments. Refer to *Table 3* to identify the attachment category. When connecting a Category I attachment to a Category II tractor hitch, install reducer bushings in the links to accommodate the smaller pins.

5.5.2 Connecting the Attachment

Follow these guidelines to connect an attachment to a three-point hitch (*Figure 28*):

Step 1 Check the operator's manual for attachment installation instructions.

Step 2 Set the rockshaft or hitch control lever to the minimum position.

> **WARNING!**
> On tractors equipped with EH hitch controls, make sure any load/depth control is set to Off or to a float position before connecting or disconnecting a three-point attachment. If optional external raise/lower controls are used, the load/depth control is automatically disabled. This prevents the hitch from suddenly and unexpectedly moving due to weight changes caused by the addition or removal of an attachment.

Step 3 Check the drawbar position to make certain it will not interfere with movement of the hitch. Some tractors have no separate drawbar. Others may require that the drawbar be removed.

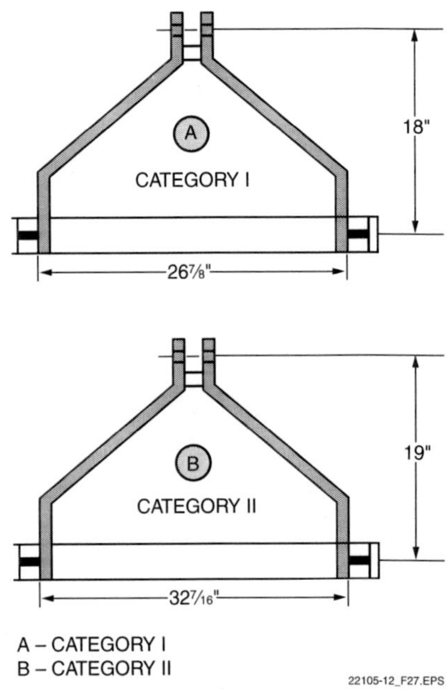

Figure 27 Category I and II attachment hitches.

Teaching Tip/Notes

30 NCCER – *Heavy Equipment Operations Level One* 22105-12

Step 4 Back the tractor so that draft links are in position to connect to attachment hitch pins.
- Make sure that the attachment is as laterally level with the tractor as possible. Use blocking if necessary.
- If the draft links are equipped with ball joints, back the tractor so that joints are over the hitch pins on the attachment.
- If the draft links are equipped with latches, align them so that latches are under the hitch pins of the attachment.

Step 5 Raise the draft links with the rockshaft or hitch control lever to the height needed to connect to attachment hitch pins.

Step 6 Place the transmission selector lever in Park or Neutral. When working on sloping ground, set the tractor wheel brakes, if provided, and block the wheels.

Table 3 Hitch Category Sizes

Category	Mast Height	Width Between Lower Pins	Lower Pin Size	Upper Pin Size
I (A)	457 mm (18 in)	682.6 mm (26⅞ in)	22 mm (⅞ in)	19 mm (¾ in)
II (B)	483 mm (19 in)	824 mm (32⁷⁄₁₆ in)	28 mm (1⅛ in)	25.4 mm (1 in)

Step 7 Connect the left draft link to the left hitch pin of the attachment. Always attach the left draft link first. If the link connection is not in line with the hitch pin on the attachment, move the attachment ahead, if possible, or rock the tractor forward and backward if it is on level ground. On tractors with extendable or telescoping draft

Figure 28 Connecting an attachment to a three-point hitch.

Classroom

Explain that it may be necessary to rock the tractor in order to line up the link connection with the hitch pin.

Teaching Tip/Notes

Audiovisual

Show Slide 36 (Figure 29). Point out the location of the center link.

! Safety

Point out that the engine must always be shut off and the PTO disengaged when attempting to connect or disconnect attachments.

links, it is not necessary to rock the tractor or pull on the attachment.

Step 8 Insert the locking pin.

Step 9 Raise or lower the right draft link to the correct height by adjusting the lift link.

Step 10 Attach the right draft link and insert the pin.

Step 11 Unhook the center link (*Figure 29*) from the transport bracket. The center link enables the operator to tilt the attachment forward or backward to level it from front to rear.

Step 12 Adjust the center link to reach the mast by turning the outer housing or by using the adjusting handle.

Step 13 Adjust the center link so that both shafts inside the outer housing are extended equally.

Step 14 Drop the center link into the attachment mast and lock into place by inserting a pin through holes in the mast and turnbuckle. Fasten the pin so that it cannot bounce out of the connecting point. If the center link is equipped with a latch assembly, depress the lever, fit the latch over the ball, and release the lever.

Step 15 Adjust sway bars or chains, or install sway blocks in a lower position with the narrow, flat edge inward. This will eliminate sway during operation and transport. Be sure draft links do not bind on sway blocks.

Step 16 Install any front end weights, as specified for the attachment, for balance.

Step 17 Raise the attachment to the transport position, using the rockshaft or hitch control lever. Make certain that it clears at all points and that the connections are secure.

5.6.0 Attaching a PTO

Use the following guidelines to attach a PTO. Be sure to have the correct rpm PTO shaft for the attachment being connected.

Step 1 Set the wheel brakes.

Step 2 Disengage the power to the PTO shaft.

Step 3 Shut off the engine.

> **WARNING!**
> Always shut off the engine and disengage the PTO before attempting to connect or disconnect attachments.

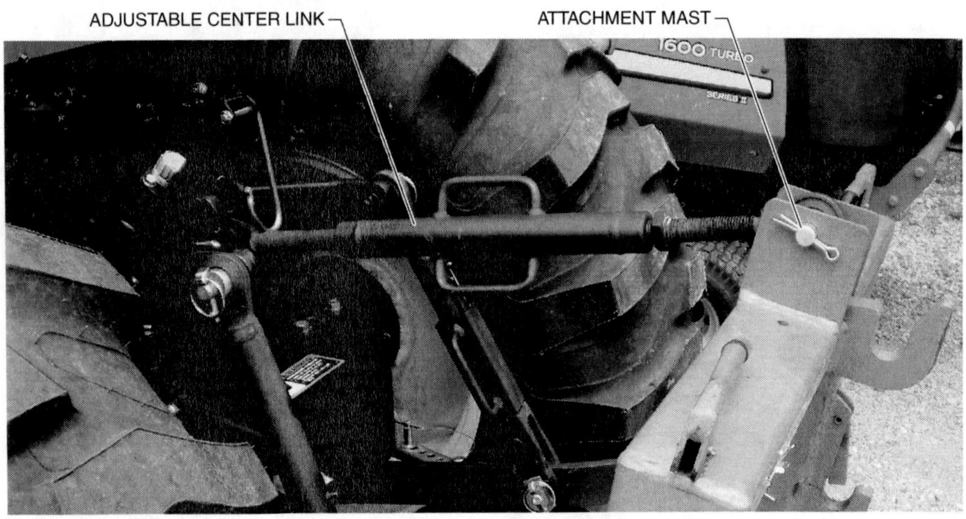

Figure 29 Location of center link.

Teaching Tip/Notes

Step 4 Remove the PTO shaft guard or master shield if the tractor is so equipped.

Step 5 Check the tractor PTO coupling and the attachment PTO stub shaft to make sure they have the same number of splines. If not, attach the correct stub shaft and manually shift the shaft speed. See the manufacturer's instructions.

Step 6 Clean the inside of the coupling of all grease and debris.

Step 7 Oil the PTO shaft spline.

Step 8 Verify that the yokes on the shaft are aligned identically as shown in *Figure 30*.

Step 9 Attach the power shaft with a spring-loaded lock (see *Figure 31*), if applicable. Other types of locking devices are common, as shown in *Figure 32*.

Step 10 Move the collar back against the stop on the yoke.

Step 11 Turn the collar to align the yoke with splines on the tractor PTO shaft. Some models require pushing the button down and turning the collar until the button stays down.

Step 12 Slide the power shaft onto the PTO shaft.

Step 13 Depress the pin or button and move the shaft forward until the collar snaps into the locked position.

Step 14 Check the locking device by pulling back on the bell or guard shaft. Do not pull on the collar; the latch will release.

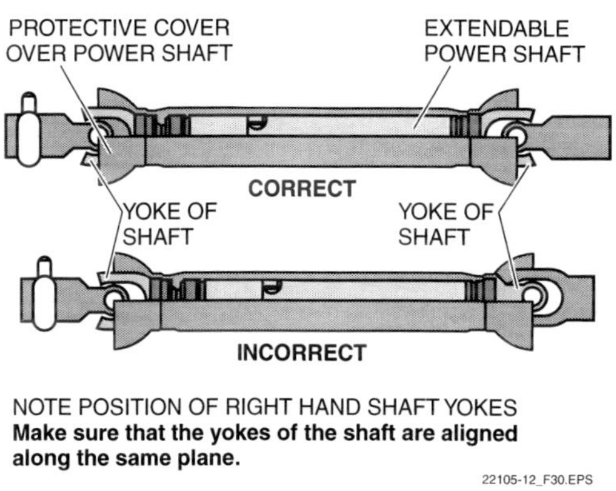

Figure 30 Positioning universal joints and shaft yokes.

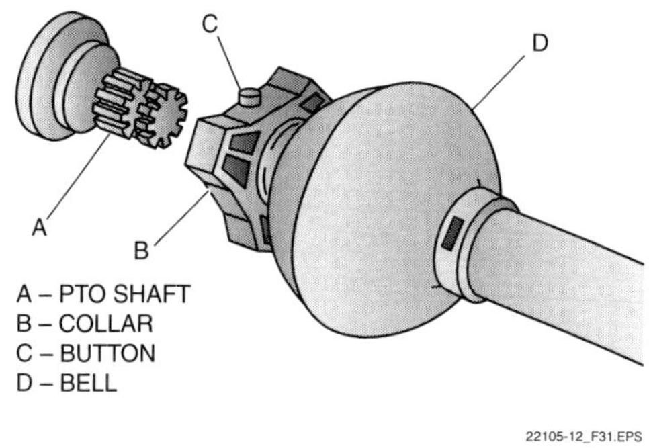

Figure 31 Power shaft with a spring-loaded lock.

Step 15 Slide the safety shield into place.

Step 16 Keep the alignment as straight as possible. Normally a PTO shaft should not be operated at more than a 15-degree angle from the U-joints.

Step 17 Maintain the proper drawbar height and length.

Step 18 Reinstall the PTO master shield.

5.7.0 Operating a PTO

The PTO provides mechanical power to the attachments. It should always be disengaged when not operating the attachment. Follow these guidelines to engage the PTO.

> **NOTE**
> PTO engagement or disengagement must be done with the tractor engine running. Newer tractors with an independent PTO do not require the use of the tractor clutch pedal; engagement and disengagement can be done with the tractor stopped or moving.

Start the tractor engine if it is not already running and, if required, depress the clutch pedal and both brakes all the way. If necessary, allow the tractor to come to a complete stop. Then, engage the PTO and, if applicable, release the brakes and engage the clutch slowly.

To shift gears and leave the PTO running, depress the clutch and, if required, both brakes. Change gears and, if used, release the brakes. Engage the clutch slowly.

To disengage the PTO and, if required, stop the tractor to do so, depress the clutch and both brakes. When the tractor stops, disengage the PTO.

Audiovisual

Show Slide 37 (Figure 30). Describe how to align yokes.

Show Slides 38 and 39 (Figures 31 and 32). Discuss the different types of locking devices used to attach the power shaft.

Classroom

Explain that PTO engagement or disengagement must be done with the tractor running.

Teaching Tip/Notes

! Safety

Explain that gloves and eye protection must be worn when connecting or disconnecting hydraulic hoses.

Audiovisual

Show Slide 40 (Figure 33).

! Safety

Point out the importance of properly positioning the SCV control lever when a hydraulic motor is connected to the SCV.

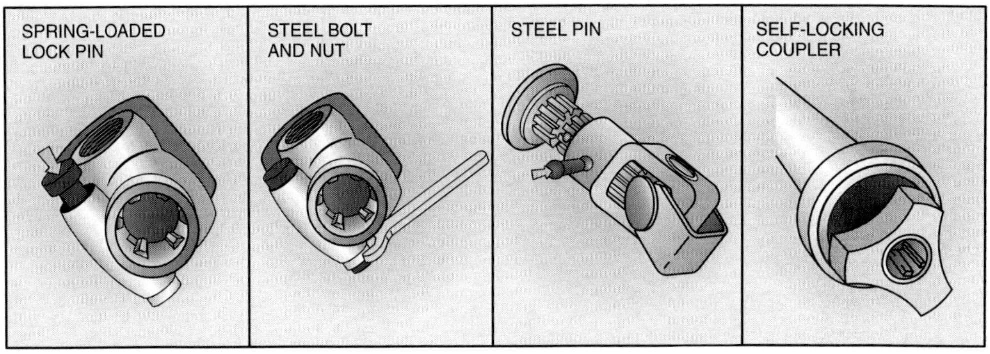

Figure 32 Locking devices.

5.8.0 Connecting and Disconnecting Attachment Hydraulic Hoses

The following guidelines cover the connection and disconnection of attachment hydraulic cylinder or motor hoses to the SCV couplers at the rear or side of a tractor.

5.8.1 Connecting Hoses

Use the following procedure to connect hoses:

> **WARNING!** Gloves and eye protection must be worn when connecting, disconnecting, or checking any high-pressure hoses.

Step 1 Stop the tractor engine.

Step 2 Cycle the SCV control levers through all positions to relieve pressure at the couplings. Then, if hydraulic motors are to be connected, set the appropriate SCV control levers in the float position. If hydraulic cylinders are to be connected, set the appropriate SCV control levers in the neutral position.

Step 3 Remove and save any attachment hose nipple caps. Then wipe the nipples clean.

Step 4 Open the covers on the SCV couplers (*Figure 33*) and verify that they are clean.

> **CAUTION** For hydraulic motors, the retract coupler must be used as the pressure source instead of the extend coupler, and a flow control valve must be present in the line either externally or internally to the valve.

Figure 33 Typical SCV coupler covers.

Step 5 Select the proper coupler for a single-acting or double-acting cylinder extend and the corresponding hose nipple. Insert the nipple in the coupler and push in firmly to **seat** the nipple.

Step 6 Repeat Step 5 for the retract coupler and hose for double-acting cylinders.

Step 7 Pull all the hoses slightly to make sure they are seated in the couplers.

> **WARNING!** Never put a SCV control lever in a partial retract or neutral position with the engine running if a hydraulic motor is connected to the SCV. The hoses may rupture or the motor may be damaged.

Teaching Tip/Notes

Step 8 Start the tractor engine and move the SCV control levers from neutral to extend and retract the cylinders slightly; or from float to full retract and back to float to start and stop hydraulic motors. Then, check for signs of any visible leaks.

Step 9 For hydraulic motors, adjust the motor speed as described previously.

5.8.2 Disconnecting Hydraulic Hoses

Use the following procedure to disconnect hoses:

> **WARNING!** Gloves and eye protection must be worn when connecting, disconnecting, or checking any high-pressure hoses.

Step 1 Lower all loads/attachments to the ground and retract all cylinders as much as possible to protect the cylinder rods from weather exposure.

> **NOTE** It is especially important to retract single-acting cylinders as much as possible to return hydraulic fluid to the tractor hydraulic system reservoir.

Step 2 Stop the tractor engine, cycle the SCV control levers through all positions to relieve pressure at the couplings, and return the levers to the Neutral position.

Step 3 Pull the hoses from the SCV couplers.

> **NOTE** Check attachment hoses and cylinders or motors for any visible signs of recent leakage.

Step 4 Replace the attachment hose nipple caps and close the SCV coupler caps.

SUMMARY

Tractors are used for a variety of operations in construction and industry, as well as in agriculture. They come in many sizes and configurations. Currently, the most common design used in the construction business is the four-wheel rubber-tired tractor. The basic tractor does not perform any operations. Instead, it is set up to accept many different types of attachments that perform the desired operations. Attachments to the front of the machine include lift buckets, fork lifts, sweepers, and scrapers. Most other attachments are mounted on the back of the machine and require some type of power source for operation. Types of attachments on the back include mowers, scraper blades, tree cutters/saws, and rollers.

Operation of a tractor requires constant attention because it can become unstable and tip over. Make sure to understand the controls and their operations before beginning to work on the machine.

Use of the hydraulic system and power takeoff require special attention and care because they can cause injury very quickly and easily if not handled properly.

Before starting the engine or operating the equipment, be sure everything is working properly by doing a walk-around inspection of the major components, such as the tires, steering, brakes, and engine.

Proper attachment of accessories that use the power takeoff is critical. The shaft connecting the tractor to the accessory must be aligned and fastened properly.

Classroom

Discuss the procedure for disconnecting hydraulic hoses.

Point out the importance of retracting single-acting cylinders.

Demonstration

Show the trainees how to perform prestart inspection and maintenance procedures and provide them with copies of inspection and maintenance checklists.

PT/Laboratory

Under your supervision, have trainees perform prestart inspection and maintenance procedures. Note the proficiency of each trainee. This task corresponds to Performance Task 1.

Demonstration

Show the trainees how to start up, warm up, and shut down a tractor.

PT/Laboratory

Under your supervision, have trainees start, warm up, and shut down a gasoline-powered and diesel-powered engine tractor. Note the proficiency of each trainee. This task corresponds to Performance Task 2.

Demonstration

Show the trainees how to perform basic maneuvering with a utility tractor.

PT/Laboratory

Under your supervision, have trainees perform basic maneuvering with a utility tractor. Note the proficiency of each trainee. This task corresponds to Performance Task 3.

Teaching Tip/Notes

Demonstration

Show the trainees how to attach implements to a drawbar and three-point hitch.

PT/Laboratory

Under your supervision, have trainees attach implements to a drawbar and three-point hitch. Note the proficiency of each trainee. This task corresponds to Performance Task 4.

Demonstration

Show the trainees how to attach and detach implements to a power takeoff.

PT/Laboratory

Under your supervision, have trainees attach and detach implements to a power takeoff. Note the proficiency of each trainee. This task corresponds to Performance Task 5.

Classroom

Summarize the major concepts presented in this module.

Have the trainees complete the Review Questions, and go over the answers prior to administering the Module Examination.

Review Questions

1. The most common type of transmission in utility tractors is the _____.
   a. sliding gear
   b. hydrostatic drive
   c. synchromesh
   d. hydraulic assist

2. The flywheel of a tractor's engine serves as what component of a basic disk clutch assembly?
   a. Drive plate
   b. Clutch plate
   c. Pressure plate
   d. Throw-out plate

3. If a tractor's charging system warning indicator shows a high negative discharge rate when the engine is at half throttle, the tractor _____.
   a. is operating properly
   b. should have its battery replaced
   c. should be run without lights or fans
   d. should be stopped and inspected to determine the cause

4. If a diesel engine has run out of fuel, have its _____.
   a. tank refilled
   b. tank refilled and engine primed
   c. fuel lines and carburetor drained
   d. fuel lines and injectors bled of air

5. What component shifts the tractor's transmission into forward or reverse?
   a. High-low shifter
   b. Range selector
   c. Gear selector
   d. Reverser

6. On a John Deere tractor, what transmission shift lever operation will change the tractor's speed up or down by ½ gear, while the tractor is in motion?
   a. High-low shifter
   b. Range selector
   c. Gear selector
   d. Speed shifter

7. At the back of most tractors, the splined shaft that provides mechanical power to connected attachments is the _____.
   a. power takeoff
   b. SVC coupler
   c. rockshaft
   d. draft link

8. A PTO shaft with 21 splines will normally be expected to turn _____.
   a. 320 rpm
   b. 540 rpm
   c. 620 rpm
   d. 1,000 rpm

9. Most selectable control valve (SCV) control levers have _____.
   a. two operating positions
   b. four operating positions
   c. six operating positions
   d. eight operating positions

10. When attempting to start a diesel engine when the air temperature is below 40°F, what ignition switch position activates the glow plugs or the intake air warmers?
    a. Accessory
    b. Warm up
    c. Start
    d. Run

11. If a person or animal approaches an operating tractor with a rotating attachment, the operator should _____.
    a. shut down the tractor
    b. carefully drive around the person or animal
    c. stop moving until the person or animal leaves
    d. slow the engine speed until the person or animal passes

12. When first moving a tractor, the operator should _____.
    a. test the brake system and steering in both directions
    b. accelerate to full speed
    c. lean out of the cab to check for clearance
    d. lock both rear wheels with the breaks

36 NCCER – *Heavy Equipment Operations Level One* 22105-12

Teaching Tip/Notes

13. When driving a tractor near the edges of deep ditches or excavations, stay at least how far away from the edges, depending upon soil conditions?
    a. 3 feet
    b. 15 feet
    c. 4 to 10 feet
    d. 6 to 12 feet

14. Always use dust caps on quick-couple disconnects when the hydraulic system is *not* in use.
    a. True
    b. False

15. Because of the increased risk of a tire bead leak on a tractor tire, inflation pressures of less than what level should be monitored regularly?
    a. 8 psi
    b. 10 psi
    c. 12 psi
    d. 20 psi

16. What device(s) are added to prevent wheel slippage when towing heavy loads?
    a. A heavier three-point hitch
    b. A heavier drawbar
    c. Front-end ballast
    d. Wheel weights

17. Before actually starting a tractor, the parking brake should be set and the transmission should be in neutral.
    a. True
    b. False

18. Before actually starting a tractor with an attachment that has a hydraulic motor connected to any selectable control valve (SCV), make sure that the control is set to the _____.
    a. neutral position
    b. extend position
    c. retract position
    d. float position

19. When shutting down a tractor with an attachment that is to be disconnected, the hydraulic cylinders are retracted as much as possible to _____.
    a. level the tractor
    b. lower the attachment
    c. re-lubricate the cylinder rods
    d. return hydraulic fluid to the reservoir

20. The tractor component that allows the operator to tilt an attachment forward or backward to level it from front to rear is the _____.
    a. draw bar
    b. center link
    c. draft link(s)
    d. sway bar(s)

Teaching Tip/Notes

Classroom

Have the trainees complete the Trade Terms Quiz, and go over the answers prior to administering the Module Examination.

? Examination

Administer the Module Examination. Record the results on Training Report Form 200, and submit the results to the Training Program Sponsor.

✓ Performance Testing

Administer the Performance Test and fill out the Performance Profile Sheets for each trainee. If desired, trainee proficiency noted during laboratory sessions may be used to complete the Performance Test. Record the results on Training Report Form 200 and submit the results to the Training Program Sponsor.

Trade Terms Quiz

Fill in the blank with the correct term that you learned from your study of this module.

1. The parallel groove that runs lengthwise on a shaft is called a(n) ________.
2. The load-pulling capacity is called the ________.
3. An exhaust-driven centrifugal air compressor that is used to increase power is called a(n) ________.
4. To ________ is to move equipment from place to place.
5. To ________ is to cause two surfaces to fit firmly together.
6. To slow down is to ________.
7. The part of the tire that fits against the rim is called the ________.
8. Controls for equipment mounted on a three-point hitch are called the ________.
9. The mechanical connection on an engine or transmission to which a cable, belt, or shaft can be connected to drive an attachment or tool is called a(n) ________.
10. Controls or transmissions that stay in neutral until the fluids in them are put into motion are said to be ________.

Trade Terms

Bead
Draft
Hydrostatic
Power takeoff (PTO)
Retard
Rockshaft unit
Seat
Spline
Travel
Turbocharger

Teaching Tip/Notes

Trade Terms Introduced in This Module

Bead: Part of tire that fits against rim.

Draft: Load-pulling capacity.

Hydrostatic: Relating to fluids at rest, or to the pressures they transmit, as in hydraulically-powered controls or transmissions that stay in a neutral condition until the hydraulic fluid in them is put into motion.

Power takeoff (PTO): Mechanical connection on an engine or transmission to which a cable, belt, or shaft can be connected to drive an attachment or tool.

Retard: To slow down.

Rockshaft unit: Controls for equipment mounted on a three-point hitch.

Seat: To cause two surfaces to fit firmly together.

Splines: Parallel grooves running lengthwise on a shaft.

Travel: Moving equipment from place to place. Can be either on the job site or over the public highway system.

Turbocharger: Exhaust-driven centrifugal air compressor used to increase power.

Teaching Tip/Notes

Additional Resources

This module presents thorough resources for task training. The following resource material is suggested further study.

Site Layout Levels 1 and *2*, Latest edition. Alachua, FL: NCCER.

Figure Credits

Deere & Company, Module opener, Figures 1 (top photo), 2, 4, 5, 7, 8, 11, 13, 18 (top, right middle, bottom photos), 24, 25, Tables 1 and 3

Mark Jones, Figures 1 (bottom photo), 15, and 18 (left middle photo)

Reprinted courtesy of Caterpillar Inc., Figure 6

Topaz Publications, Inc., Figures 10, 14, 16, 17, 28, 29, 33, and SA01

Teaching Tip/Notes

ANSWERS TO REVIEW QUESTIONS

Answer		Section Reference
1.	c	2.1.1
2.	a	2.1.2
3.	d	2.3.1
4.	d	2.3.3
5.	c	2.4.2
6.	a	2.4.2
7.	a	2.4.8
8.	d	2.4.8
9.	b	2.4.10
10.	d	2.4.11
11.	a	3.1.0
12.	a	3.2.0
13.	c	3.2.0
14.	a	3.4.0
15.	c	3.6.0
16.	d	4.4.0
17.	a	5.2.2
18.	d	5.2.2
19.	d	5.2.3
20.	b	5.5.2

ANSWERS TO TRADE TERMS QUIZ

1. Spline
2. Draft
3. Turbocharger
4. Travel
5. Seat
6. Retard
7. Bead
8. Rockshaft unit
9. Power takeoff (PTO)
10. Hydrostatic

NCCER CURRICULA — USER UPDATE

NCCER makes every effort to keep its textbooks up-to-date and free of technical errors. We appreciate your help in this process. If you find an error, a typographical mistake, or an inaccuracy in NCCER's curricula, please fill out this form (or a photocopy), or complete the online form at **www.nccer.org/olf**. Be sure to include the exact module ID number, page number, a detailed description, and your recommended correction. Your input will be brought to the attention of the Authoring Team. Thank you for your assistance.

Instructors – If you have an idea for improving this textbook, or have found that additional materials were necessary to teach this module effectively, please let us know so that we may present your suggestions to the Authoring Team.

NCCER Product Development and Revision

13614 Progress Blvd., Alachua, FL 32615

Email: curriculum@nccer.org

Online: www.nccer.org/olf

❑ Trainee Guide ❑ AIG ❑ Exam ❑ PowerPoints Other ______________________________

Craft / Level: Copyright Date:

Module ID Number / Title:

Section Number(s):

Description:

Recommended Correction:

Your Name:

Address:

Email: Phone:

22201-12

Introduction to Earthmoving

Module Six

Materials

reminds you to gather materials needed for classes, labs, and testing.

Homework

suggests that you assign reading for the next session.

Classroom

prompts you to discuss a topic, stress a point, explain a concept, or provide examples.

Audiovisual

indicates an appropriate time to show a slide or transparency.

! Safety

used to emphasize safety issues.

Demonstration

directs you to show trainees how to perform tasks.

Laboratory

used when trainees are to practice performing tasks.

✓ Performance Testing

recommends administering a performance test, or a portion thereof.

PT / Laboratory

indicates that lab practice corresponds with a performance task. If desired, you can note the proficiency of the trainees during the lab to satisfy performance testing requirements.

Teaching Tip

indicates tips for practice exercises, Quick Quizzes, field trips, or other activities designed to encourage class participation.

? Examination

recommends administering the written module exam.

NCCER Standardized Craft Training Program

NCCER provides a standardized national program of accredited craft training. Key features of the program include instructor certification, competency-based training, and performance testing. The program provides trainees, instructors, and companies with a standard form of recognition through the National Registry. The program is described in full in the *Guidelines for Accreditation*, published by NCCER. For more information on standardized craft training, contact NCCER by writing to 13614 Progress Blvd., Alachua FL 32615; calling 1-888-NCCER20; or emailing **info@nccer.org**. More information is available at **www.nccer.org**.

How to Use this Annotated Instructor's Guide

Each page presents two sections of information. The larger section displays each page exactly as it appears in the Trainee Module. The narrow column ties suggested trainee and instructor actions to each page and provides icons (detailed on the left) to call your attention to material, safety, audiovisual, or testing requirements. The bottom of each page includes space for your notes, or when available, teaching tips that suggest activities or class exercises that will help your students learn the material.

Preparation

Before teaching this module, you should review the Objectives, Performance Tasks, Materials and Equipment List, and Module Outline. Be sure to allow ample time to prepare your own training or lesson plan and gather all required materials and equipment.

Instructors purchasing a single module receive the printed exam and performance tasks with the AIG module. Perfect-bound versions of this title no longer include the printed exam booklet and performance task sheets. Log on to the Instructor Resource Center at **www.NCCERIRC.com** to download these materials. Your access code is on the card bound into the front of this AIG. This code will allow you to access exams and performance tasks for all levels of this craft. Be sure to keep all testing materials in a secure place at all times.

Introduction to Earthmoving
Annotated Instructor's Guide

Module 22201-12

Module Overview

This module introduces the trainee to the basic concepts and procedures related to the use of heavy equipment to perform earthmoving work. It identifies the most appropriate types of equipment for a given task and describes the use of the equipment to perform the work.

Objectives

Upon completion of this module, the trainee will be able to do the following:

1. Identify and explain earthmoving terms and methods.
2. Describe how to safely set up and coordinate earthmoving operations.
3. Identify and explain earthmoving operations.
4. Identify and explain soil stabilization methods.
5. Identify the best equipment for performing a given earthmoving operation.
6. List, in the correct order, the steps involved in an earthmoving operation.

Performance Tasks

Under the supervision of the instructor, the trainee should be able to do the following:

1. Draw a plan for basic earthmoving operations:
   - Clearing and grubbing
   - Excavating the foundation
   - Constructing embankments
   - Backfilling
   - Compacting
2. Lay out a basic earthmoving operation.
3. Identify and select the proper equipment for a given earthmoving operation.

Materials and Equipment

Markers/chalk
Pencils and paper
Whiteboard/chalkboard
Heavy Equipment Operations Level One PowerPoint® Presentation Slides (ISBN 978-0-13-292167-1)
Multimedia projector and screen
Computer
Appropriate personal protective equipment
Jars with different types of soil
Set of site plans including excavation drawings
Marked grade and finish stakes
Example of an NPDES stormwater permit
Trade Terms Quiz*
Module Examinations**
Performance Profile Sheets**

* Located at the back of the Trainee Guide module

** Single-module AIG purchases include the printed exam and performance task sheet. If you have purchased the perfect-bound version of this title, download these materials from the IRC using your access code.

Safety Considerations

Ensure that the trainees are equipped with appropriate personal protective equipment and know how to use it properly. Review safety guidelines associated with working on or around heavy equipment. Emphasize the importance of proper housekeeping.

Additional Resources

This module presents thorough resources for task training. The following resource material is suggested for further study.

"Avoiding Enforcement Actions: How to Effectively Manage Your Construction Site and Survive Increased Regulatory Scrutiny," *Grading and Excavation Contractor* magazine. Carol L. Forrest. July/August 2005. Santa Barbara, CA: Forester Communication.

Caterpillar Performance Handbook, Latest Edition. A CAT® Publication. Peoria, IL: Caterpillar, Inc.

"Chemical Soil Stabilization," *Erosion Control* magazine. January/February, 2003. Janis Keating.

United States Environmental Protection Agency (EPA) web site, National Pollutant Discharge Elimination System (NPDES): www.epa.gov.

Teaching Time for This Module

An outline for use in developing your lesson plan is presented below. Note that each Roman numeral in the outline equates to one session of instruction. Each session has a suggested time period of 2½ hours. This includes 10 minutes at the beginning of each session for administrative tasks and one 10-minute break during the session. Approximately 12½ hours are suggested to cover *Introduction to Earthmoving*. You will need to adjust the time required for hands-on activity and testing based on your class size and resources. Because laboratories often correspond to Performance Tasks, the proficiency of the trainees may be noted during these exercises for Performance Testing purposes.

Topic	Planned Time
Session I. Introduction; Mining; Earthmoving Operations, Part One	
A. Introduction	________
B. Mining	________
C. Earthmoving Operations	________
1. Preliminary Activities	________
2. Laying Out Slopes and Grades	________
3. Setting Up and Coordinating Operations	________
Session II. Earthmoving Operations, Part Two	
4. Production Measurement	________
5. Maintaining Haul Roads	________
6. Drainage Requirements	________
7. Site Excavation	________
8. Loading	________
Session III. Earthmoving Operations, Part Three; Stabilizing Soils; Safety Guidelines	
9. Hauling	________
10. Dumping	________
11. Fill and Embankment	________
12. Compaction	________
D. Stabilizing Soils	________
1. Choosing a Binder	________
2. Preparing the Subgrade	________
3. Spreading the Binder	________
4. Determining Logistics	________
5. Stabilizing	________
6. Compacting	________
7. Checking Quality	________

E. Safety Guidelines
   1. Personal Safety
   2. Safeguarding Property
   3. Underground Utilities

Session IV. PT/Laboratory

A. PT/Laboratory
   1. Have trainees draw a plan for basic earthmoving operations, including:
      - Clearing and grubbing
      - Excavating the foundation
      - Constructing embankments
      - Backfilling
      - Compacting

      This laboratory corresponds to Performance Task 1.
   2. Have trainees lay out a basic earthmoving operation. This laboratory corresponds to Performance Task 2.
   3. Have trainees identify and select the proper equipment for a given earthmoving operation. This laboratory corresponds to Performance Task 3.

Session V. Review and Testing

A. Review

B. Module Examination
   1. Trainees must score 70% or higher to receive recognition from NCCER.
   2. Record the testing results on Training Report Form 200 and submit the results to the Training Program Sponsor.

C. Performance Testing
   1. Trainees must perform each task to the satisfaction of the instructor to receive recognition from NCCER. If applicable, proficiency noted during laboratory exercises can be used to satisfy the Performance Testing requirements.
   2. Record the testing results on Training Report Form 200 and submit the results to the Training Program Sponsor.

22201-12

Introduction to Earthmoving

Module Six

Homework

Assign the reading of this module.

Teaching Tip/Notes

Trainees with successful module completions may be eligible for credentialing through NCCER's National Registry. To learn more, go to **www.nccer.org** or contact us at **1.888.622.3720**. Our website has information on the latest product releases and training, as well as online versions of our *Cornerstone* newsletter and Pearson's product catalog.

Your feedback is welcome. You may email your comments to **curriculum@nccer.org**, send general comments and inquiries to **info@nccer.org**, or fill in the User Update form at the back of this module.

V.1 4/12

Teaching Tip/Notes

22201-12
Introduction to Earthmoving

Objectives

When you have completed this module, you will be able to do the following:

1. Identify and explain earthmoving terms and methods.
2. Describe how to safely set up and coordinate earthmoving operations.
3. Identify and explain earthmoving operations.
4. Identify and explain soil stabilization methods.
5. Identify the best equipment for performing a given earthmoving operation.
6. List, in the correct order, the steps involved in an earthmoving operation.

Performance Tasks

Under the supervision of your instructor, you should be able to do the following:

1. Draw a plan for basic earthmoving operations:
   - Clearing and grubbing
   - Excavating the foundation
   - Constructing embankments
   - Backfilling
   - Compacting
2. Lay out a basic earthmoving operation.
3. Identify and select the proper equipment for a given earthmoving operation.

Trade Terms

Backhauling	Expansive soil	Shoring
Bedrock	Gradation	Soil testing
Center line	Groundwater	Spoils
Cohesive	Impervious	Stormwater
Consolidation	Inorganic	Test boring
Core sample	Organic	Test pit
Cycle time	Pay item	Trench
Dewatering	Pit	Water table
Dragline	Riprap	Windrow
Embankment	Select material	

Industry Recognized Credentials

If you're training through an NCCER-accredited sponsor you may be eligible for credentials from NCCER's Registry. The ID number for this module is 22201-12. Note that this module may have been used in other NCCER curricula and may apply to other level completions. Contact NCCER's Registry at 888.622.3720 or go to nccer.org for more information.

Teaching Tip/Notes

When you call on someone to answer a question, the rest of the class relaxes or even tunes out because they expect that the question and answer will take place only between you and the trainee you called on. Instead, use this technique to involve more trainees in answering questions and to keep them on their toes:

1. Ask trainees to define a term or explain a concept.
2. After one trainee has answered, ask a trainee seated nearby if the answer is right. Then ask whether a trainee in the back of the room agrees.
3. Ask trainees to explain why they think an answer is right or wrong.
4. Use the session to clear up incorrect ideas and encourage trainees to learn from their mistakes.

Materials

Ensure that you have everything required to teach the course. Check the Materials and Equipment list at the front of this module.

Classroom

Explain that terms shown in bold (blue) are defined in the Glossary at the back of this module.

Audiovisual

Show Slide 1, Objectives, and Slide 2, Performance Tasks. Review the goals of the module, and explain what will be expected of the trainees.

Contents

Topics to be presented in this module include:

Teaching Tip/Notes

Teaching Tip/Notes

Figures and Tables

Teaching Tip/Notes

1.0.0 INTRODUCTION

Earthmoving is the process of digging, loading, hauling, and dumping any material that is needed for construction or is in the way of construction. The process may also include digging and removing material from a **pit** or mine and processing the material. The primary process operations are soil or rock excavation, backfill, and **embankment** construction. Earthmoving may also include the transportation of materials, stabilization of embankments, and control of **groundwater.**

The word *earth* refers to any material in Earth's crust that is above **bedrock.** Bedrock is Earth's hard, **impervious** foundation, and includes soil and rock. Rock is a natural aggregate of minerals that are joined by strong and permanent **cohesive** forces. Soil includes **organic** and **inorganic** material, as well as large and small pieces of rock.

The methods of excavating and removing rock differ from those required for removing soil. They are also more expensive. Because of this cost, contractors must first define what makes up rock excavation, and then consider everything else to be soil excavation. The common practice is to call all excavation unclassified until the earthmoving contractor defines how much of the excavation is rock and how much is soil. This allows the contractor to determine what equipment is needed and thus the cost of the excavation.

2.0.0 Mining

Mining is a type of earthmoving work and is a form of excavation. Mining requires special equipment and methods, so it is described separately. Mining safety is governed by the Mine Safety and Health Administration (MSHA), which is charged with guarding miners' safety in order to prevent death, disease, and injury from mining. MSHA works to promote a safe, healthy work environment for miners.

The term *pit* is used to describe any open excavation that is made to obtain material of value, such as coal, mineral ore, select fill material, gravel, or quarry rock (*Figure 1*). Usually, any pit operation involves not only the use of heavy equipment, but also blasting, processing, and drainage. The size of pits varies from very small project pits, where **select material** is quarried for fill, to large open-pit mining operations such as those used for mining iron ore or bauxite. Quarries that yield construction materials, such as sand, limestone, and granite, are located all over the United States. Some are set up temporarily for the purpose of supplying material to a specific project, whereas others are set up as permanent business ventures.

22201-12_F01.EPS

Figure 1 Pit mining.

The type of heavy equipment used for a pit operation depends on the size of the pit and the material being excavated. For small pits, bulldozers, front-end loaders, and dump trucks are sufficient. For large or more permanent operations, **draglines** like the one shown in *Figure 2* may be more economical. In larger pit operations, there may still be a need for other larger specialized equipment to perform jobs, such as heavy duty hauling (*Figure 3*), the construction of haul roads, drainage work, and stockpiling.

3.0.0 EARTHMOVING OPERATIONS

Earthmoving consists of moving material, usually soil or rock, from one place to another. This can be a very short distance, as in the case of excavating a **trench,** or a longer distance, such as a cut

22201-12_F02.EPS

Figure 2 Dragline.

Classroom

Describe the activities that are part of earthmoving.

Explain what bedrock is and how it affects earthmoving operations.

Explain the differences between pit mining and surface mining.

Describe the use of draglines.

Audiovisual

Show Slides 3–5 (Figures 1–3).

Teaching Tip/Notes

Classroom

Describe the planning tasks that must be performed before earthmoving work can begin.

Discuss the different types of soils that can be encountered on a job site.

Note that extensive soil testing is usually required. Explain how this information is used to identify equipment needed for the work.

22201-12_F03.EPS

Figure 3 Rigid frame truck.

and fill operation for construction of a highway or building site.

3.1.0 Preliminary Activities

Several activities are required before any excavation and embankment work can be started. The type, quantity, and location of material on the site determine what kind and how many pieces of equipment are needed to do the job. Each project has a set of grading plans that show what the site is to look like after the earthwork is complete. Survey crews transfer the information on the plans to the actual building site using grade stakes so that workers know what areas require excavation and fill.

Projects next to public highways or other areas where people pass by require traffic and pedestrian control. The control must be put in place before construction begins. The project drawings and specifications usually show the required signs and layout.

3.1.1 Soils

The success and stability of any construction project depend upon the ground on which it is built. Most people give little thought to soil, but there is nothing more important in determining the suitability of a site for building. Site selection involves a series of complicated and rigorous tests performed by specially trained workers to determine the site's soil composition and properties.

There are many different types of soils, and each has unique characteristics. Soil composition varies by region, with most areas having a combination of two or more types. It is very rare to find a site that is composed of a single soil type. There are several classification systems for soil. A general description of the different types of soils follows:

- *Gravel* – Any rock-like material greater than 0.125 inches (½ inch) in diameter. Larger particles are called cobbles or stones, and those larger than 10 inches are called boulders. Gravel occurs naturally or it can be made by crushing rock. Natural gravel is usually rounded from the effects of water, while crushed rock is usually angular.
- *Sand (coarse and fine)* – Mineral grains measuring 0.002 to 0.125 inches. Sand is made from grinding or decaying rock. It usually contains a high amount of quartz. It is called granular material because it separates easily, giving it almost no cohesive strength. Coarse sand is frequently rounded like gravel and is often found mixed with gravel, but fine sand is usually more angular.
- *Inorganic silt* – Very fine sand with particles that are 0.002 inches or less. Silt is sand that has been ground very fine. It has a dusty appearance and powdery texture when dry, but sticks together when wet. Silt has almost no cohesive strength, so dried lumps are easily crushed. It has a tendency to absorb moisture by capillary action, which means that moisture wicks up through the soil, making it problematic in areas where the **water table** is shallow.
- *Clay* – The finest size of soil particles. Clay is very cohesive. When wet, clay-like soils feel like putty and can be easily molded and rolled into ribbons. When dry, clay is very strong and clumps are difficult to crush. A small amount of clay in the soil at a building site makes it ideal for building, but clay is an **expansive soil;** that is, it swells and shrinks with moisture changes. For that reason, pure clay is not suitable for building.
- *Organic matter and colloids* – Partly decomposed vegetable and animal material. Organic matter is usually soft and fibrous and is odorous when warm. This material is not suitable for building or as fill because it decays and it loses volume, which causes air pockets and makes the ground unstable. Colloidal clays are very fine particles that can be suspended in water and do not settle quickly. Individual particles cannot be seen with the naked eye. These materials are very susceptible to swelling and shrinking, so they are unacceptable for building.

Soil testing is required at many job sites. Exceptions are some small foundation excavation, small trenching, utility work, and landscaping. For large projects, testing is performed before any earthmoving work begins. From this information,

Teaching Tip/Notes

This exercise will help trainees become familiar with different types of soils. Trainees will need pencils and paper. You will need to set up several workstations and divide the class into small groups. Allow 15 to 20 minutes for this exercise.

1. Set up several workstations. Each station should have a jar of a different type of soil.
2. Divide the trainees into small groups. Have each group circulate among the tables and identify each type of soil. If you wish, also have trainees write one fact about each type of soil.
3. At the last rotation, have one person in the group stand up and identify the soil at their station.
4. Answer any questions the trainees may have.

the earthmoving contractor has a good idea of what kinds of equipment are needed and how to plan the work.

On some jobs, it may be necessary to examine the soils or determine rock surfaces by digging a **test pit.** Although the pit furnishes the best information on the different layers of soil and rock, digging pits is slow and costly work. Pits are usually dug only where special foundation conditions need investigation. **Test boring** is the preferred method where any depth is required. **Core samples** taken from the borings provide the needed information about the various soil types.

Site plans are developed based in part on the information gathered during soil testing, so it is important for equipment operators to understand and accurately follow the project plans and specifications. When in doubt, consult the supervisor for guidance.

3.1.2 Review Plans and Specifications

Construction personnel need to understand what is included in project plans and specifications and how that information is transformed into directions for doing the work.

Most new construction work requires a single set of plans. Other work, such as major reconstruction or rehabilitation of roads and vertical structures, has two sets of plans: the as-built plans documenting the existing roads or structures, and a set of plans for the new work. As-built plans can be very helpful in determining what kind of problems the operator might have in excavating, **dewatering,** or demolishing the existing site.

Project specifications are important documents that provide directions, provisions, and requirements for performing the work illustrated and described in the plans. The items in the specifications cover the methods of performing the work or describe the quality and quantity of materials and labor to be furnished under the contract. Addenda are issued as approved additions and revisions to the original specifications.

A number of other documents are normally incorporated by name into the specifications. These include standards for material specifications, test procedures, safety regulations, and traffic control devices.

Plans and specifications should be accurate, complete, and leave little room for assumptions or interpretation. They should also define specific responsibilities under the contract. If there is a discrepancy, the authority hierarchy is usually established in the contract documents. When in doubt, consult the supervisor.

A typical set of plans is made up of pages of drawings and notes that provide information about the dimensions, materials, and construction sequence of the project. Information in plans varies, depending on the type of project being built. Plans for horizontal construction projects, such as roads, canals, and airport runways, normally contain many drawings showing cross-section details and plan and profile sheets. For vertical construction, such as warehouses, office buildings, and houses, the plan set has foundation drawings and any special **shoring** details for excavating the foundation, as well as the normal building design drawings.

Heavy equipment operators are often involved in the cut and fill work that needs to be done in preparing a job site for construction. The plans that guide cut and fill operations are part of the drawing set that defines a project. *Figure 4* shows a three-dimensional view of a job site. The coloring on the plan has significance: green represents the site as it exists; blue indicates fill required; and red/yellow shows where cuts are needed.

Figure 5 shows cross-sections of the same site. The red lines on the charts represent locations of rock; the blue lines are finish grade; and the green lines are existing grade. *Figure 6* shows how much cut and fill work is needed to complete the project. F represents fill, and each F bubble indicates how much fill is needed. For example, F1+72 means fill 1 foot, 7 tenths, and 2 hundredths. The C (cut) bubbles are interpreted the same way.

Figure 7 is an example of a site plan for part of a subdivision. It shows property line boundaries, existing roadways, and building foundations. In addition, it shows the existing contours of the ground in dashed lines, and the planned grades in solid lines.

3.1.3 Emergency Services

If the job is expected to last for any length of time or is separate from an existing facility or site, certain preparations are required for emergency services. Most jurisdictions have strict requirements for emergency vehicle access to construction sites. Usually the first step is to set up a 911 (or local emergency number) address. This is a unique address given to a site that allows police, fire, and other emergency vehicles to find the location easily. Many areas require that the number portion of the 911 address be clearly visible from the road. The numbers are usually a specified height and are made of reflective material.

In some cases, local authorities may call for construction of special roads that allow emergency vehicles easy access to the site. For example, Cali-

Classroom

Define the terms *test pit, test boring,* and *core sample.*

Explain the importance of plans and specifications to the earthmoving process.

Describe the types of plans used.

Audiovisual

Shows Slides 6–8 (Figures 4–6). Describe how each of these drawings is used in cut and fill operations and explain how to interpret cut and fill instructions. Show how the information from the cut and fill plan is transferred to stakes that the operator will use.

Show Slide 9 (Figure 7). Explain how to read a site plan.

Classroom

Describe the requirements for emergency access to a job site.

Explain why site access roads may need to be specially designed for emergency vehicle access.

Teaching Tip/Notes

Obtain a set of site and excavation plans from a local contractor or engineering firm and allow the trainees to examine them. Explain the purpose of each plan.

Audiovisual

Show Slide 10 (Figure 8). Explain how to interpret a cut and fill stake.

Classroom

Point out the importance of respecting property lines.

Describe the use of stakes to show grading requirements.

Explain that slope stakes are used to maintain grade control during rough grading.

Describe how final slopes are checked.

Figure 4 Three-dimensional view of a job site.

fornia, which is subject to wildfires, requires an access road at least 20 feet wide with a minimum 13½-foot vertical clearance and a suitable load-bearing capability to ensure easy access for fire department equipment. Some jurisdictions may require standpipe or fire hydrant installations to supply water for firefighting equipment.

3.2.0 Laying Out Slopes and Grades

The layout of construction sites varies according to the kind of construction required, the terrain, and the location. The layout is normally done using wooden and metal stakes placed in the ground. Spray paint and flagging attached to some solid object in the construction area may also be used. Right-of-way stakes determine the limits of highway and road projects.

> **NOTE**
> The area beyond the limit of construction may be private property and should not be entered unless noted on the plans.

Grade requirements for the site are written on the stakes set by the survey crew. These stakes tell workers about various cut and fill requirements (*Figure 8*). It is essential for all workers to understand the importance of stakes and to avoid damaging or removing stakes as work is performed. Resetting stakes is a costly effort that can delay the completion of a job and cause cost overruns.

3.2.1 Layout Control

Layout controls are established by accurately setting specific stakes to use as permanent reference points. These are called benchmarks. In turn, the benchmarks are used for placing other stakes, such as slope and grade control stakes. All of these points are shown on the site construction plans.

3.2.2 Slope Control

During excavation and fill activities, slope control is handled with offset stakes that are placed at the edge of the construction limits by the survey crew. These stakes are the principal guides for grading operations. It is important that they not be disturbed by construction equipment.

After rough grading work, slopes must be checked against the requirements written on the slope stakes. Although this is usually the responsibility of the contractor, the owner's representa-

Teaching Tip/Notes

Provide marked grade and finish stakes for the trainees to examine.

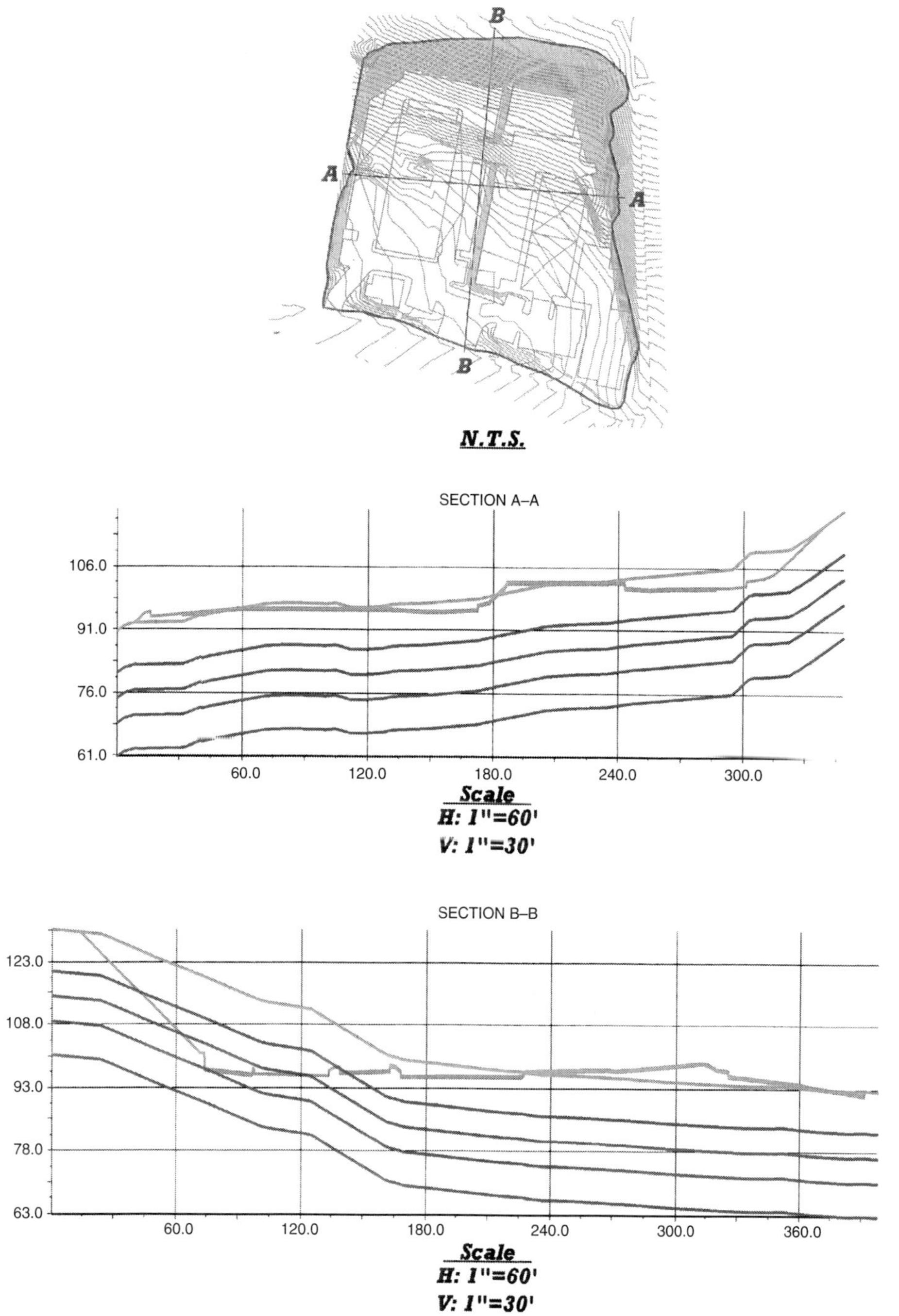

Figure 5 Job site cross-section.

Teaching Tip/Notes

22201-12 Introduction to Earthmoving Module Six 5

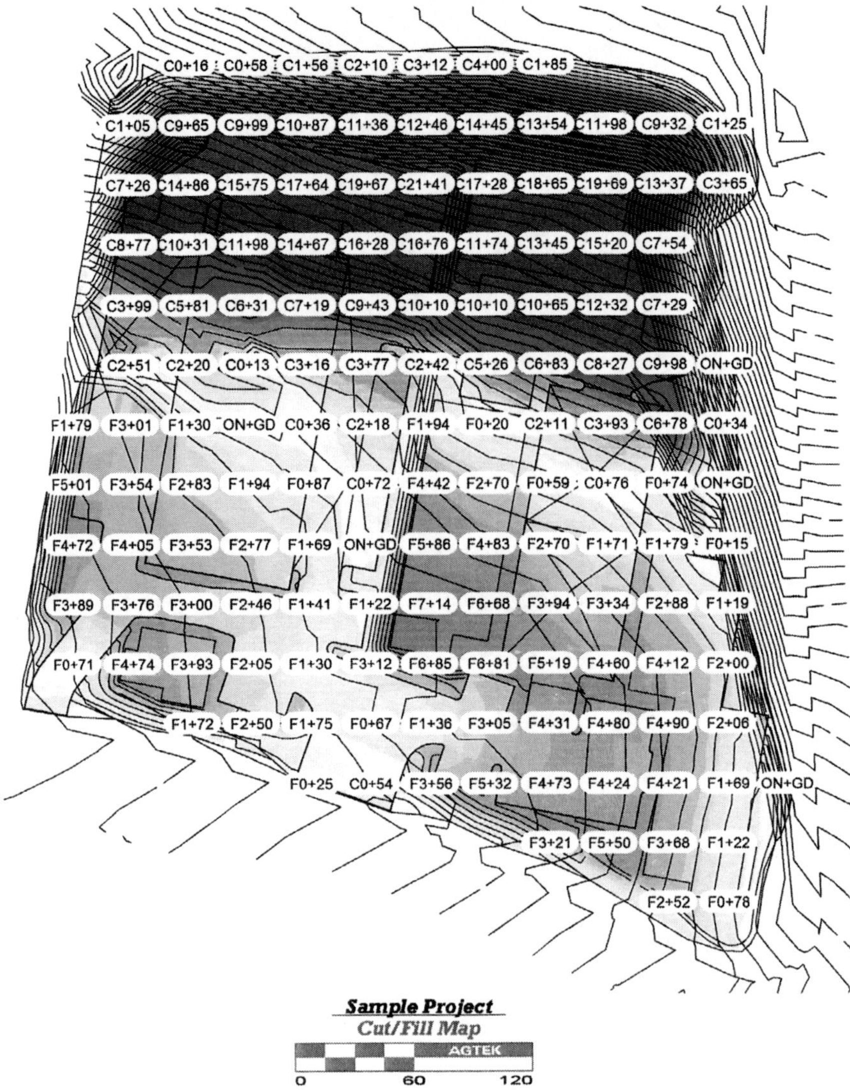

22201-12_F06.EPS

Figure 6 Cut and fill map.

tive also checks them to be sure the slope is built properly. Slopes should be checked when:

- The equipment operation is complete but before it is moved out of the area.
- There is an indication that the slope is flatter or steeper than indicated on the plans, even though the slope agrees with the stakes. The possibility of human error should be considered when checking the slopes.
- The equipment is finishing the backslope or inslope. These are checked as they are being built to prevent extra work later.

Teaching Tip/Notes

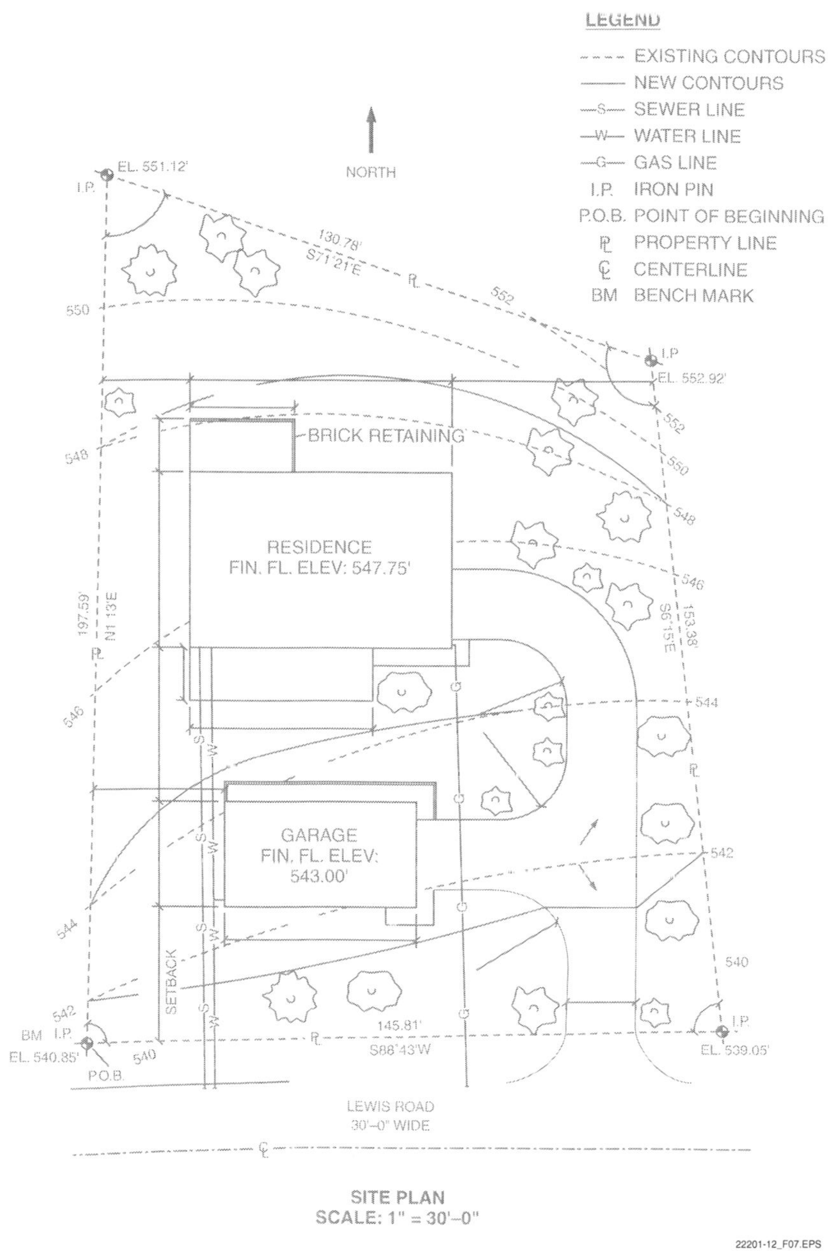

Figure 7 Site layout for part of a subdivision.

Teaching Tip/Notes

22201-12 Introduction to Earthmoving Module Six 7

Classroom

Explain how to interpret a finish stake.

Describe setup procedures for highway and construction site work.

Explain the precautions that must be taken when working in populated areas.

Discuss planning issues such as trench protection and topsoil management.

Discuss the issues that are typically covered in a preconstruction conference.

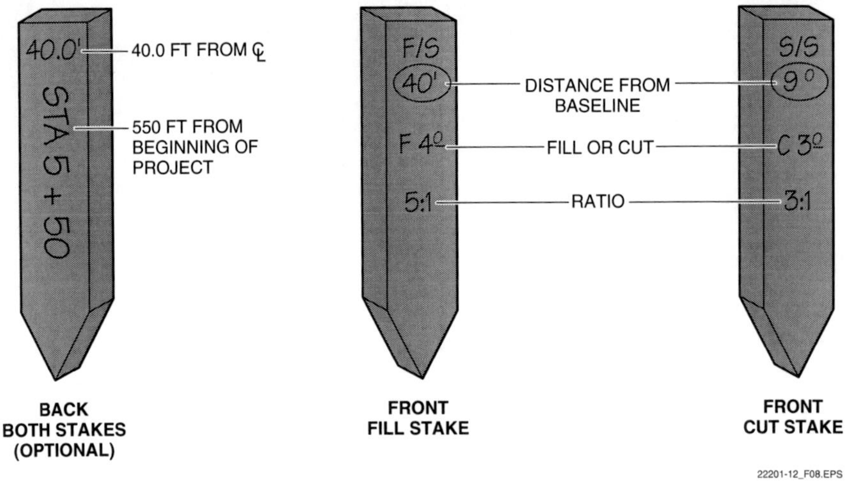

Figure 8 Examples of cut and fill stakes.

3.2.3 Grade Control

The grade is the rise and fall of the constructed surface along the horizontal plane. The basic grade requirements are shown on the profile sheets in the project plans. Slope stakes provide the primary information needed to control longitudinal grades for roadway projects. These stakes normally show the vertical distance from the stakes to the subgrade shoulder points, along with cut and fill information. This information is used to maintain grade control during rough grading.

Finish stakes, called blue tops, are placed by the survey crew to guide finishing work and trimming of the subgrade. These blue tops are driven so that their tops are flush with the proposed grade lines along each edge of pavement and the **center line.**

Heavy equipment operators are responsible for cutting or filling areas to the prescribed grade as stated on the grade stake.

3.3.0 Setting Up and Coordinating Operations

Before beginning any earthwork, the contractor must set up the job. The setup depends on the type of project. A highway project is characterized by a long, narrow area where access may only be available from one or both ends. To get to intermediate points along the route, access roads may have to be cut so that equipment can be carried to the job site. Setting up for a commercial or industrial site is a little different from setting up for highway work. Because these sites can be accessed from existing roads, equipment may be easily transported to the site and unloaded. These areas are often located in populated or high traffic areas, so security and site access may be a more significant issue than for highway work. In addition, these sites may have added complications of excavating for parking areas, landscape areas, and drainage. Also, there may be different requirements for excavating and hauling that affect the type of equipment to use. For example:

- Are there requirements to fill planters and other landscape areas with topsoil?
- How much contour grading is required?
- Is there any topsoil that must be stripped and saved, or does it have to be hauled away? If there is any vertical excavation, how much shoring is required?
- Have all underground utilities been accounted for?

Regardless of the type of project, preparations for construction usually include a preconstruction conference that is held after the contract is negotiated and before any work begins. At the preconstruction conference, the owner's representative, the designer, the construction manager, and the contractor discuss each item of work. The main topics may include the following:

- Foreseeable problems and unusual conditions
- Structures designated for removal
- Erosion, sediment, and **stormwater** runoff control
- Traffic control
- Utility locations

Teaching Tip/Notes

- Emergency plan setup
- Debris disposal
- Signing requirements
- The contractor's work plan and schedule of operation
- Project requirements and specifications
- Equal Employment Opportunity Commission (EEOC) regulations
- OSHA regulations and MSHA regulations
- Any applicable local, state, and federal rules or regulations
- Permits required before work can begin

The purpose of the meeting is to bring the contractor and the owner to the same understanding of the project requirements.

3.3.1 Staging Areas, Equipment Parking Areas, and Site Access

During the preconstruction conference, requirements for equipment staging and parking areas for servicing may be reviewed. On smaller jobs, this may not be a problem because there may be some space that is not being excavated or filled. On highway projects where larger equipment such as scrapers and bulldozers are used, there may be a problem with keeping idle equipment out of the way. Most contractors are very concerned about the condition and security of their equipment. They prefer to move it to an area around the work site or to a fenced location when it is not in use. This may not be practical for highway jobs because of the distance a machine must be transported to get it to a secured area. For highway work, the equipment is often shut down and parked on some level spot close to where the machine is working.

Site access is a concern for many contractors because of the liability that can result from accidents. Security of the equipment must be considered because of vandalism and theft. For highway projects where the work is strung out over several miles, it is difficult to secure the area with any kind of fencing or enclosures. Depending on local conditions, the contractor may decide to park equipment in only one or two locations that can be guarded. For commercial and industrial construction sites, the area is usually smaller and the boundaries are well defined.

Equipment operators should know the policy and procedures for securing their equipment during non-working periods. This includes what can be left in and on the equipment after shutdown, as well as where the equipment is to be parked.

Access to the site should be limited to authorized persons at all times, so operators should be alert to any unauthorized persons on the job site. This includes people who wander onto the site by accident, people who are trying to take shortcuts, and children who are curious about the equipment. Normally, requirements for being on the job site include the use of hard hats and safety vests, so it is relatively easy to spot someone who does not belong there. If anyone looks out of place or looks as if they do not belong on the site, call it to the attention of the supervisor.

3.3.2 Construction Signs

In some cases, it may be necessary to post construction signs to warn the public of the construction activity. Flaggers or other methods of traffic control may be required, especially where construction equipment must cross the road or when the work is very close to a public area such as a commercial building site in an urban area. Work may not begin until all signs and warning devices are installed according to the project plans and applicable government regulations.

3.3.3 Clearing and Grubbing

The work site must first be stripped of all vegetation, obstructions, and unwanted structures before excavation or other construction activity. The removal of this material is referred to as clearing and grubbing, or right-of-way preparation. It also includes the preservation of trees and other objects that are to be saved. Generally, the distinction between the two operations is as follows:

- Clearing is the removal of objects above the original ground.
- Grubbing is the removal of objects below the original ground.

The contractor is generally required to clear and grub all areas of the site. In the cross-section shown in *Figure 9*, the area within the property lines is to be cleared and grubbed of all trees, roots, stumps, and other protruding objects not designated to remain. The depth of cover for objects usually determines whether they need to be removed.

Clearing and grubbing may include the removal of structures and obstructions, including buildings, fences, old pavements, and abandoned pipe or culverts. After the area has been cleared and grubbed, it is scalped of brush, roots, and grass. The top soil is generally removed and stockpiled during this phase.

If the area is undeveloped, some vegetation may need to be cleared. Usually the vegetation is removed by scraping it off with earthmoving

Classroom

Discuss equipment staging for different types of projects.

Explain site access considerations such as equipment parking, site security, equipment security, and safety.

Discuss the types of signs used on construction sites.

Explain the difference between clearing and grubbing.

Audiovisual

Show Slide 11 (Figure 9).

Classroom

Explain how vegetation is removed.

Teaching Tip/Notes

Ask the trainees to give reasons why top soil is removed and stockpiled.

Classroom

Discuss removal and disposal of debris from the clearing and grubbing process.

Homework

Have trainees review Sections 3.4.0–3.8.2.

Materials

Ensure that you have everything required for teaching this session.

Classroom

Explain the purpose of a mass haul diagram.

Describe how cycle times are calculated for different types of equipment and processes.

Explain why cycle times are important for planning jobs.

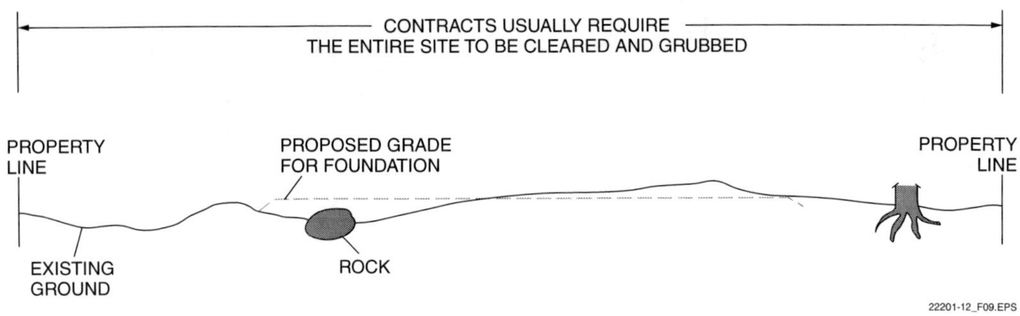

Figure 9 Clearing and grubbing.

equipment and placing it in a designated area. When clearing areas, remember the following points:

- When using a scraper to remove vegetation material, remove shanks and teeth from the scraper's cutting edge to reduce the amount of soil taken with the vegetation.
- Do not leave any vegetation in the construction area. Anything that grows is unsuitable for compaction because it loses volume and causes a depression to form on the excavation surface as it decays.
- Use the proper equipment for grubbing. There may be large trees or boulders that need to be removed before a scraper or grader can start pushing the soil around. Depending on the size, an excavator, bulldozer, or track loader can be used to remove such obstructions.

There are several ways to dispose of debris. It may be hauled away from the construction area, burned (if allowed), buried, or mulched. High-quality top soil may be stockpiled to be used during landscaping at the end of the project. Heavy equipment operators should be briefed on the desired disposal method to avoid costly errors.

3.4.0 Production Measurement

Although equipment operators usually do not get involved in the details of planning the overall project work, they need to understand how it is done. At some point, they may also provide suggestions or input based on their experience with different types of soils and the handling of different pieces of equipment under various conditions.

3.4.1 Mass Haul Diagram

The mass haul diagram is a graphic method of analyzing the movement of soil from one loading area to the dumping site. It is mainly used in highway, large commercial, or industrial projects where the cut and fill work is spread out in a long narrow band. The mass haul diagram relates the proposed grade to the existing grade and is useful in determining the location and volume of excavations and the haul distances to points in the fill.

This analysis identifies any areas where it would be better to obtain material from a place outside the project limits rather than haul material a very long distance. Also, it shows where to dump surplus excavated material in a **spoils** area rather than haul it to a fill area that is further away. These decisions depend on the cost of the equipment and the materials being used.

3.4.2 Cycle Times

Cycle time is the measurement of how long it takes to perform a specific operation, such as the time for a piece of equipment to load, haul, dump, and return to the loading point to start the next cycle. This allows the contractor to estimate how long it will take to complete a specific activity, like rough grading or excavating a ditch.

Cycle times vary for different types of equipment. For scrapers that haul and dump as well as load, the cycle time includes picking up the material from the ground, hauling a full bowl to the dump site, spreading the material, and then driving back to the excavation area to scrape another load. In the case of a front-end loader, the cycle time is measured from the time the loader begins digging with the bucket, through maneuvering to load the truck or stockpile, loading (or dumping the bucket), and then returning to the point of digging.

Most equipment manufacturers provide information about average cycle times, which can be used to estimate the time and cost of doing the work. Contractors also use cycle time measurements to determine the operator's efficiency.

Teaching Tip/Notes

Obtain a copy of a mass haul diagram and allow the trainees to examine it.

3.5.0 Maintaining Haul Roads

Well-maintained haul roads are important for efficient operation. On long haul roads, a grader and water truck should make a pass over the road periodically to keep it in good condition. This reduces the wear and tear on equipment as well as cycle time.

On small jobs, a grader may not be available for use on the haul road. In this case, the scraper operators can maintain the road. If the road is being watered to keep the dust down, the scraper operators must be careful. Wet soil can cause a scraper to skid out of control.

3.6.0 Drainage Requirements

Excavation descriptions in this module are based on the movement of dry material. However, most materials are not completely dry in any natural environment. Even in a desert, the first couple of feet of soil are dry, but as the digging goes deeper, the earth feels cool and moist to the touch. Millions of yards of material are moved every year without major problems with water. Rain may slow or temporarily halt operations because of its effect on the surface, but usually this does not produce impossible excavating conditions. Many construction sites have plans to control the flow of stormwater over the site to prevent erosion and water pollution. The federal government requires these plans, so it is important to understand and follow them. Other drainage problems on construction sites that are related to earthmoving are the control of groundwater and the allowance for drainage from embankment materials.

3.6.1 Control of Stormwater

The Environmental Protection Agency (EPA) requires that contractors performing any activity that disturbs one acre or more of land obtain a National Pollutant Discharge Elimination System (NPDES) stormwater permit before work is started at the site. The purpose of the permit is to ensure that the contractor has a workable plan to prevent pollution and control stormwater runoff from the site, as well as to prevent erosion.

> **NOTE**
> Construction sites that are smaller than one acre, but are part of a larger development, also need to have this permit.

Contractors need to use adequate measures to prevent erosion and sedimentation and to control stormwater runoff. Erosion is the eating away of soil by water or wind. Sediment refers to soil that has been moved from its original place by wind, water, or other means. Stormwater runoff is water from rain and snow that is shed from the ground rather than being absorbed. Heavy equipment operators need to know and understand the site's erosion, sediment, and runoff prevention plans.

Every site is different, so each has its own erosion and sediment control plan. The best way to prevent erosion is to avoid disturbing the surface of the soil, so it is important to follow plans closely and disturb only the earth that needs to be graded. Other measures include reducing the steepness of slopes, covering disturbed soil with mulch, erosion cloth, or **riprap,** and contouring the grade so that water is absorbed rather than shed. A slope graded with a rough surface slows the flow of water and promotes absorption of water rather than runoff, thus helping to also prevent erosion.

Because sediment is soil that has been moved from its original place, the best way to avoid sediment is to prevent erosion. In earthmoving, some erosion is unavoidable, so most sites use devices such as silt fences to trap displaced soil. Some ways to help prevent sedimentation are to avoid walking or driving across disturbed soil; washing soil off truck tires before leaving the site; and confining vehicle operation to approved haul roads.

As rain washes over a construction site, it picks up loose soil and debris, as well as gasoline, oil, and other chemicals from soil or paved areas. These chemicals can pollute water and destroy the wildlife habitat. The first step in preventing such pollution is to prevent the release of pollutants into the environment. One way to do this is by being very careful not to spill solvents, chemicals, oil, grease, paints, and gasoline onto the ground. Employers usually have designated areas for equipment maintenance and wash-downs and a protected storage facility for chemicals, paints, solvents, and other potentially toxic materials. Workers can also help prevent pollution by placing trash in the proper receptacle and using the sanitary facilities provided on the site.

3.6.2 Control of Groundwater

Water existing below the ground surface is typically divided into two zones. The zone of aeration contains varying amounts of moisture and air.

Classroom

Discuss the importance of properly maintaining haul roads.

Explain how drainage and stormwater control are affected by excavation work.

Explain the purpose of an NPDES stormwater permit.

Discuss soil erosion and its causes.

Describe the different methods used to control soil erosion.

Explain the importance of avoiding chemical spills on a site. Describe what can happen as a result of such spills.

Teaching Tip/Notes

Provide a copy of an NPDES permit and allow the trainees to examine it.

Classroom

Define the terms *groundwater* and *water table*.

Describe how the depth of the water table is determined.

Audiovisual

Show Slide 12 (Figure 10).

Classroom

Explain what a sump is and how it is used.

Discuss how water that drains from stockpiles is controlled.

Audiovisual

Show Slide 13 (Table 1). Review the color codes for buried utilities and stress the importance of locating these utilities before starting excavation.

Beneath it, water is continuously present in the zone of saturation. The water in this lower area is called groundwater. The top of this layer is called the water table. *Figure 10* shows a cross-section of the zones of aeration and saturation.

The depth of the water table varies widely. It depends on the amount of annual precipitation, the terrain, and the amount of extraction due to irrigation and other uses of the water.

The depth of the water table and the amount of groundwater may be determined by soil borings. Borings are made at specific locations over the job site to plot the profile of the water table. From this profile, the need for any type of special dewatering activity, such as sumps or various types of drains, can be established.

Sumps are open pits or holes built to collect groundwater. Sumps are then pumped out mechanically. The requirements for sumps vary with the depth and amount of flow rate of the groundwater.

Drains are various types and sizes of ditches that collect and channel the flow of water away from the excavation area. They should be limited in width and depth to the minimum requirements and should intercept the groundwater as close to the bottom of the excavation as possible. The layout of these temporary drains is determined by soil conditions and the location, direction, and quantity of flow.

3.6.3 Drainage of Stockpiles

A typical problem with stockpiles is the storage of wet materials. Wet material does not stack well and tends to spread out or flow. The ease of outflow of water from the stockpile depends on the type of material.

Water that is drained off the stockpile must be controlled. When building a large stockpile of select material, a shallow channel should be cut on the stockpile to collect the runoff and drain it away from the storage area to a holding area. Silt fences must be used to control water flow and discharge of sediment.

3.7.0 Site Excavation

Planning is important in excavation work. Work should be planned so that the equipment never stands idle while grades are being checked or while a decision is being made on where to start the next cut or fill. The operator should study the plans and stakes carefully before starting to work, taking time to analyze the type of equipment needed. For example, is a scraper sufficient to do the cutting of the soil, or is the soil hard enough that some type of ripping is needed to loosen it before using the scraper?

Before breaking ground, it is important to determine if there are any utilities or other structures buried on the site. In most areas a special 8-1-1 number has been established so that contractors can arrange to have a survey conducted to locate and mark buried utilities. This service requires advance notice. Markers such as flags are commonly used to mark the locations and types of utilities. These markers are color-coded using the standard underground color code found in *Table 1*. When workers see these markers, they should act accordingly. The plans should be checked to make sure no utilities were missed. If in doubt about the proper actions to take, workers should ask the supervisor.

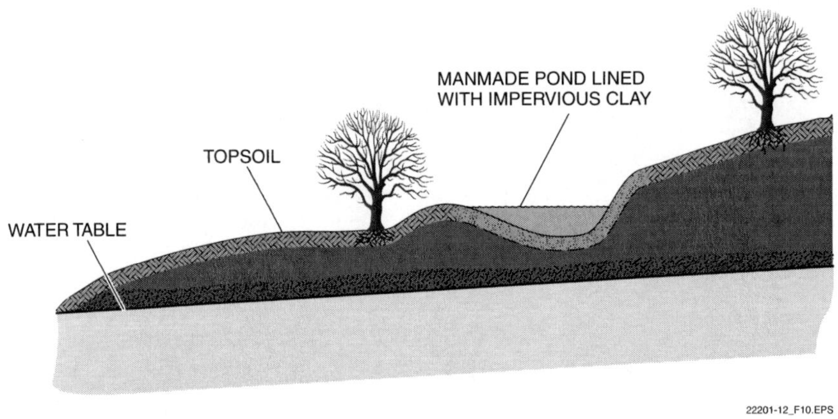

Figure 10 Cross-section of a water table.

Teaching Tip/Notes

Table 1 APWA Underground Utility Color Codes

Color	Meaning
Red	Electric power lines, cables, conduit and lighting cables
Yellow	Gas, oil, steam, petroleum, or gaseous material
Orange	Communication, alarm or signal lines, cables or conduits
Blue	Potable water
Green	Sewers and drain lines
Purple	Reclaimed water, irrigation and slurry lines
White	Proposed excavation
Pink	Temporary survey markings

Several types of equipment are used for excavating. The machine that is chosen for a particular job must fit the requirements of the work and have a cycle time that allows the work to be done cost effectively. Typically, the equipment is matched to the type of material to be excavated. The equipment must be capable of operating in the environment without being unstable or inefficient. In general, the following types of heavy equipment are used for excavation:

- Scrapers (*Figure 11*)
- Excavators (*Figure 12*)
- Motor graders (*Figure 13*)
- Bulldozers (*Figure 14*)
- Wheel and track loaders (*Figure 15*)
- Backhoe loaders (*Figure 16*)

Each type of excavation presents its own challenges. Some jobs require work at several different excavations sites, including the following:

- Highway/roadway excavation
- Several types of bulk excavations
- Limited-area vertical excavation
- Trenching

22201-12_F11.EPS

Figure 11 Scraper.

22201-12_F12.EPS

Figure 12 Excavator loading a rigid-frame, off-road dump truck.

22201-12_F13.EPS

Figure 13 Motor grader.

22201-12_F14.EPS

Figure 14 Bulldozer.

Audiovisual

Shows Slides 14–19 (Figures 11–16). Explain how each machine is likely to be used on an excavation job.

Teaching Tip/Notes

22201-12 Introduction to Earthmoving Module Six 13

Audiovisual

Show Slide 20 (Figure 17). Explain what a balanced job is.

Classroom

Explain when a bulk-pit excavation is needed.

Audiovisual

Show Slide 21 (Figure 18). Explain the importance of cycle time and how it can be minimized.

Show Slide 22 (Figure 19). Discuss the pros and cons of rearranging the entry and exit points.

Classroom

Discuss gradation of material in a bulk-wide area excavation.

22201-12_F15.EPS

Figure 15 Wheel loader.

22201-12_F16.EPS

Figure 16 Backhoe loader.

3.7.1 Roadway Excavation

Roadway excavation is a leveling process used in highway construction and airfield grading. It often involves the movement of material from high ground to low areas. This is called a balanced job because the soil needed for fill areas is obtained from cut areas and there is no need to haul material in from outside the project (*Figure 17*). In other cases, cut material may be wasted because it is not needed or is unsuitable for fill. Material moved from a borrow area to fill low spots for embankment construction may also be included.

3.7.2 Bulk-Pit Excavation

Bulk-pit excavation is, as the name implies, the digging of a pit, or a large, deep excavation with vertical sides. It involves digging and loading material of great depth and volume and then hauling the excavated material to another site. The pit may be at the project site, such as the construction of a building with a basement, or it may be outside the work site, such as at a borrow pit.

On Site

Potholing

A method known as potholing is used when it is necessary to locate the exact position of a buried utility by exposing it. Potholes can be dug with mechanical equipment, such as air blasters or excavation vacuums, or digging by hand.

As covered earlier in this module, the cycle time of any equipment is the measurement of how long it takes to perform a specific function. When excavated material is loaded onto a truck for removal from the site, it is important to minimize the movement of the excavator or loader for the shortest cycle time. *Figure 18* shows methods of loading trucks. In Method 1, the truck is positioned so that the excavator boom swings between the excavation face and the truck bed without having to move the excavator. In Method 2, the loader must move to the excavation face for loading, than back up and move into position to dump its load into the truck bed. In order to keep the cycle time at a minimum, the loader picks up a load, backs up a short distance, turns 45 degrees, and raises its bucket. The truck then backs under the raised bucket.

There is a third method of moving excavations away from the excavation site, and it is used in special conditions where it is not necessary to do more than push the excavated material aside. In this method, a crawler-mounted loader or bulldozer uses its bucket or blade to dig and collect material on the front, and then pushes beyond the access ramp to the excavation site.

In the scenario in *Figure 19*, both the entering and exiting traffic are using the paved road below the site. If the paved road above the site were used for entry and the one below the site used for exiting vehicles, it would permit more efficient traffic flow. Even when workers are fortunate enough to experience this on large jobs, local permitting may limit the construction traffic flow to a single point. In addition, the construction of two access roads would increase the cost of the job.

3.7.3 Bulk Wide-Area Excavation

Bulk wide-area excavation is similar to bulk-pit excavation, but this type of excavation permits access to the area from many directions, making it easier to access and leave the excavation site. Bulk wide-area excavation is usually shallower in depth but larger in area than bulk-pit excavation.

Teaching Tip/Notes

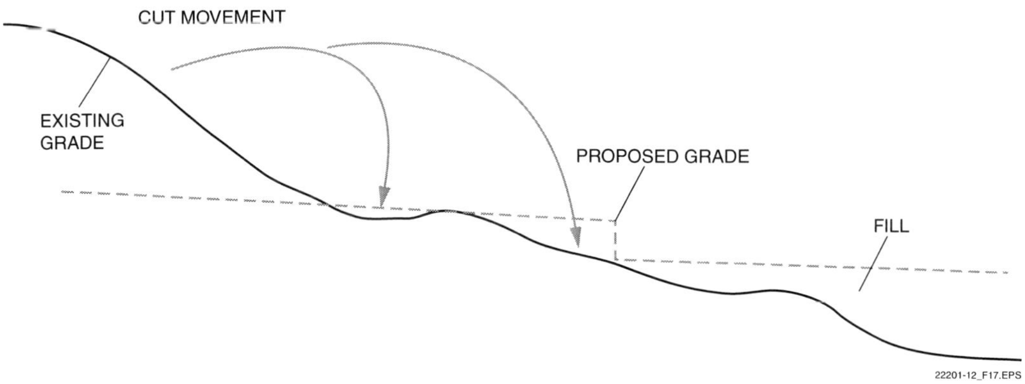

Figure 17 Balanced excavation.

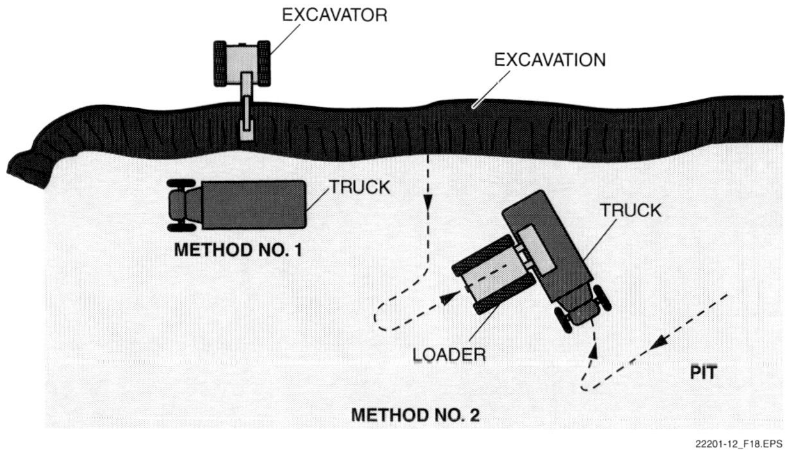

Figure 18 Loading a truck at the digging face.

Bulk wide-area excavation is used in highway construction, airfield runway grading, the removal of top layers of soil (as in quarrying or strip mining), and the building of earthen dams. In some cases, it only involves moving material from high ground to low areas. Other times, it involves wasting material that is not needed or that is unsuitable for fill. Material moved from a borrow area to make up deficiencies in embankment construction may also be included.

Equipment for this type of operation usually involves scrapers, bulldozers, and possibly front-end loaders. Selection of the specific pieces of equipment depends on the size of the job and the characteristics of the soils being moved. Scrapers have difficulty in loading loose, dry sands and rock, even if it has been crushed. They also have trouble unloading wet sticky clays. The **gradation** of the material is also an important consideration when choosing the type of equipment.

In this type of excavation, the ideal location ensures short haul distances. For short distances, bulldozers and graders can be used to scrape the material, push it the required distance, and then level it as required. If haul distances are longer, scrapers should be used when the material can be loaded efficiently. Scrapers can cut a layer of material while moving forward, and they can travel at speeds up to about 30 miles per hour (mph) when fully loaded. At the dump area, they can slow down and dump the material by spreading it in layers at the specified thickness. *Figure 20* shows effective haul distances for various types of hauling equipment.

Classroom

Describe the types of locations where bulk-wide area excavation might be used.

Audiovisual

Show Slide 23 (Figure 20). Discuss the effective haul distances for different machines.

Teaching Tip/Notes

When this type of excavation is located far from the dump site, material is loaded onto trucks or other carriers and carried to the dump site. If this is the case, the loading area is set up to permit the smooth flow of traffic and quick loading of material.

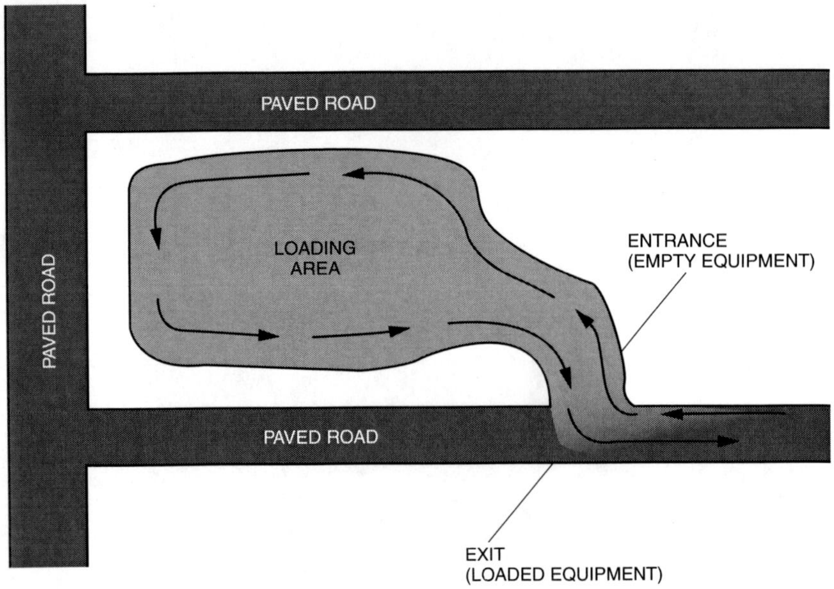

Figure 19 Bulk-pit excavation.

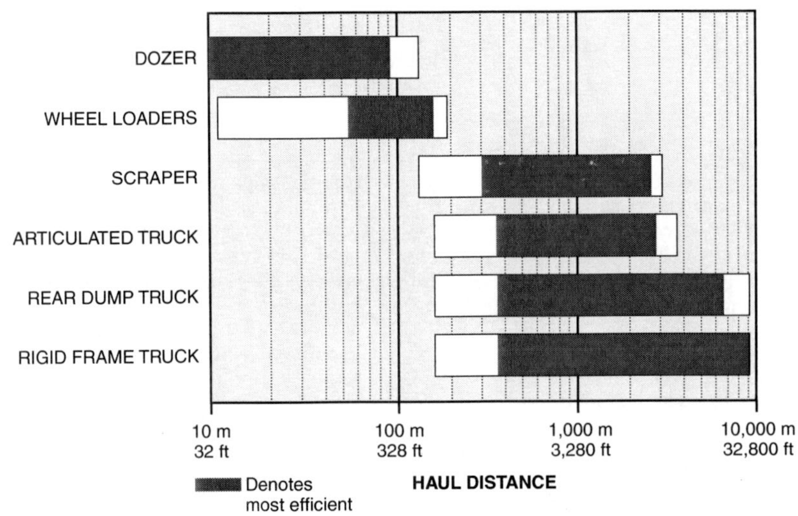

Figure 20 Effective haul distances for various types of equipment.

Teaching Tip/Notes

16 NCCER – *Heavy Equipment Operations Level One* 22201-12

3.7.4 Channel Excavation

Channel excavation is the excavation of loose, unconsolidated materials, of material lying under water (such as a drainage channel), or of saturated soils that prevent the equipment from traveling over the excavated area. This type of excavation is usually performed by equipment standing on solid ground at the same level or higher than the material being worked.

3.7.5 Limited-Area Vertical Excavation

Some excavation must be done by lifting the material vertically out of the pit because the sides of the excavation require shoring or shielding. This type of excavation is known as limited-area vertical excavation. *Figure 21* shows an example.

3.7.6 Trench Excavation

A trench is a temporary opening in which something, such as a pipe or box culvert, is placed and covered. The trench width, and usually the depth, are limited to 15 feet. One type of trench excavation is known as cut and cover. This sometimes includes excavating a large and sometimes deep trench, laying the specified pipe or culvert, and then covering everything over or building on top of it. This option is frequently used for highway and transit system tunneling when the excavation of the trench would cost less than boring a tunnel.

Trench excavation is different from pit or wide-area excavation. Pit or wide-area excavation focuses on moving large quantities of material. With trench excavation, the focus is on the rate at which the pipe or culvert can be placed. This factor affects the type and size of equipment selected for the trenching job. The trench should not be excavated any wider than is necessary for bedding and setting the structure. Shoring can be used to reinforce the sides of the trench.

> **NOTE**
> Do not confuse a trench with a ditch. A ditch is a narrow slot cut in the ground and left open. A trench is an opening in which something is placed and then covered.

The three main classes of equipment used in trenching are backhoes, excavators, and trenching machines. The trenching machine is specifically designed for trench excavation. *Figure 22* shows different types of trenchers. *Figure 23* shows a

22201-12_F21.EPS

Figure 21 Limited-area vertical excavation.

22201-12_F22.EPS

Figure 22 Trenchers

Classroom

Point out that channel excavation is typically performed using an excavator working from dry ground.

Audiovisual

Show Slide 24 (Figure 21). Describe limited-area vertical excavations.

Classroom

Explain the difference between a trench and a ditch.

Audiovisual

Show Slides 25 and 26 (Figures 22 and 23). Identify situations in which a trencher might be used as opposed to a backhoe or excavator.

Teaching Tip/Notes

Classroom

Explain that loading is the first step in the process of moving material from one location to another.

Describe the situations in which a loader, excavator, or scraper might be selected for loading.

Describe the difference between loading from a stockpile and loading from an excavation.

skid steer digging a trench with a trenching attachment.

3.8.0 Loading

Regardless of the type of project, if the material is to be moved from one area to another, it must be loaded, hauled, and dumped. The equipment that transports the earth must return to the loading point for another load. Selection of the right type and size of equipment to do these basic operations is important for efficiency and cost. Usually, there is more than one way to perform the work, but the most efficient and cost-effective method should be selected.

Loading is the first step in the earthmoving cycle. Loading is performed as a basic function, such as loading material from a stockpile or spoils, or as a secondary function when some other work is being performed. This could be some type of excavation, such as trenching or cutting with a scraper for cut and fill operations.

The specific job and the type of material encountered has a great influence on both the selection of equipment and the method used.

3.8.1 Selection of Loading Equipment

Selection of loading equipment depends on factors such as cycle time, type of material, and equipment efficiency. For operations that move large amounts of material, the type of hauling equipment must also be considered when choosing the loading unit. Another important factor is the overall cost of the operation, which includes the hourly cost of using a certain type of equipment.

Figure 23 Trenching attachment in use.

The options available for mobile loading equipment include the following:

- Front-end loaders (and variations)
- Excavators
- Scrapers
- Shovels

Front-end loaders can be used for loading from stockpiles, **windrows,** and spoils piles. They can also be used to excavate and load in the same operation. Configurations include wheel and track loaders. The size range and tight turning radii of front-end loaders make them well suited for loading operations.

Excavators come in many sizes and configurations. The term *excavator* includes backhoes and telescoping excavators. Although they are primarily used for excavation of material below the grade or into banks, they can also load material into dump trucks and other haul equipment. When the job requires only one operation, such as clearing or excavating ditches, an excavator can be used to do the excavation and load the truck.

Scrapers are special pieces of equipment that typically are used where large quantities of earth need to be moved a long distance. They are built to pick up the material through a scraping action, load it into the bowl, and transport it to the dump site. Scrapers also can be used as haul units, with other pieces of equipment loading material into the scraper bowl.

Shovels are mainly used to excavate material in pits and quarries. Because they are track mounted, they do not move very quickly. However, shovels can be manufactured in large sizes that can dig and load large quantities of material in one cycle.

Other types of loading equipment include conveyors, overhead buckets, and cribs.

3.8.2 Methods of Loading

Equipment unit size and maneuverability establish the method of loading. As described earlier, some equipment can load as well as excavate. Loading consists of two basic methods: loading from piled material and loading from an excavation.

When loading from stockpiles and spoils, the front-end loader is the primary piece of equipment used. It is able to maneuver well and work with many different configurations of haul units. Loaders usually follow an I or Y pattern when loading from a bank of material. The configura-

Teaching Tip/Notes

tion used depends on the amount of available work area. Generally, if the loader has to travel more than one and a half wheel revolutions, the operation is inefficient. *Figure 24* shows a front-end loader working against a bank while loading trucks using the Y pattern.

Excavators are most efficient if they remain stationary while they work. Their method of loading is limited to loading directly from the excavation or first dumping the material in a pile next to their position and then loading from the pile into the haul unit. This method is not suitable if the area is too narrow to permit the excavator and haul unit access to the site at the same time.

3.9.0 Hauling

Hauling is the movement of material from one point to another. This may be a very short distance within the construction site or a long haul to dispose of material, such as that from clearing and grubbing operations. Hauling is normally broken into two types:

- Over-the-road hauling is hauling over public highways using a haul unit such as a dump truck.
- Off-road hauling is done within a job site and not on public highways. Equipment used for off-road hauling includes large trucks that are not allowed on public highways, as well as scrapers, front-end loaders, and regular trucks.

3.9.1 Selection of Hauling Equipment

Equipment used for the two types of hauling is very different in design and construction. The specific operation determines which type to use.

Over-the-road haul units have very distinct limitations because highways are designed for specific load conditions, and truck load limits are established according to their loadbearing capacity. These units are also restricted in width and height. These limitations are enforced by state or federal law.

Off-road hauling units may include scrapers, front-end loaders, and trucks of various types. Off-road equipment is usually wider and higher than over-the-road dump trucks. This allows for an overall increase in the size and capacity of the equipment. For example, a tandem-axle over-the-road dump truck can carry a load of about 12 cubic yards of material; a large scraper can carry up to 44 cubic yards.

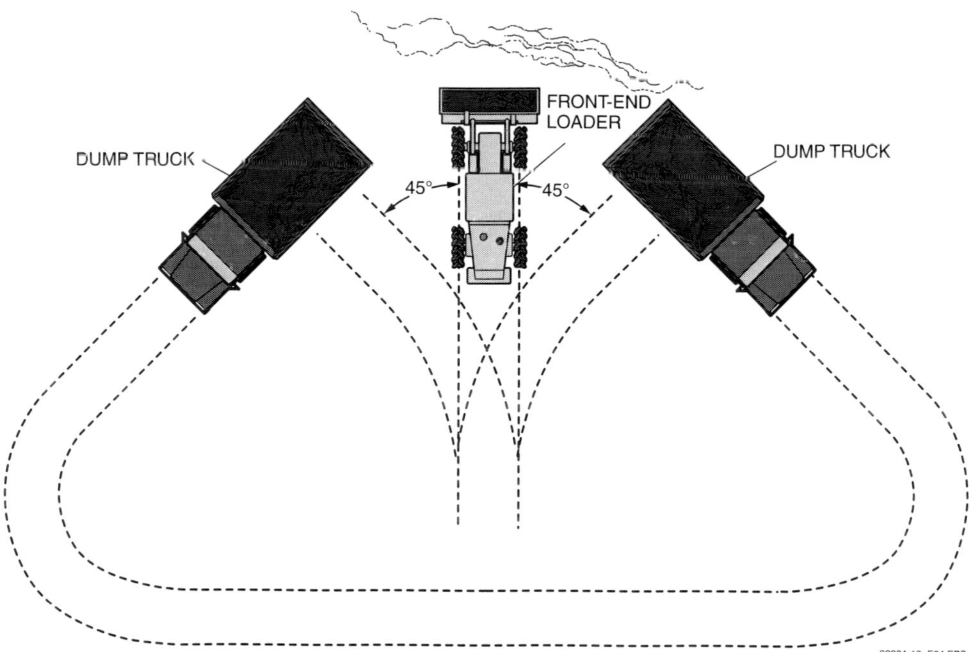

Figure 24 Loading dump trucks using the Y pattern.

Audiovisual

Show Slide 27 (Figure 24). Discuss why this loading method is more efficient than others.

Homework

Have trainees review Sections 3.9.0–5.2.0.

Materials

Ensure that you have everything required for teaching this session.

Classroom

Explain the difference between off-road and over-the-road hauling.

Describe the criteria considered in selecting haul units for a particular project.

Audiovisual

Refer back to the haul distances chart (Slide 23/Figure 20).

Classroom

Discuss the different types of off-road hauling equipment.

Teaching Tip/Notes

Classroom

Describe the use of haul roads and other methods that can improve hauling efficiency.

Explain the differences between dumping spoils and building an embankment.

Describe how stockpiles are used and explain what a pay item is.

There are several considerations in the selection of the most appropriate haul unit. The first is whether any portion of the haul will be over public highways. If so, equipment-use restrictions must be checked. The person operating the equipment is responsible. If there is a violation, the operator gets the citation. Also, the haul route needs to be checked for potential problems, such as bridge height, width, and weight limitations. Hauling over narrow, curving roads with high crowns can be hazardous for larger trucks. High road crowns that can tilt the truck may limit the allowable heaping when the unit is loaded.

Individual job conditions dictate off-the-road types of hauling equipment. There are basically three configurations to be considered.

- *End-dump (rear dump)* – These types of trucks should be used if material consists of rock fragments or masses of shale that are too large for other units to unload, or if dumping is to be done over the edge of a waste bank or fill. Rear dumps provide maximum flexibility for a variety of job conditions. They can be straight body or articulated trucks.
- *Bottom-dump* – These units are good for transporting free-flowing materials over reasonably level haul routes that permit a high travel speed. They are best for dumping in windrows over a wide area.
- *Scrapers* – Scrapers can haul a large amount of material at relatively high speeds for short distances. They have an added advantage of being able to load the material directly by cutting or scraping a thin layer of soil while moving forward at a slow speed. When the bowl is full, the scraper can travel at a higher speed to the dump site and spread the material in a thin layer. Because of their large size, scrapers are not normally used in small or confined areas. They are good for cut and fill work on highways and airports, as well as for general contouring of the ground.

Another type of truck that may be used under special circumstances is a side-dump truck. The trailer on this type of truck is built to pivot to the right or left and quickly dump its load. It is good for road-widening applications.

3.9.2 Hauling Methods

Hauling is the largest single factor in cycle times for earthmoving operations. Assuming that the material is transported by some type of haul unit instead of a cable or conveyor system, the synchronization of the hauling operation is the main concern. This includes the timing of the hauling as well as the traffic pattern. If the job requires a considerable number of haul units, a lot of time can be lost if work operations are not timed properly. Also, for jobs requiring the placement of large amounts of cut and fill or the removal of excavated material, it may be necessary to set up special haul roads and lay out traffic patterns.

A traffic control supervisor may be required for operations using a substantial number of haul units. The supervisor controls operating conditions at both the loading and dumping sites and keeps track of the haul units in transit.

3.10.0 Dumping

Material may be dumped either to build up an area to make an embankment or to dispose of spoils. The dumping must be closely controlled when building up an embankment or when dumping at a spoils site.

3.10.1 Dump Sites

Spoils areas are used for the disposal of surplus excavation materials or materials that are unsuitable for fills. The location of the spoils area may depend on the type of material and the haul distance involved. Soft materials going to a spoil area may require mixing with firmer soils. Organic debris may be burned or spread out in thin layers. Rocks and boulders may require placement in a specific way to reduce environmental concerns, such as erosion. Dumping of spoils outside the project site may be more convenient, but the cost of each option must be considered.

If the material to be dumped is for filling, backfilling, or making an embankment, these areas are shown on the plans and indicated with the survey and slope stakes. Material is sometimes dumped alongside the area to be filled and then moved into place with bulldozers or graders. The project engineer or activity supervisor marks these areas with stakes or paint so the equipment operator knows where to dump each load.

Another type of dump site is a stockpile. Stockpiles are used to store material for later use in the project. A good example of this is the topsoil that is removed at the beginning of a project. Because good topsoil is an expensive **pay item,** it is saved for landscaping and contouring at the end of the project.

3.10.2 Dump Patterns

Dump patterns depend on the material being dumped and the type of haul equipment being used. For spoils dumping, the material may be

Teaching Tip/Notes

dumped into another excavation, such as an abandoned pit or over an embankment, or it may be dumped in a level, open area. Often, spoils are hauled in a rear-dump truck. Depending on the amount of spoils, dozers or front-end loaders may be needed at the dump site to move the material around and blend it in with the surrounding terrain.

The haul equipment for fill material usually includes bottom- or rear-dump trucks or scrapers. When backfilling, the material is dumped directly into the excavation or beside the excavation, and then another piece of equipment, such as a loader or dozer, is used to place the material.

Embankment construction fill material can be placed by either trucks or scrapers. *Figure 25* shows several different dumping patterns.

A dumped fill pattern for level ground is indicated in *Figure 25A*. The material is placed in many individual piles so it is easier to spread. This

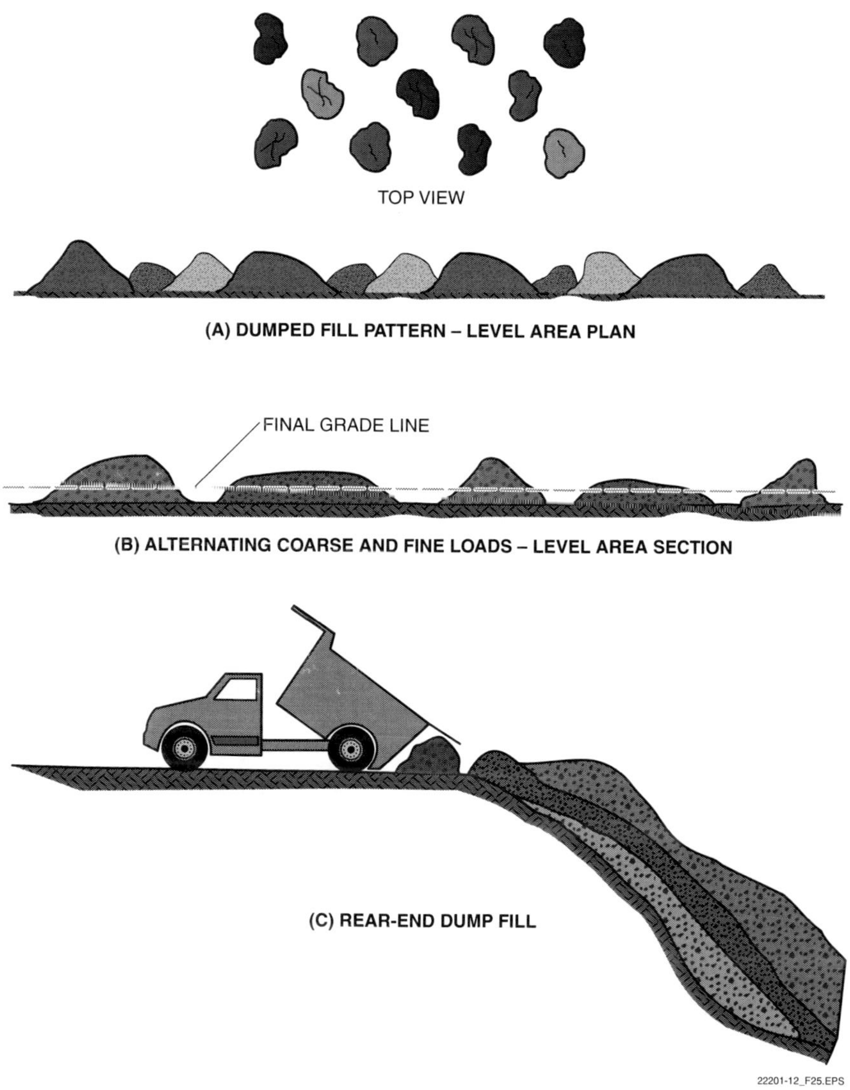

Figure 25 Dumping patterns.

Audiovisual

Show Slide 28 (Figure 25). Discuss different methods of dumping.

Teaching Tip/Notes

Classroom

Describe how materials may be dumped when they are to be blended.

! Safety

Point out the importance of having a berm in place when dumping down a hill.

Classroom

Describe methods that can be used to safely reduce the time required for return runs.

Audiovisual

Show Slide 29 (Figure 26).

Classroom

Point out the need to test fill material and to make sure it does not violate government regulations.

Define the term *common borrow* and explain the ways in which common borrow is used.

method reduces the amount of work required to move material over longer distances.

Figures 25B and *25C* show dumping alternating coarse and fine material along an embankment in order to provide for thorough mixing of the material. In *Figure 25B*, the material is dumped as the truck is moving, in order to spread the material. The method shown in *Figure 25B* is used for consolidated materials that will blend well. Note the safety berm behind the truck in *Figure 25C*. Once the loads have been dumped, a dozer or motor grader is used to blend the coarse and fine material together. Another method used to place two different types of material prior to blending is to completely spread one type of material before the second type is dumped and spread on top. This method is used for materials such as clay and sand that do not blend easily. A scarifier is then used to mix the coarse and fine material before compacting.

In highway construction, material used to build an embankment is usually dumped in layers. This can be accomplished with scrapers, which can spread the material as they dump, or with trucks, which dump in piles or windrows. These piles and windrows are then leveled out with a dozer or motor grader.

3.10.3 Return Run

The success or failure of an earthmoving operation is based in part on how quickly a truck can safely return to the excavation site for another load. Equipment operators can help make the operation a success by the following behaviors:

- Driving cautiously at all times.
- Using the prescribed return route from the dump site. It is usually set based on the distance, traffic, and number of turns between the excavation area and the dump site. If haul roads are used, the pattern for the return trip is set and must be followed by all operators to avoid disrupting traffic flow.
- If there are no haul roads, then the truck must be driven through the construction site. In this case, the project engineer should establish a pattern for the hauling and **backhauling.** Operators follow markers laid down for this purpose, and ground personnel guide the equipment out of the area once the dump is made.
- Staying in line and following the directions of the traffic supervisor.
- Thinking ahead about how to maneuver the machine when approaching the loading site for the next cycle.

For scraper operations, the dozer is positioned where the scraper can come up to it. (*Figure 26*). A good scraper operator, working with the dozer operator, can perform this operation smoothly and quickly. Otherwise, much time is lost in starting, stopping, and lining up the two machines.

Dump truck operators can help loader operators by getting into position quickly and keeping an eye on the loader. If a spotting marker has been placed, the operator should know where it is and how to position in relation to it.

3.11.0 Fill and Embankment

Excavated materials can often be used as fill, backfill, or to build an embankment, but it may be subject to one or more soil tests before it can be used as backfill or embankment. The number of tests required depends on the intended use of the material. Before filling, backfilling, or creating an embankment, always check federal and local regulations for the types of materials and the materials content allowed.

3.11.1 Fill Material

Some fill may be material stockpiled from excavations within the site. Other fill may be brought into the site from another area as needed. Fill material can be classified as three types: common borrow fill, waste fill, and select fill. In general, all fill should be kept free of organic matter if any future use of the filled area is planned.

Common borrow fill meets a pre-established project requirement. It is usually brought onto the construction site and dumped in piles. It is then spread uniformly over an area and compacted according to the job specifications.

Common borrow fill can be used to build up an area where little or no compaction is required.

22201-12_F26.EPS

Figure 26 Dozer pushing a scraper.

Teaching Tip/Notes

This exercise will familiarize trainees with excavation practices. Trainees will need appropriate personal protective equipment, pencils, and paper. You will need to arrange for an opportunity to tour an excavation construction site. Alternatively, invite a safety expert to give a presentation. Allow 1 to 2 hours for this exercise.

1. Discuss excavation procedures. Review the excavation sequence, equipment used, and the site safety plan.

The fill is simply dumped into the area, such as embankments, scraped area, or around some types of buildings. When fill is dumped into a deep fill area and allowed to roll down the slope, it tends to separate because the larger fragments roll to the bottom and the fine material stays close to the top. Dumped fills should be kept free of tree stumps, organic matter, trash, and sod if any future use of the filled area is planned. An exception to this process is the special case of sanitary landfills, where garbage and trash are spread several feet thick over the ground surface, covered with additional clean earth, and left to compost.

Waste fill is recycled, treated material approved for use in certain landfill areas. This could include embankment, landscaped areas, or backfill around some types of structures. Compaction requirements are based on where the material is being placed. When rock is used, it should be spread over the area rather than piled at a single point.

Select fill is material with specific properties that can be obtained only by selective excavation or manufacture. This category includes selected gravel or sand; selected rock used for stabilizing slopes; and dumped riprap. Again, the equipment used to place fill material includes the dump truck, scraper, dozer, and sometimes the front-end loader. Spreading of the material is accomplished by the scraper, dozer, and motor grader. Compaction can be accomplished by wheel or tractor compaction or by a mechanical roller of some type. Select fill is usually more expensive than other types of fill and must be used, placed, and compacted as specified by the project plans.

3.11.2 Backfill

Backfill is the returning of excavated material to an excavation site. Before the excavations, the ground was fairly stable because the soil had settled. This is called **consolidation.** It is important to backfill and compact excavations so that a minimal amount of future settling occurs at the site. This may seem simple, but because of compaction issues, some backfills are poorly made and cause damage to the buried structure and the surrounding area.

The main concern with backfilling is choosing the best method for compacting the fill material. Although the specifications for some work require that all excavated material be disposed of and specifically selected materials be hauled in for backfilling, most contracts allow the use of excavated or dumped fill for backfill. If dumped material is used, compaction criteria must be varied to suit the soil encountered. While poorly compacted material does ultimately settle, overcompacted material does not settle at the same rate or under the same loads as adjacent natural soils, and this effect can be just as damaging.

For backfilling around small structures and buried utilities (*Figure 27*), a backhoe is the ideal piece of equipment. The backhoe operator needs to understand the process of compacting backfill material so that he or she can place material at points where the workers need it to build up the backfill. By compacting too much in one area, pressure builds up against the structure and causes it to bend out of alignment, or even break.

3.11.3 Embankment Construction

Embankments are fills that are placed for specific purposes under carefully controlled conditions. Embankments usually consist of soils of a prescribed particle size placed with careful selection, compaction, moisture control, and mixing.

Embankments are used for highways and roads, earth dams, levees, canals, and runways and are classified according to method of compaction:

- *Equipment-compacted fill* – This is generally a fill, often of select material, that is compacted by the wheels of the haul units. Moisture control may be required. This type of embankment construction normally does not have a compaction requirement.
- *Rolled-earth fill* – This type of compaction is used for roads, building foundations, dams, canals, and other sites where highly impervious fills are required. The degree of compaction is normally specified for rolled-earth fills. The

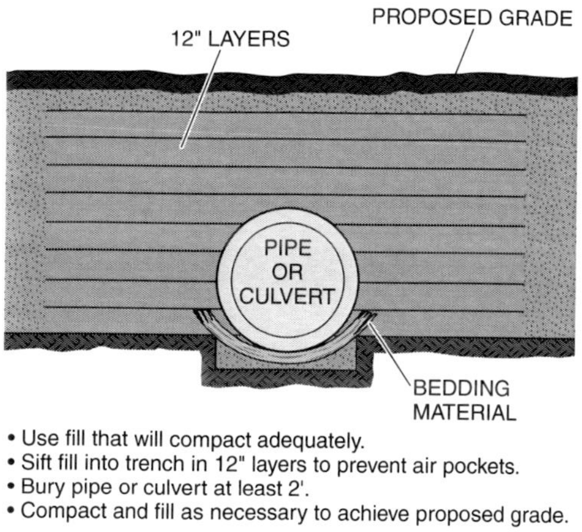

Figure 27 Backfill.

Classroom

Explain that material used as fill must be free of stumps and other organic debris.

Describe the difference between waste fill and select fill.

Explain that backfill must be properly compacted.

Point out that a backhoe is commonly used for backfilling.

Audiovisual

Show Slide 30 (Figure 27). Describe the proper method of backfilling.

Classroom

Explain that embankments are used for dams, roads, canals, and other purposes and that they contain soils of a prescribed particle size.

Describe the two classifications of embankments.

Teaching Tip/Notes

(continued)

2. Ask the safety coordinator to give a tour of the site. Ask the presenter to cover excavation, loading, hauling, dumping, communications, and site safety. Have trainees take notes and write down questions during the tour or presentation.
3. Ask company personnel to spend some time after the tour or presentation to answer questions. Have trainees present their questions.

Audiovisual

Show Slide 31 (Figure 28). Describe the construction of an earthen dam.

Classroom

Explain how to ensure uniformity when building a rolled embankment.

Point out that the original ground must be broken up by scarifying.

Explain that compacting soil forces out air and water from between the particles.

design is based on the properties of the soils to be used. *Figure 28* shows a rolled-earth fill.

The embankment may be placed on the natural ground, or it may require excavation to another layer under the ground. Embankment material is spread and leveled in layers or lifts.

The equipment and procedures involved in dumping and spreading the embankment material depend largely on the type of equipment the contractor has on the project and the type of material being used. Embankments are built in lifts, one on top of the other, for the full width unless otherwise specified. The contractor must have enough spreading, mixing, and compacting equipment and use the procedures necessary to meet moisture-density requirements and obtain uniform, well-mixed layers of embankment.

The scrapers usually follow patterns in picking up and depositing material that do not interfere with other equipment operations. The work should progress at a uniform rate without slowdowns caused by improper direction of equipment. It is extremely important that embankments be built uniformly. The following work activities ensure uniformity:

- Fill should be placed evenly from one side of the embankment to the other.
- Each lift should be spread to a uniform thickness that does not exceed the specified thickness before compaction. When excavated material consists predominately of rock fragments, the specification may allow the material to be placed in the embankment in layers not exceeding the thickness of the average size of the larger rocks. Each of these rock lifts must be leveled and smoothed with finer material, but the final rock lift must be several feet below the finish grade.
- Each lift must be compacted before the next lift is placed. Double dumping, or placing two lifts without compaction, is not allowed.
- Where large chunks of material are placed, extra compaction and blading are needed to spread the material uniformly.

Under embankment areas where fill is to be placed, the original ground must be broken up by scarifying it to a specified depth and then recompacting it.

3.12.0 Compaction

Compaction increases the density of soil to create a stable ground surface. It involves pressing soil particles together to force out air and water from the spaces between the particles. Compaction reduces the risk of future settling, so it provides a solid surface on which to build. After soil is disturbed, compaction occurs naturally over time as a result of wetting, drying, freezing, thawing, groundwater movement, and the weight of the top layer of soil.

In construction, it is necessary to speed the compaction process with equipment that imitates nature. Various types of equipment use pressure, kneading, vibration, and impaction, or a combination of these methods to compact the soil. The equipment provides the forces needed to quickly settle soil and allow building to begin.

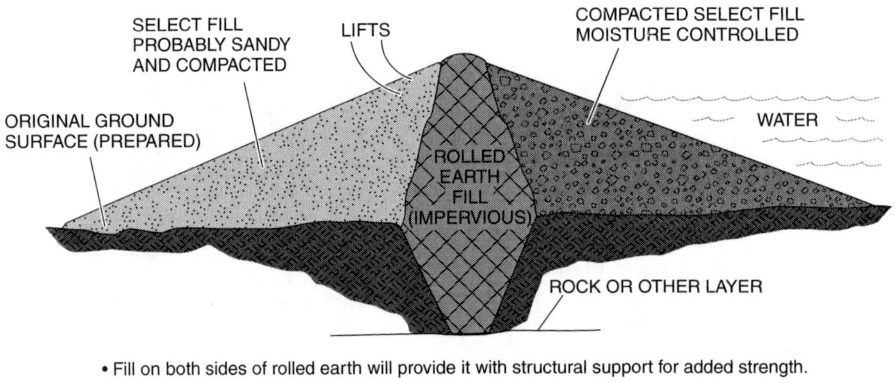

Figure 28 Rolled earthen dam.

Teaching Tip/Notes

4.0.0 Stabilizing Soils

Soil stabilization is the process of preparing subbase soils to provide a higher loadbearing capacity. The process involves pulverizing and mixing the soil thoroughly, often with binders, so that after compaction and curing, the soil is more dense, more stable, and stronger. The general requirement is to stabilize the area well beyond the locations where heavy loads occur. For a road subgrade, this means stabilizing past the road boundaries. *Figure 29* shows the subgrade stabilization limits.

Stability and loadbearing capacity depend on the internal friction and cohesion of the soil particles. The higher the internal friction of soil, the higher the cohesion and the greater the capacity to bear loads will be. Chemical binders are used to stabilize existing soils by adding internal friction and cohesion.

The first three steps in stabilization are to determine soil properties, the amount and type of binder required, and the degree of compaction needed. These are all a result of analyzing the soil or material to be stabilized. Analysis includes gradation, moisture content, density, internal friction, and cohesion. Tests and classification of soils were discussed in an earlier section.

There are many advantages to the stabilization of soils. These advantages include the following:

- Allows use of in place soil
- Eliminates need for excavating
- Eliminates the cost of replacing soil
- Reduces cost by mixing in place
- Reduces construction time

The stabilization process includes the following series of steps:

- Choosing a binder
- Preparing the subgrade
- Spreading the binder
- Determining the logistics of the stabilization job
- Stabilizing
- Compacting
- Checking quality

A soil stabilizer machine cuts, mixes, and pulverizes native, in-place soils or select material to modify and stabilize the soil for a strong base. Stabilizing agents may be mixed into the soil by the machine during this process. Stabilizers use blades on rotors or teeth on mandrels for cutting, and many have automatic depth control. Some models have a mixing chamber, which provides for consistent blending of materials. *Figure 30* shows a large stabilizer.

To carry out stabilization by mechanical methods, the entire area must be mixed thoroughly to the designated limits. The simplest type of equipment is a traveling mixing plant with a single-rotor mixer design. This machine can move rapidly and can mix an area in several passes.

The cutting depth for stabilization is usually 12 to 15 inches, and the cutting width is typically 8 feet. The production rate of the stabilizer can be calculated with these numbers and the travel speed, or it can be looked up on production charts.

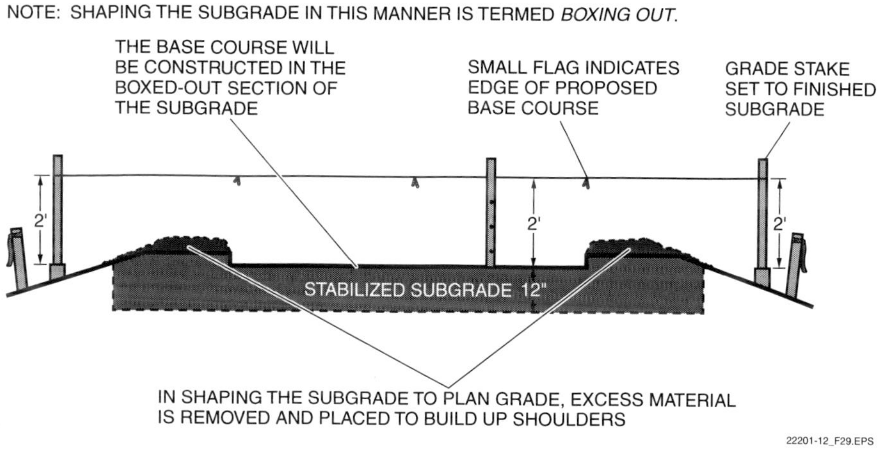

Figure 29 Subgrade stabilization limits.

Classroom

Explain that soil stabilization involves pulverizing soil and mixing it with binders.

Audiovisual

Show Slide 32 (Figure 29). Explain that subgrade stabilization limits extend beyond the areas where heavy loads occur.

Classroom

Discuss the three steps in soil stabilization. Explain the benefits of soil stabilization.

Audiovisual

Show Slide 33 (Figure 30). Explain how a stabilizer works.

Teaching Tip/Notes

Classroom

List different types of binders and explain how they are applied.

Explain how cement functions when used as a binder.

Point out that lime is an effective binder in clay-like soils.

Explain where fly ash comes from and how it is used.

22201-12_F30.EPS

Figure 30 Large soil stabilizer.

Stabilizer features may include the following:

- Mechanical rotor drive
- Microprocessor control of major machine systems
- Four steering modes with automatic rear wheel alignment
- Choice of rotors
- Roll-over protection
- Liquid additive and/or water spray systems in accordance with local environmental regulations

The machine should be able to cut up or down, since individual operations determine which type of cutting will result in more complete mixing. The ability of the machine to accomplish the specified mixing depth is critical to permit one-pass operation. Sometimes it is desirable to make an initial pass to establish proper gradation of the soil, followed by application of the binder, and then make a second pass with the soil stabilizer. Some stabilizer models offer the ability to reverse direction without turning around—a time-saving feature.

4.1.0 Choosing a Binder

Binders are selected to complement the particular soil being treated. The type and amount are dictated by the soil type, the size or gradation of its particles, its moisture content, and the loadbearing specifications of the project.

Binders may be dry, such as lime or cement, or wet, such as lime slurry. Binders can be spread by dump trucks, spreader trucks, or tanker trucks ahead of a soil stabilizer. A soil stabilizer has a flat-bottomed mixing chamber attachment. The soil is picked up and can be mixed, blended, or pulverized in the chamber. The soil is then put back down and flattened on the roadbed or surface. A soil stabilizer can stabilize a road base to approximately a 12-inch depth in a single pass. If such machines are used, binders can also be added directly to the soil in the stabilizer mixing chamber.

The method of adding binders depends on the type of stabilizer equipment used, the size of the project, and the types of additives. The binder must be thoroughly mixed with the base material and the mixture pulverized for uniform color and size. Common binders for roads include the following:

- Cement
- Lime
- Fly ash
- Calcium chloride
- Polymer

4.1.1 Cement

Cement is a fine, powdered hydraulic binder made from calcified lime compounds mixed with silica, alumina, and iron oxide. When mixed with water, cement sets and hardens in air or under water. Soil cement is made by mixing dry portland cement and water with the soil. The soil is compacted and paved.

Cement binder provides the highest loadbearing capacity of all binders. It can be used in gravel, sand, or reclaimed asphalt. Silt and clay may also be stabilized with cement.

4.1.2 Lime

Lime is especially effective as a binder in clay-bearing soils and aggregates because it reacts both chemically and physically with the soil. Lime used for stabilization includes quicklime (calcium oxide) and hydrated lime (calcium hydroxide). Both are burned forms of limestone.

Lime is applied as a dry powder, as a slurry, or in pellets. Lime reacts chemically with clay soils to produce a cementing action. Lime reacts favorably with soils in the higher plasticity index ranges. Lime treatment is used for sub-base stabilization, base stabilization, and lime modification.

4.1.3 Fly Ash

If lime is used on low plasticity index soil, a second additive may be used to produce the required reaction to the lime. This is called fly ash treatment. Fly ash is a residue recovered from smoke produced by coal-fired burners. Fly ash consists of minute spherical particles of glass, crystalline matter, and carbon. Its chemical analysis re-

Teaching Tip/Notes

sembles portland cement without lime. It reacts chemically with lime to enhance the strength of the soil, giving it cement-like properties. Fly ash combined with lime also produces an excellent road base when mixed with aggregate.

4.1.4 Calcium Chloride

Calcium chloride is another material used as a binder. It increases soil density like other binders and is particularly valuable in colder climates. Calcium chloride lowers the freezing point of water. The roads that use it for stabilization of the base material are better protected against frost damage. The use of calcium chloride for stabilization may require less compaction. It can be applied dry or in a solution.

4.1.5 Polymers

Widespread use of polymers as soil stabilizers is a fairly recent occurrence. A polymer and one or more other materials are bonded together by chemical reaction. In soil stabilization, polymers bind with soil particles to create a durable surface that permits motorized traffic. Application requirements vary with each product, so it is important to follow the manufacturer's instructions. Most products are applied as a viscous liquid that dries quickly (in a matter of minutes to hours, depending on the weather). Some polymers can be used in a weaker solution to suppress dust for erosion control.

On Site

Admixtures

Mineral admixtures are used primarily to reduce the amount of cement used in making concrete. These admixtures are either hydraulic or pozzolanic. Pozzolanic materials contain reactive silicates or alumino-silicates. When finely ground and used in a concrete mix, they react with other materials in portland cement to produce cement-like compounds. The Romans used a mixture of hydraulic lime and fine volcanic ash (a pozzolan) in the Appian Way, which connected Rome with southeastern Italy. The long road, still standing today, was started in 312 BC. Fly ash, a byproduct of combustion of ground coal in coal-burning electric utilities, is a frequently used pozzolanic material.

4.2.0 Preparing the Subgrade

To prepare the subgrade, top soil and large stones are removed from the proposed site and the subgrade is carefully shaped to grade. The desired site profile and the mix uniformity depend on accurate grading. The grading is done by a motor grader or by a fine grade trimmer. A trimmer controls profile with extreme accuracy and can accomplish the job in one pass.

When grading is complete, scarifying, or disc harrowing to the depth of stabilization is sometimes used. However, a stabilizing machine capable of mixing the binder and pulverizing the soil in one operation can eliminate this step.

4.3.0 Spreading the Binder

Dry binders must be evenly distributed over the area to be stabilized to assure that the specified soil-to-binder ratios are met. A box-type spreader attached to a dump truck or a rear-mounted spreader on a pneumatic truck is used for this purpose. The spreader meters out the material as the load is discharged ahead of the soil stabilizer.

Lime slurry or another liquid binder is best distributed by tanker truck, which spreads the product to precalculated depths as called for in the specifications.

4.4.0 Determining the Logistics of the Stabilization Job

When planning the stabilization job, logistics play an important part. Planning considerations include the following:

- *Atmospheric conditions* – The temperature should be mild with little or no precipitation expected. The initial mixing process should not be done in heavy rain. After the initial course is compacted and set, moderate rain is no problem.
- *Binder material supply* – Binder material must be available as needed to avoid delays in application.
- *Water supply* – When cement is being used for a binder, it is critical to maintain the optimum moisture content. Equipment must be available to supply water for the mix as needed. Lime is usually added to soils with high moisture, so water is not an issue for mixing. However, sufficient water should be available to control dust when using lime.

Classroom

Explain that calcium chloride, when used as a binder, provides protection against frost damage.

Explain that polymers are widely used on surfaces subjected to motorized traffic.

Describe how the subgrade is prepared for application of a binder.

Explain that a box-type spreader attached to a dump truck is often used to spread a binder evenly and that tanker trucks are used with slurries or liquid binders.

Describe the factors that must be considered when planning for soil stabilization.

Teaching Tip/Notes

Classroom

Explain the importance of soil compaction.

Audiovisual

Show Slide 34 (Figure 31). Point out that when a cement binder is used, it must be compacted immediately with a vibratory roller.

Classroom

List the quality control checks that must be performed on each compacted layer of a road bed.

Explain the importance of walking an excavation site to look for problems such as contamination, uneven ground, stinging insects, and other hazards.

- *Compaction equipment* – Appropriate equipment should be available for compaction to begin immediately after the stabilizer finishes the mixing and grading process.
- *Trimmer equipment* – After compaction, the surface should be trimmed to accurate grade and crown specifications.

4.5.0 Compacting

The stabilized area must be compacted after the stabilizer has finished. Only adequate compaction produces the required density. If cement is used, the compaction must be done immediately after mixing before the cement in the mixture begins to set. A cement mixture also requires that a vibratory roller compaction machine be used to obtain the needed results. *Figure 31* shows a vibratory roller.

When compacting a lime-stabilized area, the compaction must be done when the moisture content has dropped to a point within the specified range. This may occur immediately or after a short delay.

The number of passes made by the compactor depends on the amount of compaction to be achieved. This will, in turn, depend on the material moisture content, the layer thickness, the compactor type, and the degree of compaction called for in the specifications. With proper compaction, a final curing stage may be eliminated before the next course is applied.

22201-12_F31.EPS

Figure 31 Vibratory roller.

4.6.0 Checking Quality

Each layer of the roadbed must meet design specifications to ensure that the final product stands up under projected loads and meets smoothness requirements. Each stabilized layer must be checked for:

- Compaction or density
- Thickness
- Mix uniformity
- Gradation
- Loadbearing capacity
- Moisture content
- Binder content

5.0.0 Safety Guidelines

Once workers are comfortable reading the construction plans for the project, it is important that they relate them to the physical building site. Safety is everyone's first priority on the job, and workers must take responsibility for their own safety. Once workers are familiar with the site plans, they should walk the construction area to get a firm idea of the work to be performed and be able to identify potential safety hazards. Report any potential problems to the supervisor. Never assume that the project engineer or supervisor is aware of a potentially unsafe condition.

5.1.0 Personal Safety

Before starting any earthwork, walk the site if possible. Be alert for large depressions or other obstacles that are not shown on the plans. This can be especially important in undeveloped areas where large animal burrows, boulders, and decayed stumps are present. Driving over unexpectedly rough terrain can throw the equipment off balance and place the operator in danger of tipping over. These areas may need to be corrected during clearing and grubbing.

In the warm weather, look for swarming insects that could mean a hornet's nest or beehive in the area. Many people are highly allergic to insect stings and need to carry special medication to counteract the sting effects. Notify the supervisor of such findings. Removal of the nest may not be possible, but other workers should be notified of its existence.

Teaching Tip/Notes

When it is not possible to walk the site before starting work, be especially careful during the first drive over a section of land. Always scan the terrain looking for potential hazards. Drive slowly and be ready to stop.

5.1.1 Site Contamination

Whether working at a new construction site, an industrial site, or performing demolition on a building, there is always the possibility of uncovering a contaminated site. While walking the site, look for signs of illegal dumping and other contamination. Contaminants can be in the soil, buried underground, or stored in some container such as a barrel. Some warning signs of potential contamination are as follows:

- Puddles of fluid on the ground, especially fluid with an unusual color or odor
- Any fluid seeping out of the ground
- Unusual or foul odors
- Unmarked barrels or tanks—often buried or otherwise camouflaged
- Unexplained and remarkably bare patch of land in an area that is otherwise heavy with vegetation

Identifying contaminants is a complicated and time-consuming procedure that requires advanced education and training. Heavy equipment operators are not expected to identify contaminants but are required to report any potential contaminations to the supervisor.

Workers can limit exposure to soil contaminants by wearing proper PPE; showering and changing into clean clothes at the job site when possible; avoiding eating or drinking in the dusty area; and periodically spraying the area with water.

5.1.2 Sewage

Construction workers at an excavation site may accidentally uncover live sewers or septic tanks. Not only is this unpleasant, but contact with raw sewage can place workers at risk for disease. One of the most common diseases is Hepatitis A. Hepatitis is a viral disease that attacks the liver. It is contracted when some of the contaminant is swallowed and can be passed on by person-to-person contact. To prevent infection, wear latex or similar gloves when contacting sewage, wash hands frequently, avoid touching the face, and do not eat or drink in the area.

WARNING!

Some people are very allergic to latex. Some symptoms can be mild, such as an itchy red rash where the latex contacts the skin. Other symptoms are life threatening, such as anaphylactic shock, which is characterized by swelling of the airway, causing shortness of breath. If any of these symptoms occur while or after wearing latex gloves, remove them immediately, wash the affected area, and seek medical assistance if necessary. Then use non-latex gloves.

If illness occurs after working on a job that included exposure to sewage, tell the doctor. Hepatitis A is not often fatal, but there have been cases where healthy individuals have died from this disease. If a job requires frequent contact with sewage, talk to the supervisor about being vaccinated against this disease.

5.2.0 Safeguarding Property

As employees, workers are responsible to their company for their actions on the job. When workers think about safeguarding property, operating equipment usually comes to mind. The operator's duties include operating the equipment responsibly, reporting defects, and securing the equipment when not in use. In addition, workers must treat the building site responsibly because the site owners have contracted with employers to construct a project to certain specifications that are shown on the plans.

Workers should treat the job site responsibly by following the plans and being careful not to unduly damage the site. This is especially important on small jobs where a designated equipment staging and maintenance site has not been selected. It may be up to the worker to select a site; it should be done with the construction plans in mind. For example, it would be hard to maintain a flowerbed if oil and hydraulic fluid dripped into the soil. If the job site has no designated areas for loading, unloading, parking, and maintenance, the worker should select a site well away from the building.

Know the general plans well enough to avoid damage that heavy equipment can cause. Locate any existing structures, roads, or walkways to ensure that the operator can avoid driving over them. Know the location of underground structures, such as culverts, pipes, and septic systems at the site. Driving over these structures with heavy equipment can crush them and cause the

Classroom

Describe how to recognize contamination.

Describe the hazards associated with raw sewage and explain how to avoid contamination.

Safety !

Point out that some people are allergic to latex gloves.

Classroom

Discuss the importance of protecting a job site from damage and respecting neighboring property.

Teaching Tip/Notes

Audiovisual

Show Slide 35 (Figure 32). Point out that survey markers may be hidden and must never be moved or destroyed.

PT/Laboratory

1. Under your supervision, have trainees draw a plan for basic earthmoving operations, including:
   - Clearing and grubbing
   - Excavating the foundation
   - Constructing embankments
   - Backfilling
   - Compacting

   This laboratory corresponds to Performance Task 1.
2. Under your supervision, have trainees lay out a basic earthmoving operation. This laboratory corresponds to Performance Task 2.
3. Under your supervision, have trainees identify and select the proper equipment for a given earthmoving operation. This laboratory corresponds to Performance Task 3.

Classroom

Summarize the major concepts presented in this module.

equipment to be thrown off balance, causing an accident.

Never cross property lines without permission. Not only is this trespassing, but heavy equipment can also damage the terrain and cause the property owner to have bad feelings toward the employer. Further, there may be underground structures on the adjacent property that the equipment could damage.

When the plans call for certain trees to be preserved, workers must know their locations and avoid guesswork. The supervisor is available to help when needed. Trees and bushes are good places to look for survey markers or benchmarks (*Figure 32*). Never destroy or move these markers. Notify the project engineer or supervisor of the locations of any markers not shown on the site plan.

22201-12_F32.EPS

Figure 32 Survey marker.

Summary

Earthmoving is the process of digging, loading, hauling, and dumping any material that is needed for, or in the way of, construction. Types of earthmoving include clearing and grubbing, embankment construction, excavation, and backfilling and compaction.

Before construction can begin, other work needs to be done to determine what type and how many pieces of equipment are needed. This includes analyzing the characteristics of the soil, planning haul routes, putting up temporary signing, and staking.

Clearing and grubbing work is the first activity to be performed. It serves to remove all the vegetation, unwanted structures, and obstructions from the site. Excavation is the digging of the earth to remove material from a site. Fills, backfills, and embankment construction involve the filling in of pits or trenches, or building up of ground in order to raise the grade.

Staking is very important for the equipment operator because it guides the operations. Different kinds of stakes have different information. Types of stakes include right-of-way, slope, grade finish, center line, and ditch line.

The selection of equipment for any type of excavation job depends on the soil characteristics, the volume of material to be excavated, and distance the material must be hauled. There are many different types of equipment with different sizes and configurations for almost every type. In order to be efficient, workers must select the right type and size of equipment to meet the job at hand.

Soil stabilization and compaction may be necessary if the soil is to carry the required weight. Binders such as cement and lime may be needed to strengthen the soil.

It is very important to always operate the machine in a safe manner. Be aware of hazardous terrain on the work site and always check for buried utilities, such as gas, electric, water, and sewer lines. If the machine damages buried utilities, it could result in serious injury or death, as well as costly project delays.

Teaching Tip/Notes

30 NCCER – *Heavy Equipment Operations Level One* 22201-12

Review Questions

1. The unique address given to a site that allows police, fire, and other emergency vehicles to find the location easily is called a(n) _____.
   a. fire road
   b. 911 address
   c. police locator
   d. emergency access

2. During excavation and fill activities, slope control is handled with grade stakes that are placed by _____.
   a. the survey crew
   b. a competent person
   c. the inspector
   d. the contractor

3. When preparing a job site for excavation or construction, the removal of objects below the original ground is called _____.
   a. clearing
   b. stripping
   c. grubbing
   d. scalping

4. Cycle time is defined as the length of time it takes to perform an operation, such as the time it takes for a piece of equipment to _____.
   a. load, haul, and dump its load and return for another load
   b. return to the excavation site after it has dumped its load
   c. haul its load to the dump site and return for another load
   d. load fill in all of the vehicles needed for the day's work

5. The EPA requires that operators of construction sites that are one acre or larger establish plans to reduce sedimentation, pollution, and _____.
   a. traffic
   b. riprap
   c. erosion
   d. dust

6. Open pits or holes constructed to collect groundwater that are then pumped out mechanically are called _____.
   a. sumps
   b. borings
   c. holding ponds
   d. drainage ditches

7. A temporary opening in which something, such as a pipe or box culvert, is placed and then covered is a _____.
   a. channel
   b. trench
   c. ditch
   d. pit

8. Backhoes, excavators, and trenching machines are the three main classes of equipment used in _____.
   a. trenching
   b. excavating
   c. slope control
   d. erosion control

9. The type of dump site used to store material for later use is called a(n) _____.
   a. spoils pile
   b. embankment
   c. stockpile
   d. landfill

10. Waste fill is recycled, treated material approved for use in certain _____.
    a. landscaping projects
    b. outdoor products
    c. playgrounds
    d. landfill areas

11. The binder that provides the highest load-bearing capacity is _____.
    a. lime
    b. fly ash
    c. cement
    d. crushed rock

Classroom

Have the trainees complete the Review Questions, and go over the answers prior to administering the Module Examination.

Have the trainees complete the Trade Terms Quiz, and go over the answers prior to administering the Module Examination.

Examination ?

Administer the Module Examination. Record the results on Training Report Form 200, and submit the results to the Training Program Sponsor.

Performance Testing ✓

Administer the Performance Test and fill out the Performance Profile Sheets for each trainee. If desired, trainee proficiency noted during laboratory sessions may be used to complete the Performance Test. Record the results on Training Report Form 200 and submit the results to the Training Program Sponsor.

Teaching Tip/Notes

22201-12 Introduction to Earthmoving Module Six 31

12. Compaction of a lime-stabilized area requires _____.
    a. that compaction be performed immediately
    b. that a vibratory roller be used
    c. that the moisture content be reduced
    d. the addition of water to the soil

13. When working on a job, the first priority is to perform the job _____.
    a. efficiently
    b. quickly
    c. safely
    d. well

14. If a foul-smelling fluid seeping out of the ground is noticed while walking the new site before starting work the operator should know that it _____.
    a. could be contamination
    b. is probably an underground spring
    c. needs to be filled before work begins
    d. needs to be shown on the site plans

15. What should an operator know or do if the site plans show that a particularly large tree on the building site needs to be preserved?
    a. Find the tree before starting any work.
    b. Try to look for the tree while working.
    c. Avoid working around any large tree.
    d. Ignore it.

Teaching Tip/Notes

Trade Terms Quiz

Fill in the blank with the correct term that you learned from your study of this module.

1. Substances that have the ability to bond together are defined as ________.
2. A line that marks the center of a roadway is known as a ________.
3. The time it takes for an equipment to complete an operation is referred to as ________.
4. The solid level of rock under Earth's surface is ________.
5. A(n) ________ is the term used to describe a machine with a bucket that is dragged toward the machine by a cable.
6. The classification of soils into different particle sizes is called ________.
7. Material that is derived from something other than living organisms is known as ________.
8. Loose pieces of rock that are placed on an embankment to prevent erosion are called ________.
9. Material that is excavated and stockpiled for future use is known as .
10. A(n) ________ is a temporary long, narrow excavation that is to be filled in
11. The return of a piece of equipment after it has dumped its load is called ________.
12. The process of compacting particles so they are close together is called ________.
13. A(n) ________ is used to describe material piled in a uniform manner to build up an area.
14. If a material will not allow water to pass through it, the material is considered ________.
15. A(n) ________ is a defined piece of work that a contractor is paid for.
16. The material used to brace the sides of a trench is known as ________.
17. A hole drilled in order to take a sample of material for testing is a(n) ________.
18. The ________ is the depth at which the soil is saturated with water.
19. A sample of earth taken from a test boring is called a(n) ________.
20. Removing water from an area using a pump or drain is known as ________.
21. A clay-like soil that swells with an increase in moisture is referred to as ________.
22. Water below the ground surface is called ________.
23. A material derived from living organisms is referred to as ________ material.
24. An open excavation that usually does not require shoring is classified as a(n) ________.
25. Soil or other material that meets a specification for size, shape, or other characteristic is called ________.
26. A mechanical or electronic test of soil to determine its density and moisture content is known as ________.
27. A long, straight pile of placed material used for the purpose of mixing is called ________.
28. Water that is derived from rain or snow is referred to as ________.

Trade Terms

Backhauling	Cycle time	Groundwater	Riprap	Test boring
Bedrock	Dewatering	Impervious	Select material	Test pit
Center line	Dragline	Inorganic	Shoring	Trench
Cohesive	Embankment	Organic	Soil testing	Water table
Consolidation	Expansive soil	Pay item	Spoils	Windrow
Core sample	Gradation	Pit	Stormwater	

Teaching Tip/Notes

22201-12 Introduction to Earthmoving Module Six 33

Trade Terms Introduced in This Module

Backhauling: The return trip of a piece of equipment after it has completed dumping its load.

Bedrock: The solid layer of rock under Earth's surface. Solid rock, as distinguished from boulders.

Center line: The line that marks the center of a roadway. This is marked on the plans by a line and on the ground by stakes.

Cohesive: The ability to bond together in a permanent or semi-permanent state. To stick together.

Consolidation: To become firm by compacting the particles so they are closer together.

Core sample: A sample of earth taken from a test boring.

Cycle time: The time it takes for a piece of equipment to complete an operation. This normally would include loading, hauling, dumping, and then returning to the starting point.

Dewatering: Removing water from an area using a drain or pump.

Dragline: An excavating machine having a bucket that is dropped from a boom and dragged toward the machine by a cable.

Embankment: Material piled in a uniform manner so as to build up the elevation of an area. Usually, the material is in long narrow strips.

Expansive soil: A clay-like soil that swells with an increase in moisture and shrinks with a decrease in moisture.

Gradation: The classification of soils into different particle sizes.

Groundwater: Water beneath the surface of the ground.

Impervious: Not allowing entrance or passage through; for example, soil that will not allow water to pass through it.

Inorganic: Derived from other than living organisms.

Organic: Derived from living organisms such as plants and animals.

Pay item: A defined piece of material or work that the contractor is paid for. Pay items are usually expressed as unit costs.

Pit: An open excavation that usually does not require vertical shoring or bracing.

Riprap: Loose pieces of rock that are placed on the slope of an embankment in order to stabilize the soil.

Select material: Soil or manufactured material that meets a predetermined specification as to some physical property such as size, shape, or hardness.

Shoring: Material used to brace the side of a trench or the vertical face of any excavation.

Soil testing: A mechanical or electronic test used to determine the density and moisture of the soil, and therefore the amount of compaction required.

Spoils: Material that has been excavated and stockpiled for future use.

Stormwater: Water from rain or snow.

Test boring: To drill or excavate a hole in order to take a sample of the material that rests in different layers beneath the surface.

Test pit: See *test boring.*

Trench: A temporary long, narrow excavation that will be covered over when work is completed.

Water table: The depth below the ground's surface at which the soil is saturated with water.

Windrow: A long, straight pile of placed material for the purpose of mixing or scraping up.

Teaching Tip/Notes

Additional Resources

This module presents thorough resources for task training. The following resource material is suggested for further study.

"Avoiding Enforcement Actions: How to Effectively Manage Your Construction Site and Survive Increased Regulatory Scrutiny," *Grading and Excavation Contractor* magazine. Carol L. Forrest. July/August 2005. Santa Barbara, CA: Forester Communication.

Caterpillar Performance Handbook, Latest Edition. A CAT® Publication. Peoria, IL: Caterpillar, Inc.

"Chemical Soil Stabilization," *Erosion Control* magazine. January/February, 2003. Janis Keating.

United States Environmental Protection Agency (EPA) web site, National Pollutant Discharge Elimination System (NPDES): www.epa.gov.

Figure Credits

Reprinted courtesy of Caterpillar Inc., Module opener, Figures 3, 11, 13–16, 23, 26, 30, and 31

Dale Chadwick, Figure 1

P&H Mining Equipment, Figure 2

Mark Jones, Figures 4–6 and 12

John Hoerlein, Figure 7

Kundel Industries, Figure 21

Topaz Publications, Inc., Figures 22 (top photo) and 32

The Charles Machine Works, Inc., Manufacturer of Ditch Witch products, Figure 22 (bottom photo)

Teaching Tip/Notes

ANSWERS TO REVIEW QUESTIONS

Answer		Section Reference
1.	b	3.1.3
2.	a	3.2.2
3.	c	3.3.3
4.	a	3.4.2
5.	c	3.6.1
6.	a	3.6.2
7.	b	3.7.6
8.	a	3.7.6
9.	c	3.10.1
10.	d	3.11.1
11.	c	4.1.1
12.	c	4.5.0
13.	c	5.0.0
14.	a	5.1.1
15.	a	5.2.0

ANSWERS TO TRADE TERMS QUIZ

1. Cohesive
2. Center line
3. Cycle time
4. Bedrock
5. Dragline
6. Gradation
7. Inorganic
8. Riprap
9. Spoils
10. Trench
11. Backhauling
12. Consolidation
13. Embankment
14. Impervious
15. Pay item
16. Shoring
17. Test boring or test pit
18. Water table
19. Core sample
20. Dewatering
21. Expansive soil
22. Ground water
23. Organic
24. Pit
25. Select material
26. Soil testing
27. Windrow
28. Stormwater

NCCER CURRICULA — USER UPDATE

NCCER makes every effort to keep its textbooks up-to-date and free of technical errors. We appreciate your help in this process. If you find an error, a typographical mistake, or an inaccuracy in NCCER's curricula, please fill out this form (or a photocopy), or complete the online form at **www.nccer.org/olf**. Be sure to include the exact module ID number, page number, a detailed description, and your recommended correction. Your input will be brought to the attention of the Authoring Team. Thank you for your assistance.

Instructors – If you have an idea for improving this textbook, or have found that additional materials were necessary to teach this module effectively, please let us know so that we may present your suggestions to the Authoring Team.

NCCER Product Development and Revision

13614 Progress Blvd., Alachua, FL 32615

Email: curriculum@nccer.org
Online: www.nccer.org/olf

❑ Trainee Guide ❑ AIG ❑ Exam ❑ PowerPoints Other ______________________

Craft / Level: Copyright Date:

Module ID Number / Title:

Section Number(s):

Description:

Recommended Correction:

Your Name:

Address:

Email: Phone:

22106-12

Grades

Module Seven

Materials

reminds you to gather materials needed for classes, labs, and testing.

Homework

suggests that you assign reading for the next session.

Classroom

prompts you to discuss a topic, stress a point, explain a concept, or provide examples.

Audiovisual

indicates an appropriate time to show a slide or transparency.

! Safety

used to emphasize safety issues.

Demonstration

directs you to show trainees how to perform tasks.

Laboratory

used when trainees are to practice performing tasks.

✓ Performance Testing

recommends administering a performance test, or a portion thereof.

PT / Laboratory

indicates that lab practice corresponds with a performance task. If desired, you can note the proficiency of the trainees during the lab to satisfy performance testing requirements.

Teaching Tip

indicates tips for practice exercises, Quick Quizzes, field trips, or other activities designed to encourage class participation.

? Examination

recommends administering the written module exam.

NCCER Standardized Craft Training Program

NCCER provides a standardized national program of accredited craft training. Key features of the program include instructor certification, competency-based training, and performance testing. The program provides trainees, instructors, and companies with a standard form of recognition through the National Registry. The program is described in full in the *Guidelines for Accreditation*, published by NCCER. For more information on standardized craft training, contact NCCER by writing to 13614 Progress Blvd., Alachua FL 32615; calling 1-888-NCCER20; or emailing **info@nccer.org**. More information is available at **www.nccer.org**.

How to Use this Annotated Instructor's Guide

Each page presents two sections of information. The larger section displays each page exactly as it appears in the Trainee Module. The narrow column ties suggested trainee and instructor actions to each page and provides icons (detailed on the left) to call your attention to material, safety, audiovisual, or testing requirements. The bottom of each page includes space for your notes, or when available, teaching tips that suggest activities or class exercises that will help your students learn the material.

Preparation

Before teaching this module, you should review the Objectives, Performance Tasks, Materials and Equipment List, and Module Outline. Be sure to allow ample time to prepare your own training or lesson plan and gather all required materials and equipment.

Instructors purchasing a single module receive the printed exam and performance tasks with the AIG module. Perfect-bound versions of this title no longer include the printed exam booklet and performance task sheets. Log on to the Instructor Resource Center at **www.NCCERIRC.com** to download these materials. Your access code is on the card bound into the front of this AIG. This code will allow you to access exams and performance tasks for all levels of this craft. Be sure to keep all testing materials in a secure place at all times.

Grades
Annotated Instructor's Guide

Module Overview

This module introduces the concept of preparing graded surfaces using heavy equipment. It also covers identification and interpretation of construction stakes and describes the methods for grading slopes.

Objectives

Upon completion of this module, the trainee will be able to do the following:

1. Explain the terms used in grade work.
2. Identify types of stakes and explain markings on grade stakes and benchmark (BM) stakes.
3. Identify equipment used by operators to check stakes.
4. Explain different types of slopes and slope ratio.
5. Check horizontal and vertical distance of cut and fill slope stakes.
6. Check finish subgrade on a cross slope.

Performance Tasks

Under the supervision of the instructor, the trainee should be able to do the following:

1. Identify types of stakes and markings on stakes.
2. Check horizontal and vertical distances of cut and fill slope stakes.
3. Check finish subgrade on a cross slope.

Materials and Equipment

Markers/chalk
Pencils and scratch paper
Whiteboard/chalkboard
Heavy Equipment Operations Level One PowerPoint® Presentation Slides (ISBN 978-0-13-292167-1)
Multimedia projector and screen
Computer
Appropriate personal protective equipment
Samples of marked surveyor stakes:
- Center line stakes
- Offset stakes
- Grade stakes
- Slope stakes
- Edge of pavement stakes

Construction plans including topographical and grade information
Leveling rods
Sight level
Automatic level
Laser level
Tripod
Pocket leveling rod
Hand level
Measuring tape
18" rulers
String
Trade Terms Quiz*
Module Examinations**
Performance Profile Sheets**

* Located at the back of the Trainee Guide module

** Single-module AIG purchases include the printed exam and performance task sheet. If you have purchased the perfect-bound version of this title, download these materials from the IRC using your access code.

Safety Considerations

Ensure that the trainees are equipped with appropriate personal protective equipment and know how to use it properly. Review safety guidelines associated with working on or around heavy equipment. Emphasize the importance of proper housekeeping.

Additional Resources

This module presents thorough resources for task training. The following resource material is suggested for further study.

Site Layout Levels 1 and 2, Latest Edition. NCCER. Upper Saddle River, NJ: Pearson Education Inc.

Teaching Time for This Module

An outline for use in developing your lesson plan is presented below. Note that each Roman numeral in the outline equates to one session of instruction. Each session has a suggested time period of 2½ hours. This includes 10 minutes at the beginning of each session for administrative tasks and one 10-minute break during the session. Approximately 15 hours are suggested to cover *Grades*. You will need to adjust the time required for hands-on activity and testing based on your class size and resources. Because laboratories often correspond to Performance Tasks, the proficiency of the trainees may be noted during these exercises for Performance Testing purposes.

Topic	Planned Time
Sessions I and II. Introduction; Planning Grades; Site Layout	
A. Introduction	________
B. Planning Grades	________
1. Profiles	________
2. Erosion Controls	________
3. Roadways	________
4. Buildings	________
5. Grade Computation	________
C. Site Layout	________
1. Common Stake Markings	________
2. Setting Stakes	________
3. Slope Stake Interpretation	________
D. PT/Laboratory	________
Have trainees identify types of stakes and markings on stakes. This laboratory corresponds to Performance Task 1.	
Sessions III and IV. Performing Site Measurements	
A. Performing Site Measurements	________
1. Manual Equipment	________
2. Electronic Measuring Devices	________
3. Global Positioning System Survey Devices	________
4. Stakeless Systems	________
B. PT/Laboratory	________
Have trainees check horizontal and vertical distances of cut and fill slope stakes. This laboratory corresponds to Performance Task 2.	
Session V. Finish Grades	
A. Finish Grades	________
1. Finish Grade Stakes	________
2. Checking Grades	________
B. PT/Laboratory	________
Have trainees check finish subgrade on a cross slope. This laboratory corresponds to Performance Task 3.	

Session VI. Review and Testing

A. Review

B. Module Examination

1. Trainees must score 70% or higher to receive recognition from NCCER.
2. Record the testing results on Training Report Form 200 and submit the results to the Training Program Sponsor.

C. Performance Testing

1. Trainees must perform each task to the satisfaction of the instructor to receive recognition from NCCER. If applicable, proficiency noted during laboratory exercises can be used to satisfy the Performance Testing requirements.
2. Record the testing results on Training Report Form 200 and submit the results to the Training Program Sponsor.

22106-12

Grades

Module Seven

Homework

Assign the reading of this module.

Teaching Tip/Notes

Trainees with successful module completions may be eligible for credentialing through NCCER's National Registry. To learn more, go to **www.nccer.org** or contact us at **1.888.622.3720**. Our website has information on the latest product releases and training, as well as online versions of our *Cornerstone* newsletter and Pearson's product catalog.

Your feedback is welcome. You may email your comments to **curriculum@nccer.org**, send general comments and inquiries to **info@nccer.org**, or fill in the User Update form at the back of this module.

V.1 4/12

Teaching Tip/Notes

22106-12 GRADES

Objectives

When you have completed this module, you will be able to do the following:

1. Explain the terms used in grade work.
2. Identify types of stakes and explain markings on grade stakes and benchmark (BM) stakes.
3. Identify equipment used by the operator to check stakes.
4. Explain different types of slopes and slope ratio.
5. Check horizontal and vertical distance of cut and fill slope stakes.
6. Check finish subgrade on a cross slope.

Performance Tasks

Under the supervision of your instructor, you should be able to do the following:

1. Identify types of stakes and markings on stakes.
2. Check horizontal and vertical distances of cut and fill slope stakes.
3. Check finish subgrade on a cross slope.

Trade Terms

Aggregate
Backsight (BS)
Backslope
Baseline
Benchmark (BM)
Center line
Crow's foot
Elevation
Erosion
Finished grade (FG)
Foresight (FS)
Global positioning system (GPS)
Grade
Grade work
Height of instrument (HI)
Hinge point (HP)
Hubs
Inslope
Line of sight
Natural ground
Profile
Ratio
Reference stake (RS)
Slope
Stake
Station number
Subgrade
Temporary benchmark (TBM)
Topographical

Industry Recognized Credentials

If you're training through an NCCER-accredited sponsor you may be eligible for credentials from NCCER's Registry. The ID number for this module is 22106-12. Note that this module may have been used in other NCCER curricula and may apply to other level completions. Contact NCCER's Registry at 888.622.3720 or go to nccer.org for more information.

Materials

Ensure that you have everything required to teach the course. Check the Materials and Equipment list at the front of this module.

Classroom

Explain that terms shown in bold (blue) are defined in the Glossary at the back of this module.

Audiovisual

Show Slide 1, Objectives, and Slide 2, Performance Tasks. Review the goals of the module, and explain what will be expected of the trainees.

Teaching Tip/Notes

When you call on someone to answer a question, the rest of the class relaxes or even tunes out because they expect that the question and answer will take place only between you and the trainee you called on. Instead, use this technique to involve more trainees in answering questions and to keep them on their toes:

1. Ask trainees to define a term or explain a concept.
2. After one trainee has answered, ask a trainee seated nearby if the answer is right. Then ask whether a trainee in the back of the room agrees.
3. Ask trainees to explain why they think an answer is right or wrong.
4. Use the session to clear up incorrect ideas and encourage trainees to learn from their mistakes.

Contents

Topics to be presented in this module include:

Figures and Tables

Teaching Tip/Notes

Teaching Tip/Notes

1.0.0 Introduction

A **grade** (*Figure 1*) is a rise or fall in the road, channel, or **natural ground.** It can be expressed as a **ratio** of vertical distance (rise) to horizontal distance (run), but it is most often expressed as a percentage above or below the horizon. Grades can rise, fall, or be level.

Heavy equipment operators are often assigned to move earth and to level the earth for the proposed construction. This is called **grade work.** Grade work is important because the stability of any building depends on the grade on which it is built. In order for the walls of the building to be level, the foundation must be level.

Heavy equipment is sometimes used to increase or decrease the **slope** of land on which a road will be built. This is important because road surfaces and curves must be safe for driving. Too sharp of a curve or too steep of a grade can be dangerous. The slope is also important because it can contribute to **erosion,** which can damage the road or the foundation of a building (*Figure 2*).

During the planning stage of the project, engineers and other professionals examine the proposed building site and determine whether earth needs to be moved to complete the project. This information is recorded on the construction plans.

A survey crew goes to the construction site and marks the site according to the construction plans. The crew marks instructions on **stakes** and drives them into the ground to indicate whether earth needs to be removed from or added to an area to change the grade.

Heavy equipment operators must learn to read the stakes and follow the directions written on them. Although they usually are not responsible for planning the grades, they may need to help set stakes on some jobs. Understanding how grades are marked is critical.

2.0.0 Planning Grades

During the planning stage of a construction project, the company in charge of the design performs a detailed **topographical** analysis of the proposed site. The analysis shows the natural features and characteristics of the site, such as hills, gullies, and other surface features.

The design plans contains information necessary for the construction crew to do its job, including information about grades. To understand how to make grades, machine operators need to know the grade names and how they affect the construction project. They also need to understand how a grade is computed.

2.1.0 Profiles

A **profile** is a side view of the proposed roadway or building. It is normally drawn on grid paper and shows the natural ground and proposed construction. For roadways, it shows grade and the **station numbers** at points along it. Station numbers measure the distance from a reference point and are usually in increments of 100 feet.

Figure 3 shows a proposed roadway. One can see that the natural ground line is irregular, while the proposed grade is smooth. To build the roadway, operators need to cut earth from some areas and fill in other areas. Also, note that the natural ground line and the proposed grade slope up. This is called a rising grade.

2.2.0 Erosion Controls

Part of the planning process involves the way construction site runoff waters are controlled. When dirt is broken up during construction, it is easily moved by water. State and federal guidelines require that erosion controls get put in place to

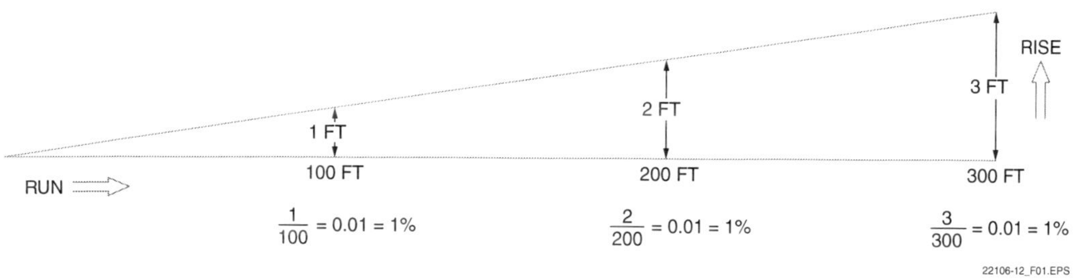

Figure 1 Grades.

Classroom

Define grades and explain how they are expressed.

Audiovisual

Show Slide 3 (Figure 1).

Classroom

Explain that heavy equipment operators often perform grade work and must be able to interpret grade stakes.

Audiovisual

Show Slide 4 (Figure 2). Explain that slopes are subject to erosion.

Classroom

Explain what a topographical analysis is.

Point out that heavy equipment operators must be able to identify types of grades.

Define the terms *profile* and *station number.*

Audiovisual

Show Slide 5 (Figure 3).

Teaching Tip/Notes

Obtain a set of site plans containing grading information for the trainees to examine.

22106-12 Grades Module Seven 1

Audiovisual

Show Slide 6 (Figure 4). Explain how silt fences are used.

Show Slide 7 (Figure 5). Describe the typical requirements for silt fencing.

22106-12_F02.EPS

Figure 2 Example of erosion.

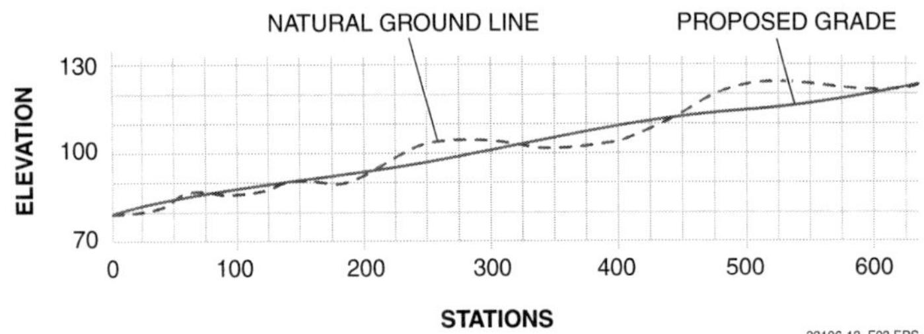

22106-12_F03.EPS

Figure 3 Roadway profile sheet.

keep any construction site sediment (dirt and debris) from flowing into nearby streams or across nearby properties. The most often used device for preventing such runoffs is a silt fence (*Figure 4*). Bales of hay may also be used in conjunction with a silt fence to control the runoff. The bales slow the water flow during heavy rains.

Different states may have different specifications on how silt fencing must be installed (*Figure 5*). In most cases, long lengths of specially woven cloth are partially buried in a trench cut across a drainage area. A backfill of dirt and rock is used on the downstream side of the fence to hold the cloth in place. The unburied portion of the cloth is then lifted and suspended between posts to form a barrier. Most specifications call for at least 12 inches of the cloth to be buried, and two to three feet of the cloth suspended to form the fence. In most cases, the posts used to support the cloth are spaced four to ten feet apart. The post will need to be closer when heavier runoffs are expected. All posts need to be driven deep enough to keep the fence from collapsing as time goes on.

Heavy equipment operators may be called upon to help install silt fencing. After it has been installed, they must make sure that they do not damage such fencing as they grade a site. If any silt fencing or other such erosion controls are disturbed or damaged during grading operations, make sure that a supervisor is informed so that the erosion control device(s) can be repaired.

GOING GREEN

Silt Fencing

In addition to protecting the environment, silt fencing also protects the companies doing the construction work. Companies can be heavily fined if erosion controls are not properly installed, or not installed at all.

Teaching Tip/Notes

Figure 4 Silt fencing.

2.3.0 Roadways

Before a roadway becomes a highway complete with blacktop and yellow lines, the grading portion of the construction must be completed.

Refer to *Figure 6*. The finished surface of the road foundation is called the **subgrade.** This is where the road foundation (called **aggregate**) and the road surface will be applied. The middle of the roadway is called the **center line.** The center line is sometimes called the **baseline** and is used as a reference point during the construction of the roadway. The center line is abbreviated with the interlocked letters CL. The subgrade is usually sloped down slightly from the center line so that water will drain from the surface of the finished roadway. This downward slope is called a falling grade, or crown.

The edges of the subgrade are called the shoulder points, the **hinge points (HP),** or the edge of road. They may also be called pavement points (E/R or E/P). These points designate the end of the road surface. On both sides of the subgrade is the **inslope.** The inslope is sloped down to move water away from the road. When the ground slopes up again, it is called a **backslope.** A backslope is used when the natural ground rises above the grade work for the roadway. Backslopes are cut so that there will be minimal erosion.

Now consider any piece of land with its natural hills and valleys. *Figure 7* shows a roadway superimposed on a piece of land. As shown in the figure, there are areas of the earth that will need to be cut away and other areas will need to be filled in so that the roadway can be built. Cutting and filling is done by heavy equipment operators.

2.4.0 Buildings

Heavy equipment operators are exposed to plans, drawings, and other site layout materials that guide them in placing stakes and earthmoving. *Figure 8* shows the foundation plan for a simple monolithic slab. Note that in this plan, the horizontal and vertical center lines are clearly shown, but this is not always the case.

To prepare a site for a monolithic slab, the area needs to be leveled (*Figure 9*). Some parts will need to be cut and other areas need to be filled. To ensure that the building pad will be level and true, corner stakes must be set and batter boards used to establish building lines. Batter boards (*Figure 10*) are often used to set reference elevations, such as the finish floor level of the building, as well as building lines. As a heavy equipment operator, it is your job to recognize and understand the purpose of these markings. If you are working on a project and see stakes that you do not understand, ask for help before beginning the grading work. Reworking a grade because of an error takes time, costs money, and it could delay the completion of the project.

2.5.0 Grade Computation

So far, it has been established that a roadway has a lengthwise grade, such as a road running up or down a mountain. Also covered was the fact that a road surface and the shoulder on the sides of the road are graded so that water moves away from the road. Engineers decide where and how much of a grade is needed. Surveyors mark the job site

Audiovisual

Show Slide 8 (Figure 6). Use the diagram to explain the terms for roadway grading.

Show Slide 9 (Figure 7). Explain the cut and fill requirements.

Show Slide 10 (Figure 8). Explain how to interpret a foundation plan.

Show Slide 11(Figure 9). Discuss the grading work that must be done to prepare a monolithic slab.

Show Slide 12 (Figure 10). Describe how batter boards are used to lay out a building foundation.

Classroom

Explain the importance of knowing how to calculate a grade.

Teaching Tip/Notes

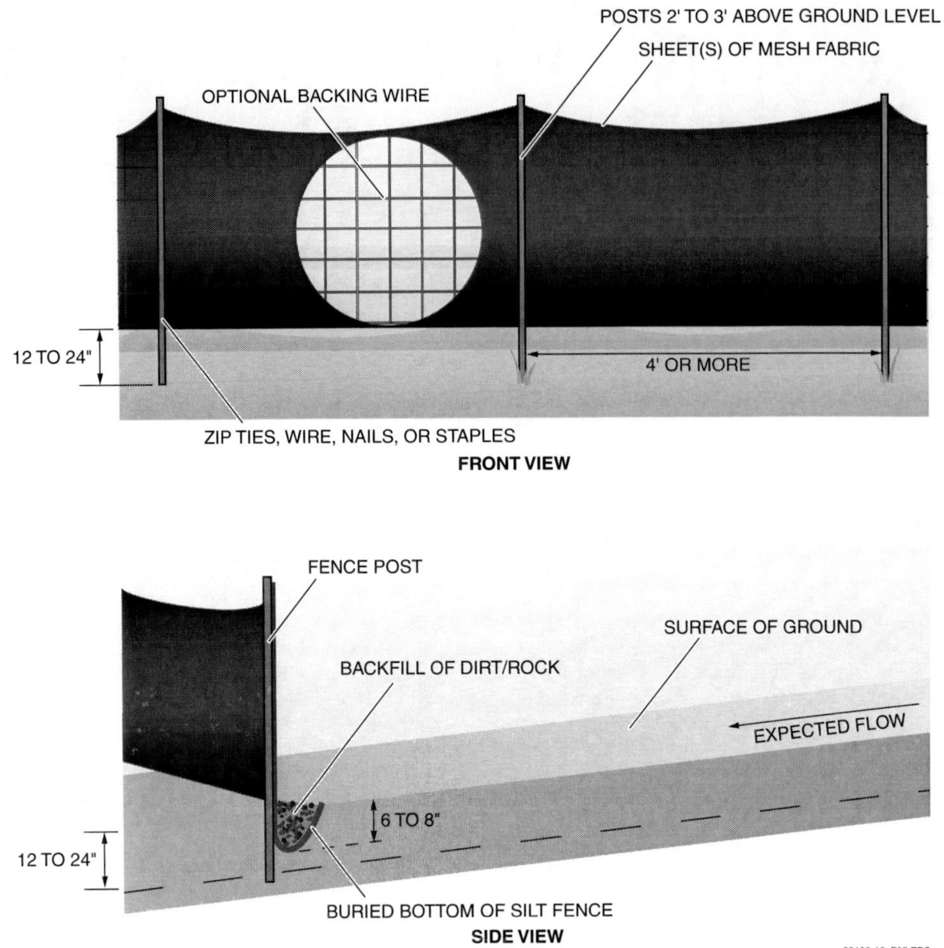

Figure 5 Side and front view of silt fencing.

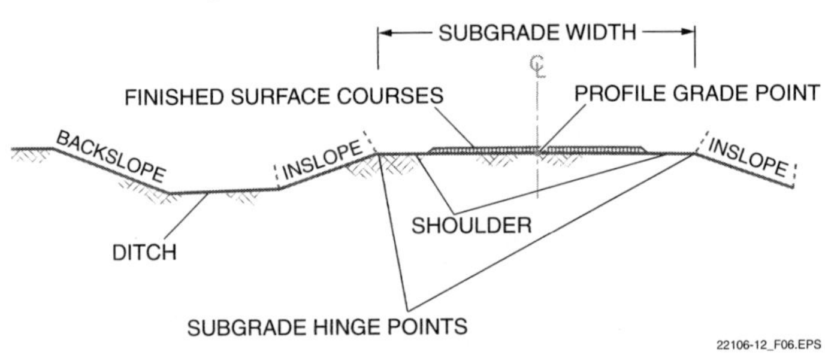

Figure 6 Parts of a roadway.

Teaching Tip/Notes

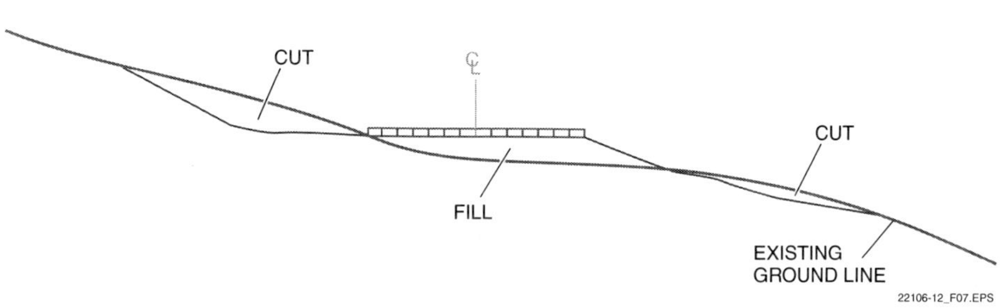

Figure 7 Excavation of a roadway.

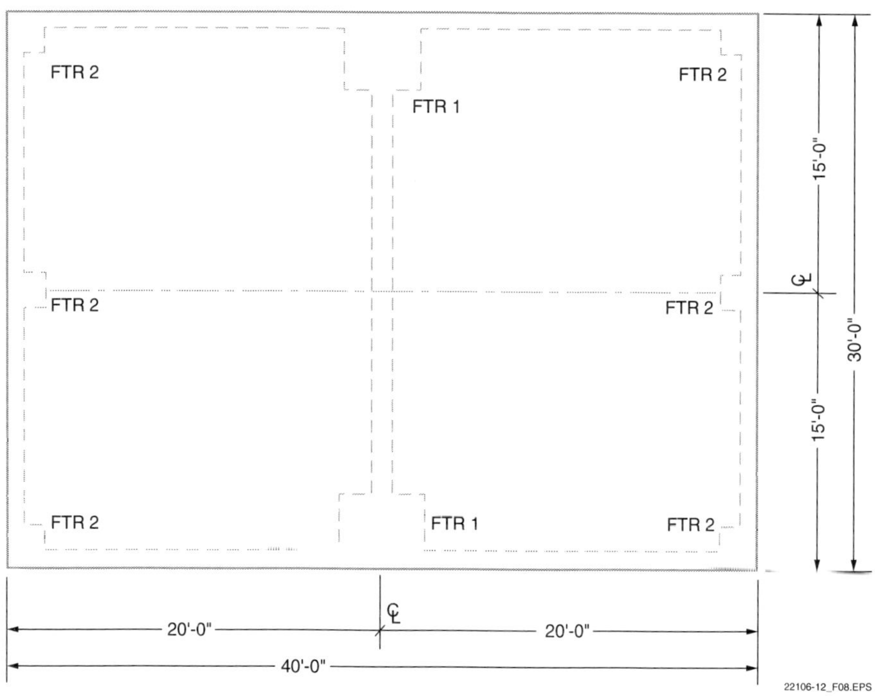

Figure 8 Foundation plan.

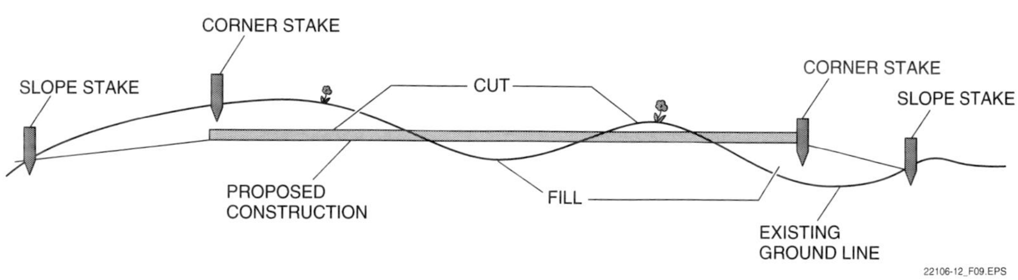

Figure 9 Excavation of a monolithic slab.

Teaching Tip/Notes

Audiovisual

Show Slide 13 (Figure 11). Demonstrate how a grade is calculated.

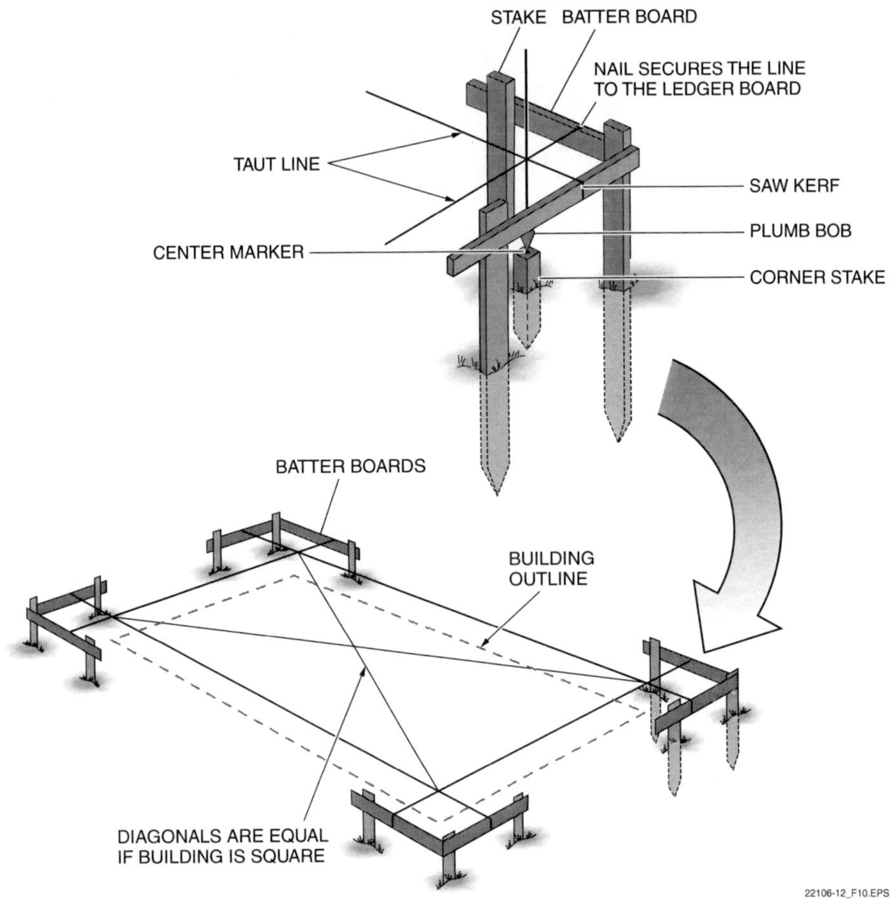

Figure 10 Typical use of batter boards.

to indicate grades. But it is the job of the equipment operator to cut the correct grade, so it is very important to understand how to compute a grade.

A grade is a rise or fall in a surface. It can be expressed as a ratio of vertical distance or **elevation** (rise) to horizontal distance (run), which is abbreviated rise:run, but it is most often expressed as a percentage above or below the horizon. Sometimes the grade is expressed as rise-over-run.

Referring to *Figure 11*, if a roadway rises 3 feet over 100 feet of travel, the rise is 3 feet and the run is 100 feet. The percentage of grade is computed by dividing the horizontal distance (run) into the vertical distance (rise):

$$\frac{\text{Vertical distance (rise)}}{\text{Horizontal distance (run)}} = \frac{3}{100}$$

or 3 ÷ 100 = 0.03 or 3 percent

If a roadway rises 3 feet over 200 feet of travel, the rise is 3 feet and the run is 200 feet. The percentage of grade is still computed by dividing the horizontal distance (run) into the vertical distance (rise):

$$\frac{\text{Vertical distance (rise)}}{\text{Horizontal distance (run)}} = \frac{3}{200}$$

or 3 ÷ 200 = 0.015 or 1.5 percent

A grade can rise, or fall, or be level. When a grade goes up, it is called a rising grade and is designated with a plus sign (+). A roadway that rises 3 feet over 100 feet of travel has a grade of +3.00 percent.

Teaching Tip/Notes

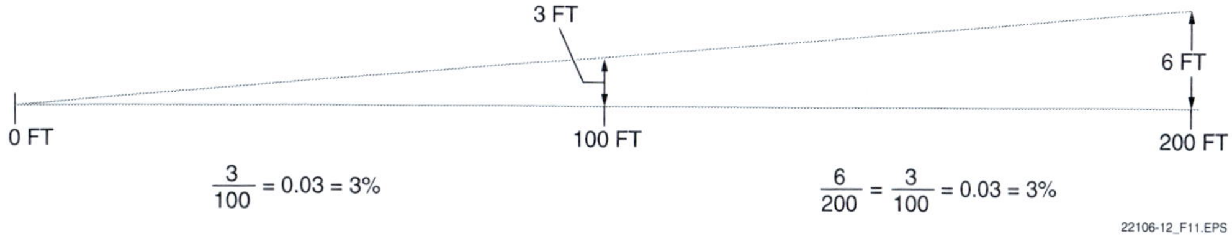

Figure 11 Calculating a grade.

When a grade goes down, it is called a falling grade and is designated with a minus sign (–). A roadway that falls 3 feet over 100 feet of travel has a grade of –3.00 percent.

When a grade neither rises nor falls, it is called level. Since there is no change in the rise, it has a zero percent grade.

3.0.0 Site Layout

Before grading work begins, the survey crew marks the area with stakes. They write letters and numbers on each stake to indicate whether the area needs be cut or be filled. Heavy equipment operators must recognize these stakes and understand the directions given on them.

> **NOTE**
> Stakes contain critical information. They must not be moved or struck by machine operators. Operators must not pull stakes unless directed to do so. If a survey stake is accidentally moved, do not reset it. Inform your supervisor.

3.1.0 Common Stake Markings

Each stake on a construction site has a purpose. Given the limited space on a stake, abbreviations are used to mark the stakes. *Table 1* shows some common abbreviations used on stakes. Operators must be able to identify each stake's purpose and to interpret the directions written on the stake. If unsure of a stake's purpose, ask a supervisor. If necessary, verify the markings with the person who made them.

> **NOTE**
> Never trust a stake that appears to have been moved and reset. If in doubt, verify the stake's position and information with your supervisor.

3.2.0 Setting Stakes

The term *stake* is used to indicate any stake, lath, or **hub** used in laying out a construction project. Stakes vary in size, shape, and material according to their uses and manufacturers. The simplest stake is made from an ordinary rough plaster lath. Stakes are usually made from lumber and can be 1 to 3 inches wide and as long as 48 inches.

> **On Site**
> **Computing Percent**
> A percent is 1/100th of something. For example, to compute the percent equivalent of one-half, convert one-half into hundredths as follows:
> $$\frac{1}{2} = \frac{1 \times 50}{2 \times 50} = \frac{50}{100} = 50 \text{ percent}$$

Table 1 General Stake Markings

Marking	Meaning	Example	Example Meaning
F	Fill	F 2.5	Fill the area with 2.5 ft of soil
C	Cut	C 2.5	Cut 2.5 ft of soil from the area
E/P or EOP	Edge of pavement		
S/S or SLOPE	Slope	S/S 3:1	For every 3 ft from the stake the ground needs to drop 1 ft.
O/S	Offset	O/S	
		(5)	5 ft offset from centerline
		(2)	2 ft offset from centerline
BM or B/M	Benchmark	B/M	

22106-12_T01.EPS

Classroom

Emphasize that grade stakes contain critical information and must be left undisturbed.

Audiovisual

Show Slide 14 (Table 1). Explain the stake marking abbreviations.

Teaching Tip/Notes

22106-12 Grades Module Seven 7

Audiovisual

Show Slides 15 and 16 (Figure 12 and Table 2). Describe a basic stake and explain the color-coded flagging system.

Classroom

Explain what a benchmark is and how it is used.

Explain the use of temporary benchmarks.

Audiovisual

Show Slides 17 and 18 (Figures 13 and 14). Explain that benchmarks are used as control points.

Classroom

Explain how stakes are used in the various stages of a project.

Audiovisual

Show Slide 19 (Figure 15).

Most have a small piece of colored flagging tied to them, as shown in *Figure 12*. Some of the stakes on the work site are placed to mark underground utilities. These markers are color-coded using the standard underground color code found in *Table 2*. When these markers are seen, act accordingly. If in doubt, ask your supervisor.

Grade stakes serve as reference points for cuts, fills, and boundaries. Each stake indicates whether to remove or add earth, and what the grade of the area should to be. The number of stakes and the information written on each stake vary according to the project and whether the stake is temporary or permanent.

Most construction projects use fixed control points as references. These points are called **benchmarks (BM)** and are used to ensure that the construction project is placed correctly. Benchmarks can be either permanent or temporary. A permanent benchmark can be some fixed object with a known position, such as the top of a fire hydrant or a spike on a telephone pole. Other permanent benchmarks are markers that are set in concrete and marked with the elevation; they are often called monuments (*Figure 13*). These markers are set by government personnel or by licensed surveyors.

Temporary benchmarks (TBM) are usually stakes, hubs, or markers that are placed by the survey crew based on the position of a permanent benchmark. Temporary benchmarks are needed because it is not always practical to start a survey at a permanent benchmark. Temporary benchmarks are protected by guard stakes that usually have some flagging tied around the top. (*Figure 14*).

Temporary benchmarks must not be disturbed or damaged. They are used throughout the project to set other stakes to indicate grading, cuts, or fills. Moving these stakes could cause part of the project to be placed incorrectly. Damaging them could cause many hours of additional work to replace them and delay the completion of the project.

For the heavy equipment operator, a construction project has three phases: initial clearing, rough grading, and final grading. When a project first begins, the survey crew sets rough boundary stakes for the clearing crew. Boundary stakes are used to permit preparation of the site and will probably be destroyed while the land is being cleared.

After the site has been cleared, the survey crew uses levels, electronic devices, and other equipment to mark the site for rough grading. On roadway projects, the survey crew sets stakes that indicate the center and shoulders of the road, and any embankments and backslopes. In some cases, the survey crew sets only center line stakes for the road, and the construction crew sets the other stakes.

Figure 15 shows a typical configuration of stakes for a roadway. Keep in mind that the stakes are set in the existing ground. Each stake contains specific information about what work needs to be done to build the roadway.

22106-12_F12.EPS

Figure 12 Grade stake.

Table 2 American Public Works Association (APWA) Underground Color Codes

Color	Meaning
Red	Electric power lines, cables, conduit, and lighting cables
Yellow	Gas, oil, steam, petroleum, or gaseous material
Orange	Communication, alarm or signal lines, cables, or conduits
Blue	Potable water
Green	Sewers and drain lines
Purple	Reclaimed water, irrigation, and slurry lines
White	Proposed excavation
Pink	Temporary survey markings

22106-12_T02.EPS

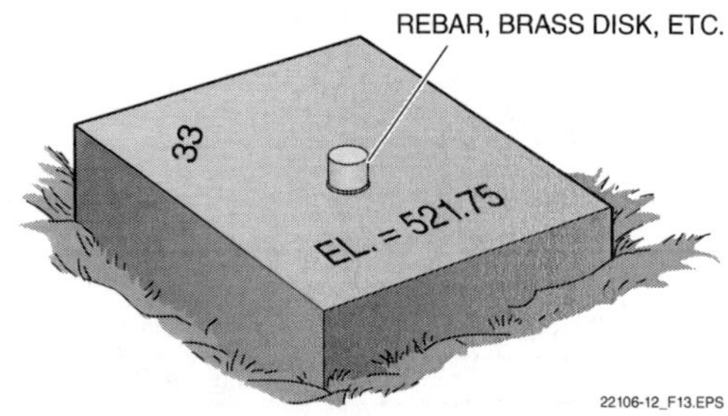

22106-12_F13.EPS

Figure 13 Permanent benchmark.

Teaching Tip/Notes

Provide grade stakes with different markings for the trainees to examine. Explain the markings.

Audiovisual

Show Slide 20 (Figure 16). Explain the crow's foot and other markings found on center line stakes.

Show Slide 21 (Figure 17). Explain what a plus station is.

Figure 14 Temporary benchmarks and guard stakes.

3.2.1 Center Line Stakes

Center line stakes mark the center of the proposed roadway and are marked with the letters (*Figure 16*) indicating cut and fill information. The station number is written on the front face of center line stakes. The station number shows the distance the stake is from a known location in 100-foot increments. The first station is marked 0 + 00, indicating 0 feet. The next station is 100 feet farther and marked 1 + 00, indicating 100 feet. Full stations are every 100 feet. Generally, the front of the stake faces the beginning point of the project.

When a culvert or the beginning of a curve is to be located between full stations, another center line stake is used. For example, if a culvert is to be located 1,857 feet from the beginning of the project, a center line stake marked with 18 + 57 is placed at the desired point (*Figure 17*). This is called a plus station because it is plus 57 feet from the last regular station (18).

Referring to *Figure 16*, the reverse side of the stake indicates cut or fill information. Cut is designated with the letter C, and fill is designated with the letter F. Cut and fill information is accompanied by the **crow's foot** symbol. The crow's foot

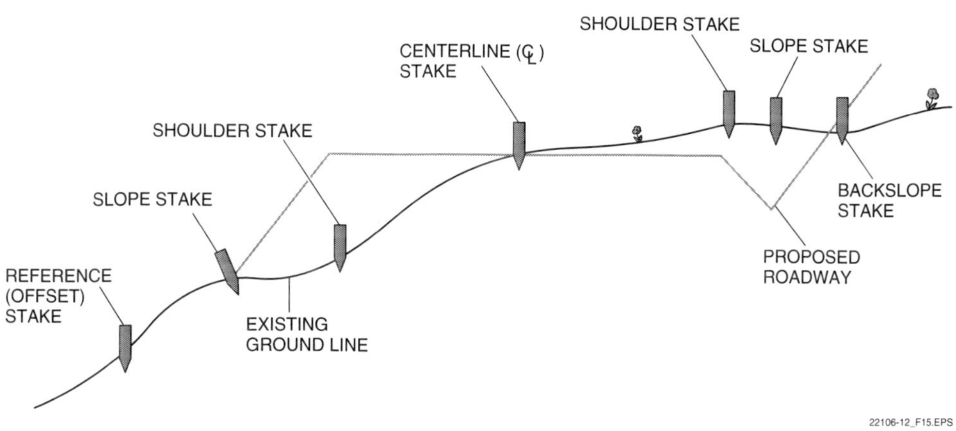

Figure 15 Stakes for rough grading.

Teaching Tip/Notes

22106-12 Grades Module Seven 9

Audiovisual

Show Slide 22 (Figure 18). Discuss the markings used on edge of pavement stakes.

Classroom

Describe the placement of slope stakes.

Explain the rise-to-run format for metric units.

Audiovisual

Show Slides 23 (Figure 19). Explain the markings used on slope stakes.

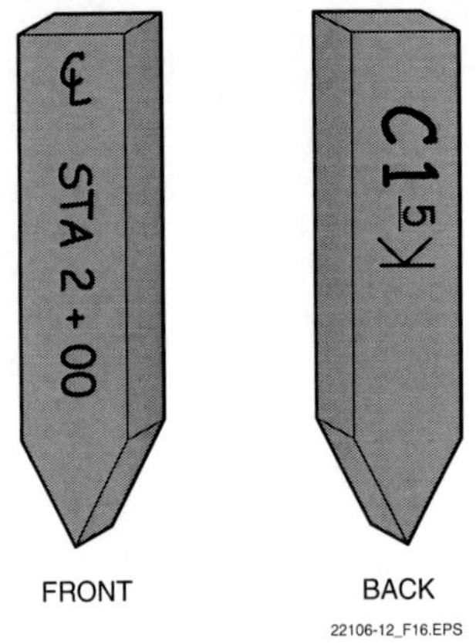

Figure 16 Center line stake.

points down (▼) for cut and up (▲) for fill. Numbers follow the letter to designate the amount of earth that needs to be removed or the amount of fill required. If no cut or fill is needed, the crow's foot is placed so that the straight line is exactly level with the ground, and the letters GR are written on the stake.

One point to keep in mind is that, in surveying, measurements are made and written in the decimal system. Cut and fill information on rough grade stakes are written in feet and tenths of feet similar to the following:

$$C\ 2^{\underline{5}} \text{ or } F\ 3^{\underline{8}}$$

Finished grade (FG) stakes are marked in hundredths of a foot. $C\ 2^5$ is a cut of 2.50 feet, or 2 feet and 6 inches. $F\ 3^8$ means a fill of 3.79 feet, or 3 feet and 9½ inches.

3.2.2 Edge of Pavement Stakes

Edge of road stakes or markers (*Figure 18*) are used to mark the edges of the paved surface. They are set on the shoulder line at right angles to the center line marker. The same station numbers written on the center line marker are also written on the back of these stakes so that they may be easily matched to a station. Cut or fill information is written on the front of the stake, which faces the center line. The distance that the stake is located from the center line may be placed below the cut or fill figure.

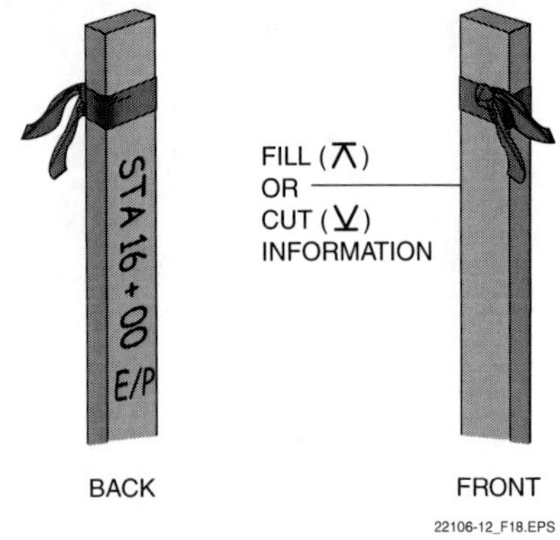

Figure 18 Edge of pavement stakes.

3.2.3 Slope Stakes

Slope stakes (*Figure 19*) are set at the lower hinge point (toe) in the slope. This indicates the limits of the earthwork. Slopes are marked on the stake as a ratio of the horizontal distance to the vertical distance (run:rise). Thus, a stake that reads SLOPE 3:1 indicates that for every three feet from the stake, the bank falls by one foot. When the slope is not indicated, it should go from the bottom of the ditch or edge of the road to the base of the stake. A slope stake located in this manner is sometimes referred to as a backslope stake. When a backslope stake is used as an offset stake, the stake indicates the amount of cut and the distance from the stake that would be the bottom of the cut.

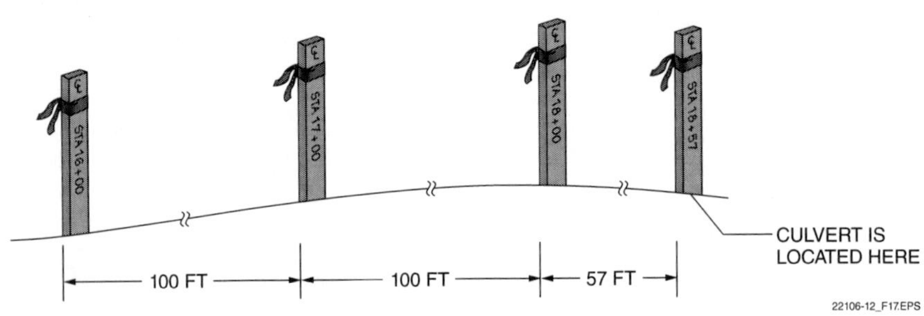

Figure 17 Plus station.

Teaching Tip/Notes

Provide center line, edge of pavement, and slope stakes for the trainees to examine. Have them compare the markings.

> **NOTE**
> For projects using metric units, slope ratios are expressed as vertical distance to horizontal distance (rise:run).

3.2.4 Offset Stakes

Once work starts on a cut or fill section, center line and slope stakes will probably be destroyed by equipment. To eliminate resurveying parts of the project to set new stakes, offset stakes are placed as **reference stakes (RS).** These stakes are located out of the actual work area so it is unlikely that they will be damaged.

Offset stakes are marked with the station number on the back and the distance that they are set from the center line on the front. The stake is designated as an offset stake by a circle around the distance from baseline number and the letters O/S. Any reference to cut or fill that is written on the stake also indicate what distance from the stake the cut or fill is desired. The offset stake in *Figure 20* indicates that it is set 40 feet from the center line with a fill of 2½ feet at 20 feet from the offset stake, and the grade is on the center line.

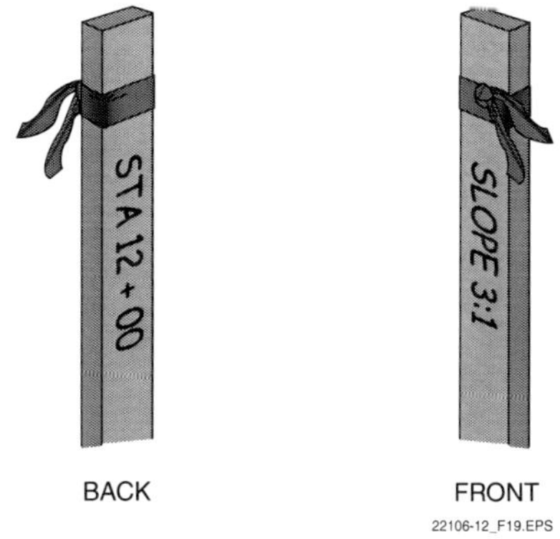

Figure 19 Slope stakes.

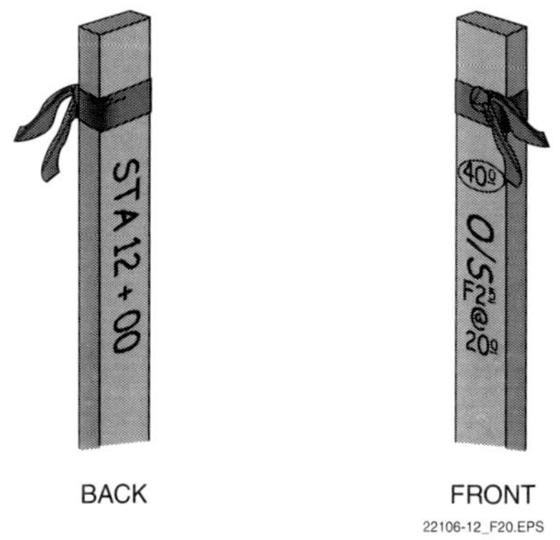

Figure 20 Offset stake.

3.3.0 Slope Stake Interpretation

Slope stakes, which are also called cut and fill stakes (*Figure 21*), are used to indicate the desired shape of a slope. A survey crew working for the engineering firm that designed the project will normally set out slope stakes. The crew calculates the information that is written on each stake as it

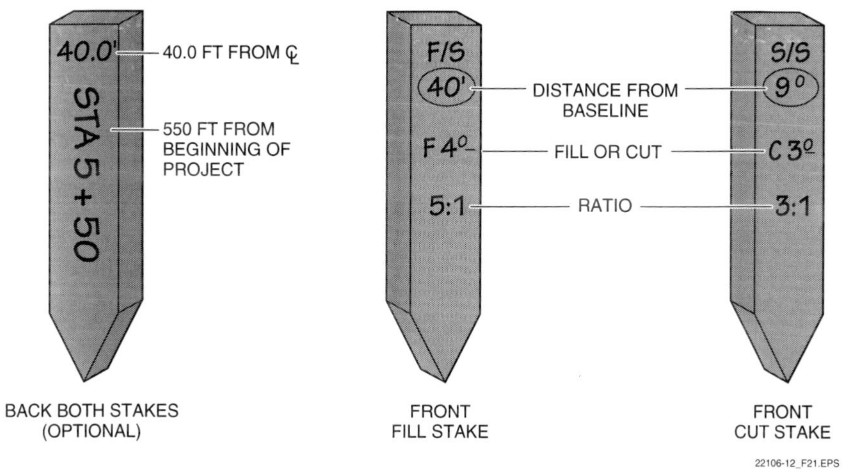

Figure 21 Cut and fill (slope) stakes.

Audiovisual

Show Slide 24 (Figure 20). Describe how offset stakes are used and how they are marked.

Classroom

Discuss the information that appears on a cut and fill stake.

Audiovisual

Show Slide 25 (Figure 21). Explain how to interpret the markings on the stake. Point out that information appears on the front and back of the stake.

Teaching Tip/Notes

Audiovisual

Show Slide 26 (Figure 22). Show how the markings on the fill stake are translated onto the diagram.

Show Slide 27 (Figure 23). Explain how grades are checked.

Show Slides 28 and 29 (Figure 24 and 25). Show how the markings on the cut stake are translated onto the diagram.

Demonstration

Show the trainees how to identify types of stakes and markings on stakes.

PT/Laboratory

Under your supervision, have trainees identify types of stakes and markings on stakes. Note the proficiency of each trainee. This laboratory corresponds to Performance Task 1.

Homework

Have trainees review Sections 4.0.0–4.4.0.

Materials

Ensure that you have everything required for teaching this session.

is placed. Operators must be able to interpret the direction on each stake. Slope stakes provide the following three items of information:

- Distance from the reference (baseline, center line, or offset stake)
- Cut or fill information
- Slope ratio

> **NOTE**
> It is always important to check both sides of a stake in order to get complete directions. The station number may be written on the back of the slope stake. The offset distance is a number inside a circle.

3.3.1 Fill Stakes

Fill stakes are placed where material must be added to build up the natural ground. This stake is designated with the letter F. In *Figure 22*, the stake shown says F 4^0 and 5:1. Remember that a grade can be expressed as a ratio of vertical distance (rise) to horizontal distance (run) or as a percentage above or below the horizon rise-over-run. So using the markings on *Figure 22*, calculate the horizontal distance (run) as follows:

$$\frac{\text{Rise}}{\text{Run}} = \frac{1}{5} = \frac{1 \times \text{fill}}{5 \times \text{fill}} = \frac{1 \times 4}{(5 \times 4)} = \frac{4}{20}$$

So this stake shows that 20 feet from the stake, the fill depth must be 4.0 feet. Operators will need to check their grade periodically as the work progresses. Refer to *Figure 23*. Remember that the desired slope ratio is 5:1, so for every 5 feet of horizontal distance (run), 1 foot of fill (rise) is needed. Thus, at 10 feet, the fill depth would be 2 feet, and at 15 feet, the fill depth would be 3 feet, as referenced from the stake.

3.3.2 Cut Stakes

In cut stakes, the ratio and distance from the reference are similar, but the cut information is interpreted differently. See *Figure 24*. This stake shows a cut of 3.0 feet and a slope ratio of 3:1. Using the rise-over-run formula, calculate the run as follows:

$$\frac{\text{Rise}}{\text{Run}} = \frac{1}{3} = \frac{1 \times \text{cut}}{3 \times \text{cut}} = \frac{1 \times 3}{(3 \times 3)} = \frac{3}{9}$$

In this case, the run is 9 feet. Remember, a ratio of 3:1 means that every 3 feet from the stake the

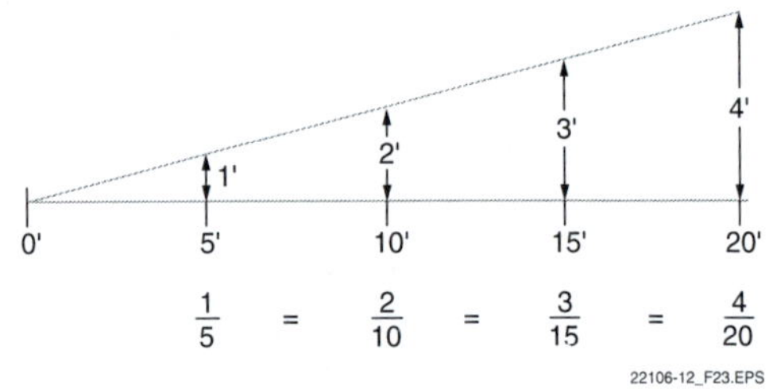

Figure 23 Checking the fill grade.

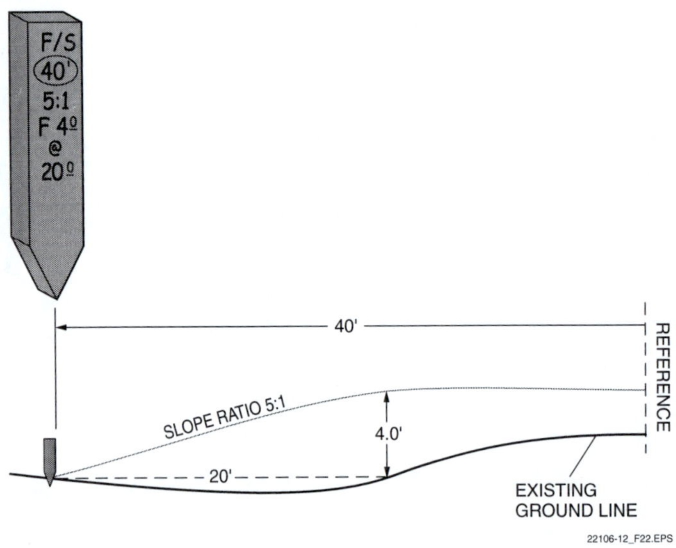

Figure 22 Cross-section of a fill.

Teaching Tip/Notes

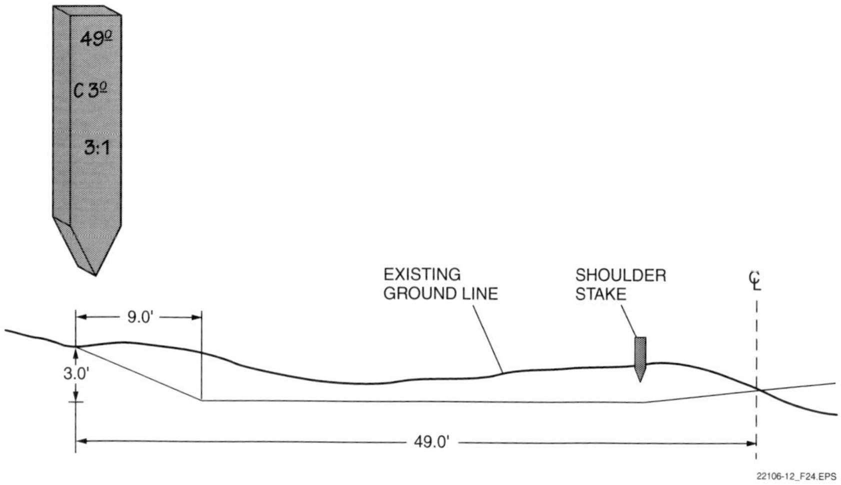

Figure 24 Cross-section of a cut.

ground must drop 1 foot, so to get to a 3.0-foot cut, the run must be 9 feet.

After the initial 3.0-foot cut, the area will probably need to be excavated to form a ditch (*Figure 25*). Sometimes the ditch will need to be a different grade than the initial cut, so the operator may need to get more information before continuing.

4.0.0 Performing Site Measurements

Horizontal and vertical distance measurements require a great deal of precision. Some contractors may use older manual measuring equipment, while others use sophisticated electronic equipment. Many modern heavy equipment machines are now equipped with some type of **global positioning system (GPS).** The GPS used on the grading equipment communicates with GPS transmitter (survey) equipment set up near the area being graded. The GPS/survey equipment is aligned to the control points positioned by the survey team.

Regardless of the positioning equipment installed on the grading equipment, heavy equipment operators must learn how to use it. They must also have a general understanding of how, why, and where markers are placed. *Figure 26* shows a grader working near a GPS transmitter.

To measure vertical distances accurately, begin with a known elevation that is used as a control point or benchmark. The benchmark may be permanent or temporary as previously discussed, or it may be assumed. Assumed benchmarks are usually used on very small jobs; it is any convenient point that is assumed to have an elevation of 100.00 feet. It is usually marked with a hub like a temporary benchmark, as shown previously in *Figure 12*. Once an assumed elevation point is chosen, all elevation readings on the site must refer to that point. Permanent and assumed benchmarks must be recorded on site plans so they may be easily located. Assumed benchmarks must be protected until the job is completed.

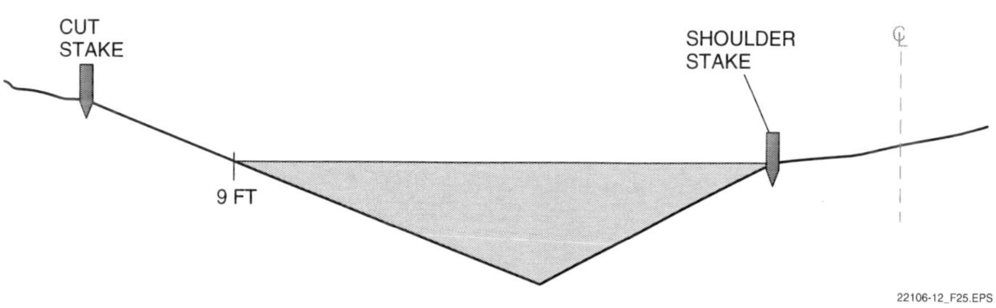

Figure 25 Ditch excavation.

Classroom

Explain the importance of precision in site distance measuring.

Describe how GPS is used in site layout work.

Audiovisual

Show Slide 30 (Figure 26).

Classroom

Describe how benchmarks are used in distance measurement and why temporary benchmarks are often necessary.

Teaching Tip/Notes

Figure 26 Grader with GPS equipment.

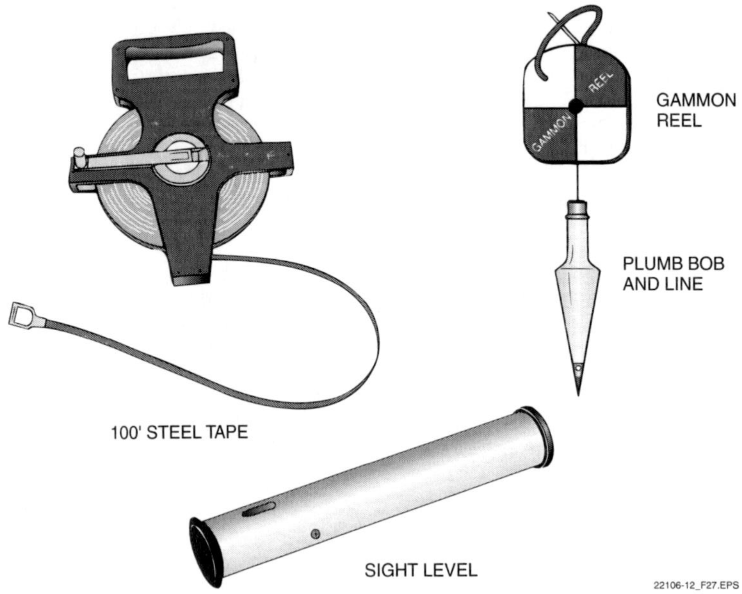

Figure 27 Basic measuring equipment.

Once the benchmark is established, temporary benchmarks may be set and marked with their elevation based on the permanent or assumed benchmark. Temporary benchmarks are necessary because some construction projects cover vast distances and elevations, such as a highway or a shopping mall, so it is not always practical to start every measurement from the permanent benchmark. Since temporary benchmarks are placed from permanent benchmarks, they may then be used as control points for other measurements.

Teaching Tip/Notes

14 NCCER – *Heavy Equipment Operations Level One* 22106-12

4.1.0 Manual Equipment

The manual equipment for measuring horizontal and vertical distances has changed very little over the centuries. While this equipment is simple in design, the accuracy of the measurements depends on the skill of the users. This type of measurement is called **line of sight** because the operator must be able to see the target.

Simple manual devices, such as a sight level, measuring tape, and plumb bob (*Figure 27*) are used to measure horizontal and vertical distances. By placing a sight level to the eye and sighting the target, the sight glass can be used to measure vertical distances. The measuring tape is used to measure horizontal distances, while the plumb bob is used to establish a vertical line. Such basic measuring devices are only used for rough grading.

Since handheld levels are a bit unstable, a more accurate way to measure vertical distances is to put an automatic level (*Figure 28*) onto a tripod that gives the device better stability. Such automatic levels usually have a good working range of up to 150 feet with an accuracy of $^1/_{16}$th of an inch at 75 feet. Automatic levels also have adjustable magnifications that allow the operator to better see the target.

The last piece of equipment needed is a leveling rod (*Figure 29*). A leveling rod is used with a sight level or an automatic level to measure vertical distances.

4.1.1 Measuring Horizontal Distances

Measuring horizontal distances across level ground is simple. Two people, a head tape person and a rear tape person, simply take the tape and measure the distance across the ground, as shown in *Figure 30*.

When measuring a horizontal distance across a steep slope, measure the distance in a series of steps while holding the tape horizontal (level) as

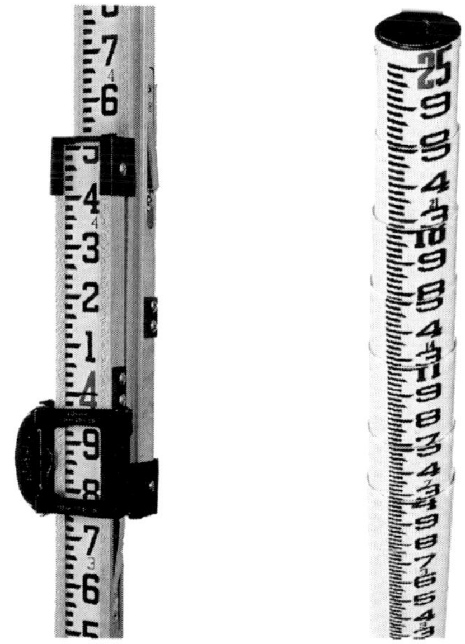
LENKER ROD (SAN FRANCISCO STYLE)

TELESCOPING ROD

22106-12_F29.EPS

Figure 29 Leveling rods.

22106-12_F28.EPS

Figure 28 An automatic level.

Audiovisual

Show Slides 31 and 32 (Figures 27 and 28). Compare the use of rough measuring equipment such as a tape with the use of an automatic level.

Show Slide 33 (Figure 29). Describe the use of a leveling rod.

Show Slide 34 (Figure 30). Describe how a tape is used to measure a horizontal distance.

Teaching Tip/Notes

Audiovisual

Show Slide 35 (Figure 31). Describe how a tape is used to measure distance across a slope.

Classroom

Explain differential leveling and give examples of how it is used.

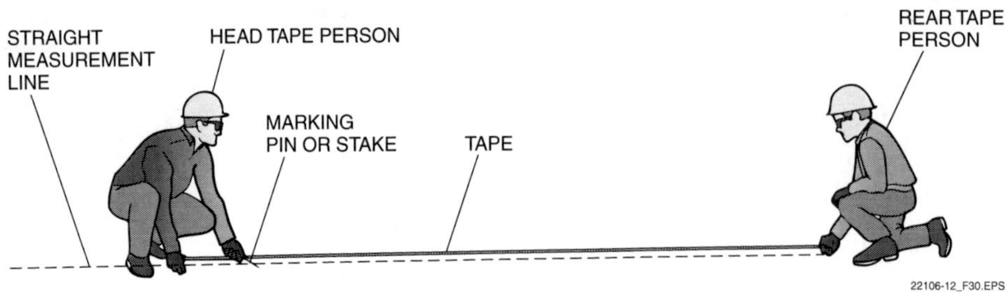

Figure 30 Measuring horizontal distances.

shown in *Figure 31*. This process is called breaking the tape. A plumb bob is used to mark the exact position on the ground where the start of the next measurement will be.

4.1.2 Measuring Vertical Distances

Measuring vertical distances or elevation is called differential leveling. Differential leveling is used to set grade stakes and to check the slopes after

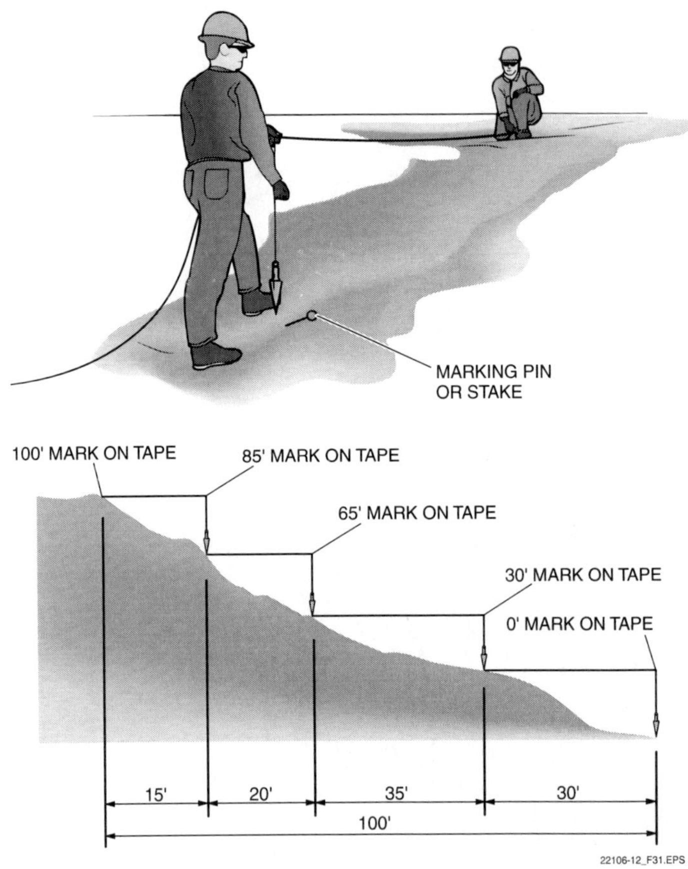

Figure 31 Measuring horizontal distance across a slope.

Teaching Tip/Notes

Did You Know?

Gunter's Chain

Surveyors in the eighteenth and nineteenth centuries did not have measuring tapes. They used a measuring device called a Gunter's chain. It was designed by British mathematician Edmund Gunter. The chain had 100 links. That is why today you will frequently hear a survey crew call a measuring tape a chain.

the grading work is done. Vertical distances may be measured using a sight or automatic level and leveling rod. It can be done with one person using additional hardware, but it is best to use two people for speed and accuracy. One person acts as the instrument person and uses the level, while the other person acts as the rod person and handles the leveling rod.

When an automatic level is used, the tripod must be set up first (*Figure 32*). The tripod must be set firmly in the ground. Any movement of the tripod during operation affects the accuracy of the measurements and requires that all measurements be repeated. Further, the tripod must be level to obtain an accurate measurement. If the tripod does not have levels attached to it, lay a small construction level across the mounting plate to ensure that it is set correctly. Note that in *Figure 32* the plumb bob is suspended from the tripod to mark the exact reference point.

The person handling the leveling rod must understand how to use it. The bottom of the rod must be placed in the proper location and then held vertically so the instrument person can see it. The instrument person usually reads the rod through the level. Always double-check the readings.

There are two types of leveling rods commonly used in the United States. One is the architect's rod, which measures in feet, inches, and eighths of an inch. The other is the engineer's rod, which measures in feet, tenths of a foot, and hundredths of a foot. See *Figure 33.*

Using a leveling rod requires that all measurements start at a known level. This may be a monument, a benchmark, or a temporary benchmark. In *Figure 34*, the leveling rod is placed on a BM, and then it is read. This reading is called the **backsight (BS)** reading. Add the backsight reading to the known elevation of the benchmark, to calculate the **height of instrument (HI).** Then the instrument operator should carefully turn the level towards the first turning point (TP1) while the rod person moves the rod to TP1. The instrument is sighted and the rod reading is taken. This

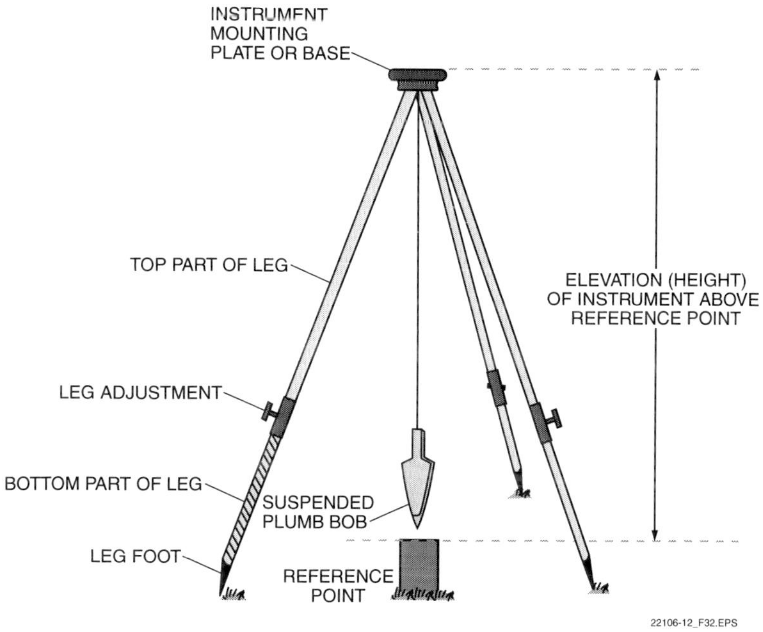

Figure 32 Automatic level tripod.

Classroom

Point out the importance of keeping the rod level.

Audiovisual

Show Slide 36 (Figure 32). Describe the purpose of the plumb bob.

Show Slide 37 (Figure 33). Demonstrate how to read a leveling rod.

Show Slide 38 and 39 (Figures 34 and 35). Explain the terms *backsight* and *foresight.*

Teaching Tip/Notes

Bring a leveling rod into the classroom and allow the trainees to practice reading it.

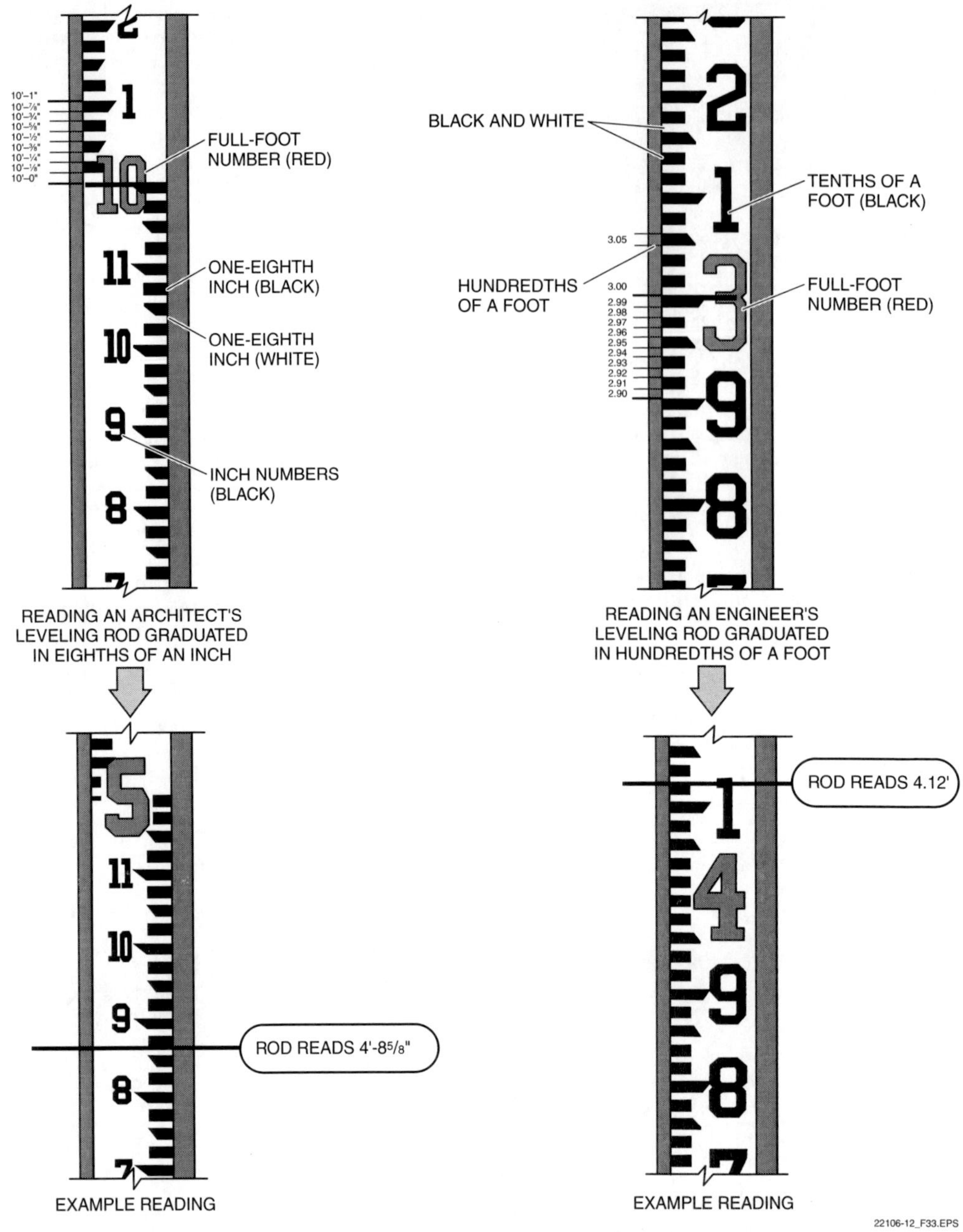

Figure 33 Reading a leveling rod.

Teaching Tip/Notes

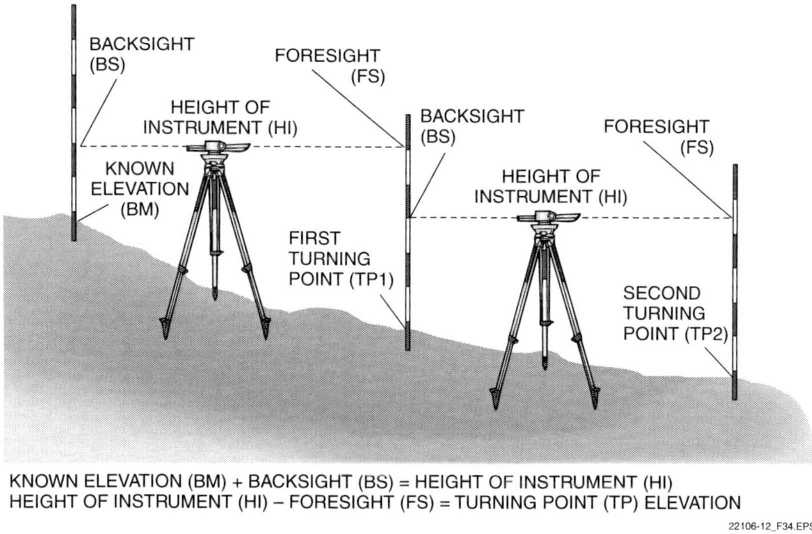

Figure 34 Using a leveling rod.

number is called the **foresight (FS)** reading. To calculate the elevation of TP1, subtract the foresight reading from the HI (*Figure 35*). Continue this process until all measurements have been taken.

4.2.0 Electronic Measuring Devices

Many of the devices discussed previously now have electronic counterparts, which provide the users with a digital readout of measurements. Laser levels (*Figure 36*) have made setting grade stakes faster and easier.

Newer laser levels are more sophisticated and now permit one-person operation (*Figure 37*). Some laser levels are self-leveling and do not turn on unless they are level. Further, some of these levels automatically turn off if they are bumped off level during use. Laser levels can speed up the task and improve the accuracy of setting stakes as well as checking grades, but only in the hands of a knowledgeable user.

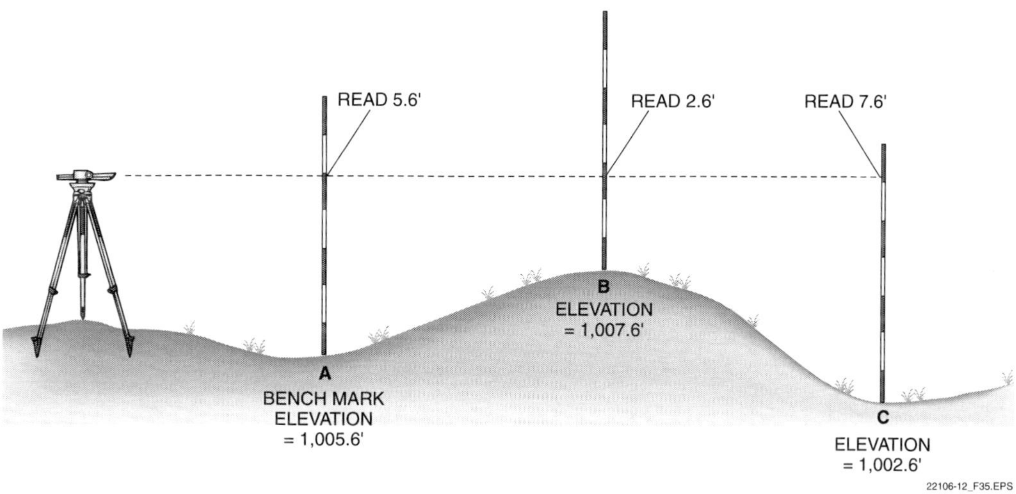

Figure 35 Using a leveling rod to calculate elevations.

Audiovisual

Show Slide 40 and 41 (Figures 36 and 37). Explain how laser levels are used.

Teaching Tip/Notes

Audiovisual

Show Slide 42 and 43 (Figures 38 and 39). Describe the operation and applications of rotating-beam laser levels.

Safety

Point out the hazards associated with the use of lasers. Explain that OSHA requires that anyone operating laser instruments be properly trained and that the training must be documented.

22106-12_F36.EPS

Figure 36 Laser transmitter and receiver rod.

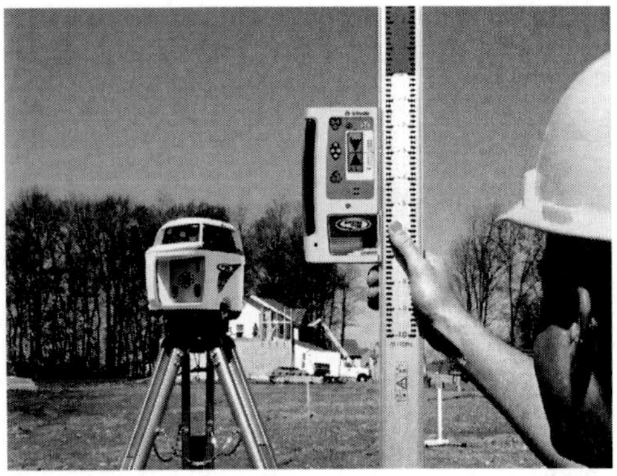

22106-12_F37.EPS

Figure 37 One-person operation.

Electronic levels work by line of sight, just as manual equipment does. They work by transmitting a visible laser beam or invisible infrared light towards a target. Some models even measure vertical distances, giving them a clear advantage over manual equipment. Some laser levels transmit a single beam, but others, like the one shown in *Figure 38*, transmit a rotating beam. The principle behind the single-beam laser is the same as the manual level; it projects a line of light (visible or invisible) towards a receiver. The receiver can be mounted on a rod or be a handheld device. If the detector is attached to a rod, the rod person moves the detector up or down until it detects the laser beam and then the rod is read. Receivers often have an audible or visible indicator to let the user know when it is receiving the beam. Rotating beam lasers sweep a beam of light over a 360-degree area so that multiple receivers can be placed around the work site (*Figure 39*). This way several points may be checked at the same time.

WARNING!

Some laser units transmit a very powerful and focused beam that can cause serious eye injury or even blindness. This can happen when the beam is reflected into the operator's eyes from a bright object, such as shiny metal or a mirror.

Because the beam from some lasers is powerful, OSHA requires that operators of laser instruments be properly trained, and the training documented, before using the equipment. Always follow manufacturer's directions regarding operation and the need for eye protection. *OSHA 29 CFR Part 1926* states that safety precautions, such as posting signs, must be used when laser instruments are in use.

4.3.0 Global Positioning System Survey Devices

The GPS surveying equipment was developed by the US military to provide precise position information for vehicle, troop, and equipment movement. It is made up of a series of satellites that circle Earth at various orbiting levels. These satellites broadcast a radio signal that contains satellite location information 24 hours a day. A GPS surveying system uses an electronic receiver for these signals. Then computers in the receiver compute the exact position.

GPS surveying equipment takes much of the guesswork out of site layout. The GPS receiver (*Figure 40*) may be placed at any convenient location on the work site and a benchmark may be quickly and accurately set. GPS surveying equipment combines satellite and laser technology so that the equipment is versatile and accurate. Many models provide on-screen graphics of construction plans (*Figure 41*). GPS systems increase the accuracy and efficiency of measurements over other methods, but they require a skilled operator to use the provided information effectively.

22106-12_F38.EPS

Figure 38 Rotating beam laser instrument.

Teaching Tip/Notes

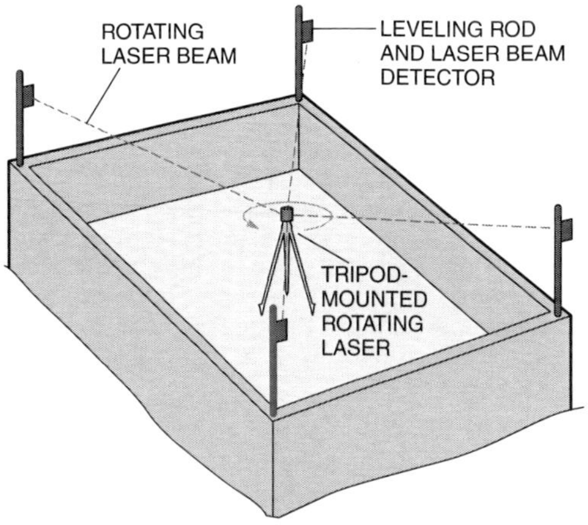

22106-12_F39.EPS

Figure 39 Rotating lasers sweep multiple sensors.

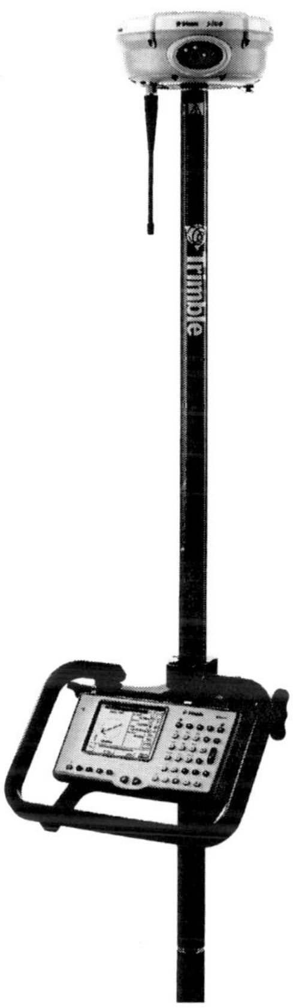

22106-12_F40.EPS

Figure 40 GPS surveying receiver.

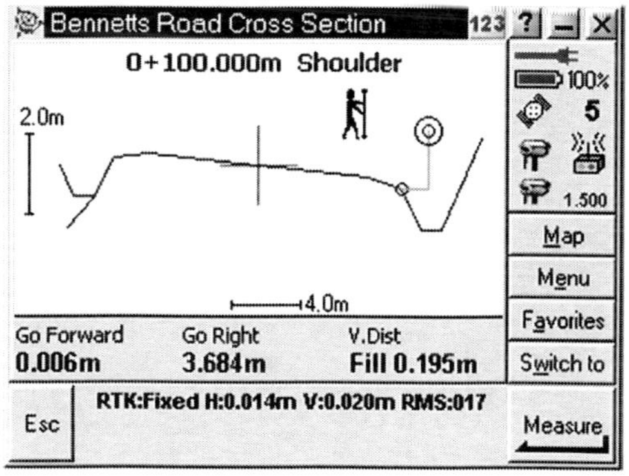

22106-12_F41.EPS

Figure 41 Graphics of construction plan on GPS surveying equipment.

Some GPS systems use two receivers. One receiver is placed in a fixed location, and the other is moved to different positions on the site. When activated, the computers calculate the horizontal and vertical difference in their positions. This method is very accurate and allows one person to set stakes.

4.4.0 Stakeless Systems

As in other industries, computers are making their mark on heavy equipment. Computer technology now allows construction sites to be laid out electronically on a computer that converts the layout to transmittable signals. Signal receivers are mounted on the heavy equipment to receive these signals. Computers calculate the location of the equipment, with or without GPS, and match it to the location on the construction plans. Based on this information the blade of the equipment is raised or lowered to perform the desired cut or fill.

On Site

Virtual Reference Stations (VRS)

To help ensure that a GPS system is being referenced to the most accurate data, cell phones can be used to connect into virtual reference station (VRS) networks that provide the most accurate, real-time, positioning data. There are now several GPS systems on the market that can be programmed to accept VRS data through a cell phone.

Audiovisual

Show Slide 44 and 45 (Figures 40 and 41). Explain how GPS is used to establish benchmarks, place stakes, and make site measurements.

Classroom

Describe how virtual reference station (VRS) technology is used with cell phones to obtain positioning information.

Describe how stakeless systems using computers and equipment-mounted receivers allow sites to be graded by electronically controlling grading equipment.

Demonstration

Demonstrate the proper method for checking horizontal and vertical distance of cut and fill slope stakes.

PT/Laboratory

Under your supervision, have trainees check horizontal and vertical distance of cut and fill slope stakes. Note the proficiency of each trainee. This laboratory corresponds to Performance Task 2.

Homework

Have trainees review Sections 5.0.0–5.2.0.

Teaching Tip/Notes

Materials

Ensure that you have everything required for teaching this session.

Classroom

Explain that finish stakes, often called red tops, are used to indicate the final grade.

Point out that finish stakes may be placed at the beginning of the job or after rough grading has been completed.

Describe the methods used to check grade.

Audiovisual

Show Slides 46 through 48 (Figures 42 through 44).

5.0.0 FINISH GRADES

On large, complex construction jobs, the survey crew is constantly working to set stakes and check grading work. After rough grading work has been completed, the survey crew sets finish grade stakes. These stakes are set at exact levels and are used to check the final level of the subgrade and grade. Remember, the subgrade is the surface of the foundation where the final surface is applied. The grade is the actual finished surface of the road.

On small jobs, the survey crew may set finish grade stakes at the beginning of the job. If this is the case, equipment operators must be aware of how they are set and ensure that the grading work is completed correctly. Reworking costs time and money, and it can delay the completion of the job.

5.1.0 Finish Grade Stakes

Finish stakes are used to indicate the exact finish of a subgrade or grade. Subgrade finish stakes are color-coded blue and are commonly called blue tops. Grade finish stakes are color-coded red and are commonly called red tops.

Finish grade stakes are set in one of two ways. Either the top of the stake is set level with the grade or subgrade, or the top of the stake is set and marked at a predetermined level above the grade or subgrade. The most common manner is to set the finish grade stakes level with the surface.

5.2.0 Checking Grades

Normally three people are needed to manually check subgrades and grades. However, one person, using various types of equipment, can complete the work. Some of these methods are crude; others are very precise. Heavy equipment operators are unlikely to be asked to check finished grades or subgrades, but they are responsible for cutting grades as close to the specified level as possible. Operators should become proficient at some measuring techniques to ensure that the grade is cut properly.

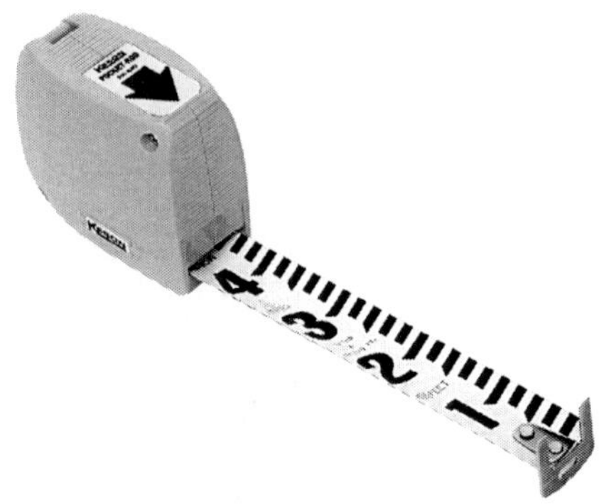

Figure 42 Pocket leveling rod.

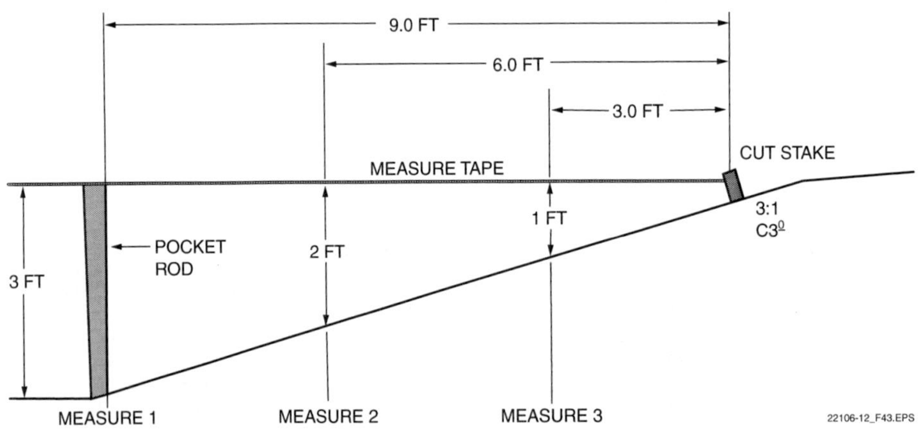

Figure 43 Measuring a cut.

Teaching Tip/Notes

The first technique may only be used to provide a rough idea of the cut. It uses a pocket leveling rod (*Figure 42*), level, and tape measure.

The tape measure is attached to the cut stake and reeled out to the point indicated on the cut stake (*Figure 43*). For example, on a cut stake marked with a 3:1 ratio and 3.0' cut, the tape is pulled out to 9 feet from the stake. The pocket leveling rod is then placed on the ground and held vertically. With the tape held horizontally against the rod, using the level to ensure that the string is level (if the distance is not too great for it to be of use), the rod should read 3 feet, indicating a three-foot cut. For a 3:1 grade, a second measurement is taken at the 6-foot mark on the tape measure, and the pocket leveling rod should read 2 feet. Finally, a third measurement is taken at the 3-foot mark on the tape measure, and the pocket leveling rod should read 1 foot. Note that this method checks both the cut depth and the grade.

A more precise method is needed to measure across a roadway. This two-step method requires three people, a length of string, and three rulers. In the first step, the level of the surface is checked. In the second step, the grade of the surface is checked.

To check the surface level, the string is stretched across the roadway and two of the people set a ruler on top of the finish stakes and hold them at the same level (*Figure 44*). The third person randomly measures points across the roadway to ensure that the surface is the same height between the subgrade and string. Each job will have its own tolerance, but every effort should be made to bring the subgrade as close as possible to specifications.

Of course, some methods are time-consuming and can be very inaccurate. Further, none of these methods tell the exact elevation of the surface. Methods that are more precise require the use of equipment such as an automatic level and rod, a laser level and detector, or a GPS device.

An automatic level or laser level and rod, or a laser level and detector, may be used to check the grade of the subgrade with greater accuracy than the previously discussed methods. However, both methods require the measurement of a known elevation (control point) to ensure that the elevation readings at the blue stakes are accurate (*Figure 45*). If a Lenker (direct elevation) rod is used, no elevation computations are needed because the elevation is read directly from the rod. However, the grade still needs to be computed manually using the rise-over-run formula.

A GPS surveying device can provide the fastest and most accurate elevation measurement. Since GPS devices provide precise elevation reading with the touch of a button, they can be used to set the finish grade stakes, as well as check the level of any point of the work site. Use GPS whenever possible, since it provides the most accurate readings.

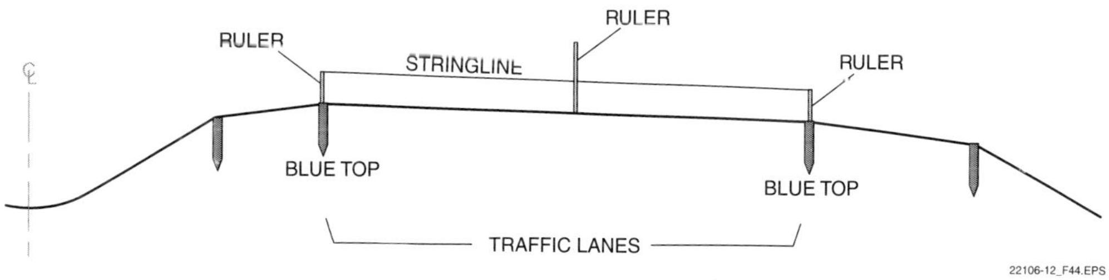

Figure 44 Checking the grade across a roadway.

Audiovisual

Show Slide 49 (Figure 45). Describe how grade can be accurately checked with an automatic level.

Demonstration

Demonstrate methods used to check subgrade on a cross slope.

PT/Laboratory

Under your supervision, have trainees check finish subgrade on a cross slope. Note the proficiency of each trainee. This laboratory corresponds to Performance Task 3.

Teaching Tip/Notes

Classroom

Summarize the major concepts presented in this module.

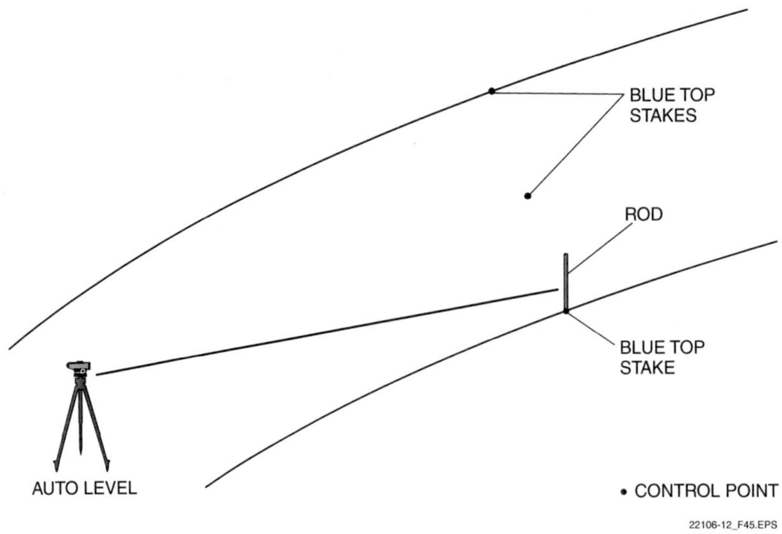

Figure 45 Checking the grade with an automatic level.

Summary

An equipment operator is expected to know about the grading requirements on construction jobs, as well as how to make a grade by reading and understanding grade stakes.

Grade is the rise or fall in a road, channel, or natural ground. It is expressed either as a ratio of vertical distance to horizontal distance (rise-over-run), or as a percent above or below the horizon.

Stakes are set at various points to provide information to workers about what should be done to the topography of the site. Special stakes called benchmarks will provide elevation information. The basic kinds of stakes normally seen on a typical job include center line, slope, offset, shoulder, and finish grade.

Center line stakes identify the center line of the roadway or construction. Shoulder stakes mark the outer edges of the pavement or hinge points. Slope stakes mark the points on either side of the construction where the planned slope intersects with the original ground. Numerical values written on each stake describe the required cut or fill amount needed to raise or lower the original ground elevation to meet grade requirements.

Offset stakes are reference stakes that are placed out of the construction area. These stakes are placed at points where they are out of the way and will not be damaged.

On occasion, an operator may be required to assist in the placement of stakes. This is done with a level rod, hand level, and measuring tape. A more modern method uses an automatic level or a laser level and a level rod, but the latest technology uses GPS survey devices.

For finish grading, the elevations are checked from one side of the grade to the other to make sure the final grade follows the prescribed elevations. Finish grades and subgrades may be checked using manual, electronic, or GPS survey equipment. Manual survey equipment requires more people and may yield inaccurate information. Electronic and GPS devices provide more precise and accurate measurements.

Teaching Tip/Notes

Review Questions

1. Station numbers are usually placed in increments of _____.
   a. 25 feet
   b. 50 feet
   c. 75 feet
   d. 100 feet

2. The center line of a roadway is sometimes called the baseline.
   a. True
   b. False

3. The letters CL on a stake indicate _____.
   a. curvature line
   b. center line
   c. culvert line
   d. cut line

4. To allow for drainage, the subgrade usually has what type of grade?
   a. Level
   b. Positive
   c. Falling
   d. Rising

5. Hinge points mark the _____.
   a. edges of the subgrade
   b. center line of the road surface
   c. boundary for the silt fence
   d. beginning of the backslope

6. Calculate the grade of a surface that rises 1 foot for every 50 feet of distance.
   a. ¼ percent
   b. ½ percent
   c. 2 percent
   d. 4 percent

7. Calculate the grade of a surface that rises 5 feet for every 200 feet of distance.
   a. ¼ percent
   b. ½ percent
   c. 2½ percent
   d. 5 percent

8. When a grade falls, it is designated with a _____.
   a. plus sign
   b. minus sign
   c. ratio sign
   d. percent sign

9. A stake with an orange flag indicates which kind of buried utility line?
   a. Communication
   b. Electric power
   c. Irrigation
   d. Sewer

10. What items are placed by the survey crew, based on the position of a permanent benchmark?
    a. Guard laths
    b. Warning flags
    c. Boundary stakes
    d. Temporary benchmarks

11. Center line stakes may contain information about cuts and fills.
    a. True
    b. False

12. If a center line stake has the number 20 + 57 written on it, those numbers represent what distance on the roadway?
    a. 2.57 feet
    b. 20.57 feet
    c. 257 feet
    d. 2,057 feet

13. The information written on the front of an edge of pavement stake is the _____.
    a. center line distance
    b. cut or fill information
    c. pavement width information
    d. pavement thickness information

14. A stake that reads SLOPE 4:1 indicates that for every _____.
    a. ¼ foot from the stake the earth falls 1 foot
    b. 1 foot from the stake the earth falls ¼ foot
    c. 1 foot from the stake the earth falls 4 feet
    d. 4 feet from the stake the earth falls 1 foot

Classroom

Have the trainees complete the Review Questions, and go over the answers prior to administering the Module Examination.

Have the trainees complete the Trade Terms Quiz, and go over the answers prior to administering the Module Examination.

Examination ?

Administer the Module Examination. Record the results on Training Report Form 200, and submit the results to the Training Program Sponsor.

Performance Testing ✓

Administer the Performance Test and fill out the Performance Profile Sheets for each trainee. If desired, trainee proficiency noted during laboratory sessions may be used to complete the Performance Test. Record the results on Training Report Form 200 and submit the results to the Training Program Sponsor.

Teaching Tip/Notes

15. To keep offset stakes from being damaged, they are placed _____.
    a. in line with the center line stakes
    b. along the side of the subgrade
    c. outside the work area
    d. with guard laths

16. If a marking on a stake is a 7 inside a circle, it means that the stake is _____.
    a. 7 feet offset from the center line
    b. 7 feet from the backslope
    c. turning point number 7
    d. station number 7

17. The letter F on a stake means that the stake is a _____.
    a. fill stake
    b. foot marker
    c. fill indicator
    d. forward stake

18. If a stake is marked with numbers and letters that say: F 1^0 and 2:1, the fill level will be _____.
    a. ½ foot at 1 foot from the stake
    b. 1 foot at ½ foot from the stake
    c. 1 foot at 2 feet from the stake
    d. 2 feet at 1 foot from the stake

19. If a stake is marked with numbers and letters that say: C 2^0 and 8:1, the cut level will be _____.
    a. 1 foot at 8 feet from the stake
    b. 1 foot at 16 feet from the stake
    c. 8 feet at 1 foot from the stake
    d. 16 feet at 1 foot from the stake

20. To measure vertical distances accurately, start at a(n) _____.
    a. known elevation
    b. boundary stake
    c. hinge point
    d. inslope

Teaching Tip/Notes

Trade Terms Quiz

Fill in the blank with the correct term that you learned from your study of this module.

1. The point indicating where the road or shoulder subgrade ends and the inslope begins is called the ________.
2. A(n) ________ is a reading taken on a leveling rod to determine an unknown elevation; used in conjunction with backsight and instrument height.
3. Ground that forms a natural or artificial incline upward or downward is called a ________.
4. The area of the roadway between the shoulder and the ditch is called the ________.
5. A ________ is a reading taken on a leveling rod held at a known elevation, which is used to determine the height of the leveling instrument.
6. A ________ is the distance above or below sea level or other reference point.
7. ________ is the surface of a road, channel, or natural ground area.
8. A ________ is a lath set with markings to indicate the final grade at a certain point.
9. The side view of a proposed construction project is called the .
10. ________ is any surface that has been cut or built to the elevation indicated for that point.
11. ________ is the area between the ditch line and a backslope stake.
12. The proportion or relationship in quantity, amount, or size between two or more things is called the ________.
13. Any type of marker that is used to mark cut, fill, and grade information, underground utilities, or survey points is called a(n) ________.
14. The system developed by the US military to provide precise information for the movement of military vehicles, equipment, and personnel is called the ________.
15. The plotting and correcting of irregularities of the ground to a definite limit of grade and alignment is called ________.
16. The ________ is the uppermost level of material placed in embankment or left at cuts in the normal grading of a road bed.
17. The ________ is the point on stakes or drawings which indicates the halfway point between two sides.
18. The description and delineation of natural and man-made features of an area are called ________ features.
19. ________ is the process in which a substance, such as soil, is slowly washed away by rain, wind, or other causes.
20. The original ground elevation before any excavation has been done is called ________.
21. A temporary point of known or assumed elevation from which surveyors can establish all their grades for a particular job is called ________.
22. ________ is the line through the leveling instrument as viewed by the eye.
23. Manufactured or natural rock or soil that has a specific size characteristic is a(n) ________.

Teaching Tip/Notes

24. ________ is the vertical distance from the benchmark to the line of sight (also called backsight) of the instrument.

25. ________ are points-of-origin stakes that identify a point on the ground.

26. The longitudinal distance of a point of a roadway from the starting or reference point is called the ________.

27. A permanent point of known or assumed elevation from which surveyors can establish their grades is called a ________.

28. ________ is a stake from which measurements and grades are established.

29. A line from which other measurements are referenced is called a(n) ________.

Trade Terms

- Aggregate
- Backsight (BS)
- Backslope
- Baseline
- Benchmark (BM)
- Center line
- Crow's foot
- Elevation
- Erosion
- Finished grade (FG)
- Foresight (FS)
- Global positioning system (GPS)
- Grade
- Grade work
- Height of instrument (HI)
- Hinge point (HP)
- Hubs
- Inslope
- Line of Sight
- Natural ground
- Profile
- Ratio
- Reference stake (RS)
- Slope
- Stake
- Station number
- Subgrade
- Temporary benchmark (TBM)
- Topographical

Teaching Tip/Notes

Trade Terms Introduced in This Module

Aggregate: Manufactured or natural rock or sand that has a specific size characteristic. Used in foundations.

Backsight (BS): A reading taken on a leveling rod held at a known elevation, which is used to determine the height of the leveling instrument.

Backslope: The area between the ditch line and a backslope stake.

Baseline: A line from which other measurements are referenced.

Benchmark (BM): Permanent point of known or assumed elevation from which surveyors can establish grades.

Center line: The point on stakes or drawings, which indicates the halfway point between two sides.

Crow's foot: Markings on a lath to indicate the final grade at a certain point.

Elevation: The distance above or below sea level or other reference point.

Erosion: The process in which a substance, such as soil, is slowly washed away by rain, wind, or other causes.

Finished grade (FG): Any surface that has been cut or filled to the final elevation indicated or designed for that point.

Foresight (FS): A reading taken on a leveling rod to determine an unknown elevation; used in conjunction with backsight and height of instrument.

Global positioning system (GPS): System developed by the US military to provide precise information for the movement of military vehicles, equipment, and personnel. This system is now available for use by civilians.

Grade: The surface of a road, channel, or natural ground area. This usually means the surface level required by the plans or specifications.

Grade work: Plotting and correction of irregularities of the ground to a definite limit of grade and alignment.

Height of instrument (HI): Vertical distance from the benchmark to the line of sight (also called backsight) of the instrument.

Hinge point (HP): A point indicating where the road or shoulder subgrade ends and the inslope begins. Sometimes called the catch point.

Hubs: Points-of-origin stakes that identify a point on the ground. The top of the hub establishes the point from which soil elevations and distances are computed; or a point to be trimmed to.

Inslope: The area of roadway between the shoulder and the ditch (natural ground and the shoulder hinge point).

Line of sight: Line through the leveling instrument as viewed by the eye; point where the vertical hair crosses the horizontal hair.

Natural ground: The original ground elevation before any excavation has been done.

Profile: The side view of a proposed construction project.

Ratio: Proportion; relationship in quantity, amount, or size between two or more things. For example, if a grade has a horizontal run of 3 feet for every 1 foot in elevation, the ratio is 3:1.

Reference stake (RS): The stake from which measurements and grades are established.

Slope: Ground that forms a natural or artificial incline upward or downward.

Stake: Any type of marker that is used to mark cut, fill, and grade information, underground utilities, or survey points.

Station number: The longitudinal (lengthwise) distance of a point of a roadway from the starting or reference point. Station numbers are usually measured in 100-foot increments.

Subgrade: The uppermost level of material placed in an embankment or left at cuts in the normal grading of a road bed. This becomes the foundation for aggregate and asphalt material.

Temporary benchmark (TBM): A temporary point of known or assumed elevation from which surveyors can establish all their grades for a particular job.

Topographical: Description and delineation of the natural and man-made features of an area.

Teaching Tip/Notes

Additional Resources

This module presents thorough resources for task training. The following resource material is suggested further study.

Site Layout Levels 1 and *2*, Latest Edition. NCCER. Upper Saddle River, NJ: Pearson Education Inc.

Figure Credits

Reprinted courtesy of Caterpillar Inc., Module opener and Figure 6

US Department of Homeland Security, Federal Emergency Management Agency, Figure 2

Topaz Publications, Inc., Figure 4

Dan Nickel, Figure 26

Robert Bosch Tool Corporation, Figures 28 and 38

Sokkia Corporation, Figure 29

Trimble Navigation Limited, Figures 36, 37, 40, and 41

Keson Industries, Inc., Figure 42

Teaching Tip/Notes

ANSWERS TO REVIEW QUESTIONS

Answer		Section Reference
1.	d	2.1.0
2.	a	2.3.0
3.	b	2.3.0
4.	c	2.3.0
5.	a	2.3.0
6.	c*	2.5.0
7.	c*	2.5.0
8.	b	2.5.0
9.	a	3.1.0; Table 2
10.	d	3.2.0
11.	a	3.2.1
12.	d	3.2.1
13.	b	3.2.2
14.	d	3.2.3
15.	c	3.2.4
16.	a	3.3.0; Figure 21
17.	a	3.3.1
18.	c*	3.3.1
19.	a*	3.3.2
20.	a	4.0.0

*Calculations for these answers appear below.

REVIEW QUESTION CALCULATIONS

6. $\frac{\text{Vertical distance (rise)}}{\text{Horizontal distance (run)}} = \frac{1}{50}$ or $1 \div 50 = 0.02$ or 2 percent

7. $\frac{\text{Vertical distance (rise)}}{\text{Horizontal distance (run)}} = \frac{5}{200}$ or $5 \div 200 = 0.025$ or 2½ percent

18. $\frac{\text{Vertical distance (rise)}}{\text{Horizontal distance (run)}} = \frac{1}{2} = \frac{1 \times 1}{2 \times 1}$ = 1 ft of rise at 2 ft out from the stake

19. $\frac{\text{Vertical distance (drop)}}{\text{Horizontal distance (run)}} = \frac{1}{8} = \frac{1 \times 2}{8 \times 2} = \frac{2}{16} = \frac{1}{8}$ = 1 ft of drop at 8 ft out from the stake

ANSWERS TO TRADE TERMS QUIZ

1. Hinge point
2. Foresight
3. Slope
4. Inslope
5. Backsight
6. Elevation
7. Grade
8. Crow's foot
9. Profile
10. Finished grade
11. Backslope
12. Ratio
13. Stake
14. Global positioning system (GPS)
15. Grade work
16. Subgrade
17. Center line
18. Topographical
19. Erosion
20. Natural ground
21. Temporary benchmark
22. Line of sight
23. Aggregate
24. Height of instrument
25. Hubs
26. Station number
27. Benchmark
28. Reference stake
29. Baseline

NCCER CURRICULA — USER UPDATE

NCCER makes every effort to keep its textbooks up-to-date and free of technical errors. We appreciate your help in this process. If you find an error, a typographical mistake, or an inaccuracy in NCCER's curricula, please fill out this form (or a photocopy), or complete the online form at **www.nccer.org/olf**. Be sure to include the exact module ID number, page number, a detailed description, and your recommended correction. Your input will be brought to the attention of the Authoring Team. Thank you for your assistance.

Instructors – If you have an idea for improving this textbook, or have found that additional materials were necessary to teach this module effectively, please let us know so that we may present your suggestions to the Authoring Team.

NCCER Product Development and Revision

13614 Progress Blvd., Alachua, FL 32615

Email: curriculum@nccer.org
Online: www.nccer.org/olf

❑ Trainee Guide ❑ AIG ❑ Exam ❑ PowerPoints Other ______________________________

Craft / Level: Copyright Date:

Module ID Number / Title:

Section Number(s):

Description:

Recommended Correction:

Your Name:

Address:

Email: Phone:

Glossary

Aggregate: Manufactured or natural rock or sand that has a specific size characteristic. Used in foundations.

Articulated: Two parts connected by a joint so as to move independently.

Backhauling: The return trip of a piece of equipment after it has completed dumping its load.

Backsight (BS): A reading taken on a leveling rod held at a known elevation, which is used to determine the height of the leveling instrument.

Backslope: The area between the ditch line and a backslope stake.

Baseline: A line from which other measurements are referenced.

Bead: Part of a tire that fits against the rim.

Bedrock: The solid layer of rock under Earth's surface. Solid rock, as distinguished from boulders.

Benchmark (BM): Permanent point of known or assumed elevation from which surveyors can establish grades.

Blade: The primary attachment on a bulldozer used to push material in front of the equipment. It is typically a concave metal plate.

Bucket: A U-shaped, closed-end scoop attached to a front-end loader or backhoe.

Center line: The line that marks the center of a roadway. This is marked on the plans by a line and on the ground by stakes.

Center line: The point on stakes or drawings, which indicates the halfway point between two sides.

Center of gravity: The point where all of an object's weight is evenly distributed.

Channeling device: Cones, barrels, or markers that are used to move traffic from its normal lane to a desired lane.

Cohesive: The ability to bond together in a permanent or semi-permanent state. To stick together.

Consolidation: To become firm by compacting the particles so they are closer together.

Core sample: A sample of earth taken from a test boring.

Crow's foot: Markings on a lath to indicate the final grade at a certain point.

Cycle time: The time it takes for a piece of equipment to complete an operation. This normally would include loading, hauling, dumping, and then returning to the starting point.

Demolition: The destruction and removal of a structure.

Dewatering: Removing water from an area using a drain or pump.

Direct current (DC): An electrical current that flows in only one direction.

Dogged: To hold or fasten tightly with a mechanical device.

Draft: Load-pulling capacity.

Dragline: An excavating machine having a bucket that is dropped from a boom and dragged toward the machine by a cable.

Elevation: The distance above or below sea level or other reference point.

Embankment: Material piled in a uniform manner so as to build up the elevation of an area. Usually, the material is in long narrow strips.

Erosion: The process in which a substance, such as soil, is slowly washed away by rain, wind, or other causes.

Expansive soil: A clay-like soil that swells with an increase in moisture and shrinks with a decrease in moisture.

Finished grade (FG): Any surface that has been cut or filled to the final elevation indicated or designed for that point.

Flagger: A person who is specially trained to direct traffic through and around a work zone.

Float: A control setting where the bucket, forks, or blade will follow the ground contour.

Foresight (FS): A reading taken on a leveling rod to determine an unknown elevation; used in conjunction with backsight and height of instrument.

Global positioning system (GPS): System developed by the US military to provide precise information for the movement of military vehicles, equipment, and personnel. This system is now available for use by civilians.

Glow plugs: Heating elements used to preheat the combustion chamber and aid in igniting fuel in a cold engine.

Gradation: The classification of soils into different particle sizes.

Grade: The surface of a road, channel, or natural ground area. This usually means the surface level required by the plans or specifications.

Grade work: Plotting and correction of irregularities of the ground to a definite limit of grade and alignment.

Groundwater: Water beneath the surface of the ground.

Haul road: A compacted dirt road used to move material and equipment on and off the site.

Height of instrument (HI): Vertical distance from the benchmark to the line of sight (also called backsight) of the instrument.

Hinge point (HP): A point indicating where the road or shoulder subgrade ends and the inslope begins. Sometimes called the catch point.

Hubs: Points-of-origin stakes that identify a point on the ground. The top of the hub establishes the point from which soil elevations and distances are computed, or a point to be trimmed to.

Hydraulic actuator: A device that converts the fluid energy of the hydraulic system into mechanical energy using a hydraulic cylinder or motor.

Hydrostatic: Relating to fluids at rest, or to the pressures they transmit, as in hydraulically-powered controls or transmissions that stay in a neutral condition until the hydraulic fluid in them is put into motion.

Impervious: Not allowing entrance or passage through; for example, soil that will not allow water to pass through it.

Infrastructure: A system of public works such as highways, bridges, and dams.

Inorganic: Derived from other than living organisms.

Inslope: The area of roadway between the shoulder and the ditch (natural ground and the shoulder hinge point).

Joystick: A control mechanism that pivots about a fixed point in four directions. It is used to control the motion of an object.

Line of sight: Line through the leveling instrument as viewed by the eye; point where the vertical hair crosses the horizontal hair.

Mine Safety and Health Administration (MSHA): A federal government agency established to ensure the safety of mining operations.

Moldboard: The cutting blade on a motor grader.

Natural ground: The original ground elevation before any excavation has been done.

Occupational Safety and Health Administration (OSHA): The federal government agency established to ensure a safe and healthy environment in the workplace.

On-the-job learning (OJL): Job-related learning acquired while working.

Organic: Derived from living organisms such as plants and animals.

Pascal's law: A scientific law stating that pressure at any one point in a closed static fluid system will be applied to all points in the system in every direction and will exert equal forces on equal areas; the basis of hydraulic theory, developed by Blaise Pascal (1623–1662), a French philosopher and mathematician.

Pay item: A defined piece of material or work that the contractor is paid for. Pay items are usually expressed as unit costs.

Pinch points: The area in which two moving equipment parts come together.

Pit: An open excavation that usually does not require vertical shoring or bracing.

Pitch: The angle of the blade and/or moldboard in relation to a vertical plane.

Pneumatic: Inflated with compressed air.

Power takeoff (PTO): A mechanism attached to a motor vehicle engine that supplies power to a non-vehicular device, such as a pump or pneumatic hammer.

Power takeoff (PTO): Mechanical connection on an engine or transmission to which a cable, belt, or shaft can be connected to drive an attachment or tool.

Profile: The side view of a proposed construction project.

Ratio: Proportion; relationship in quantity, amount, or size between two or more things. For example, if a grade has a horizontal run of 3 feet for every 1 foot in elevation, the ratio is 3:1.

Reference stake (RS): The stake from which measurements and grades are established.

Retard: To slow down.

Ripper: A towed attachment with teeth used on dozers, motor graders and other machines to loosen heavily compacted soil and soft rock.

Riprap: Loose pieces of rock that are placed on the slope of an embankment in order to stabilize the soil.

Rockshaft unit: Controls for equipment mounted on a three-point hitch.

Roll-over protective structure (ROPS): A metal structure attached to the equipment's cab designed to protect the operator should the equipment roll over.

Scarifier: An attachment with teeth used on motor graders to loosen soil.

Scarifying: To loosen the top surface of material using a set of metal shanks (teeth).

Seat: To cause two surfaces to fit firmly together.

Select material: Soil or manufactured material that meets a predetermined specification as to some physical property such as size, shape, or hardness.

Shoring: Material used to brace the side of a trench or the vertical face of any excavation.

Slope: Ground that forms a natural or artificial incline upward or downward.

Soil testing: A mechanical or electronic test used to determine the density and moisture of the soil, and therefore the amount of compaction required.

Splines: Parallel grooves running lengthwise on a shaft.

Spoils: Material that has been excavated and stockpiled for future use.

Stabilizers: Hydraulic arms on backhoes and other equipment that can be positioned on the right and left sides of the machine to give it additional stability.

Stake: Any type of marker that is used to mark cut, fill, and grade information, underground utilities, or survey points.

Station number: The longitudinal (lengthwise) distance of a point of a roadway from the starting or reference point. Station numbers are usually measured in 100-foot increments.

Stormwater: Water from rain or snow.

Subgrade: The uppermost level of material placed in an embankment or left at cuts in the normal grading of a road bed. This becomes the foundation for aggregate and asphalt material.

Temporary benchmark (TBM): A temporary point of known or assumed elevation from which surveyors can establish all their grades for a particular job.

Temporary traffic control (TTC): Device and plan used to safely divert traffic when the normal use of the road is disrupted due to constriction.

Test boring: To drill or excavate a hole in order to take a sample of the material that rests in different layers beneath the surface.

Test pit: See *test boring*.

Throttle: A lever that regulates the supply of fuel to an engine.

Topographical: Description and delineation of the natural and man-made features of an area.

Torque converter: A device that uses fluid, usually oil, to transmit torque from one shaft to another. Also known as a fluid coupler.

Travel: Moving equipment from place to place. Can be either on the job site or over the public highway system.

Trench: A temporary long, narrow excavation that will be covered over when work is completed.

Turbocharger: Exhaust-driven centrifugal air compressor used to increase power.

Water table: The depth below the ground's surface at which the soil is saturated with water.

Windrow: A long, straight pile of placed material for the purpose of mixing or scraping up.

Index

C

D

E

F

I

J

K

L

R

S

T

U

V

W

Y